Herbicides and Plant Physiology

Second Edition

Professor Andrew H. Cobb
Dr John P.H. Reade

Crop and Environment Research Centre
Harper Adams University College
Newport
Shropshire
UK

WILEY-BLACKWELL

A John Wiley & Sons, Ltd., Publication

This edition first published 2010
© 2010 A.H. Cobb and J.P.H. Reade

Blackwell Publishing was acquired by John Wiley & Sons in February 2007. Blackwell's publishing programme has been merged with Wiley's global Scientific, Technical, and Medical business to form Wiley-Blackwell.

Registered office
John Wiley & Sons Ltd, The Atrium, Southern Gate, Chichester, West Sussex, PO19 8SQ, United Kingdom

Editorial offices
9600 Garsington Road, Oxford, OX4 2DQ, United Kingdom
2121 State Avenue, Ames, Iowa 50014-8300, USA

For details of our global editorial offices, for customer services and for information about how to apply for permission to reuse the copyright material in this book please see our website at www.wiley.com/wiley-blackwell.

The right of the author to be identified as the author of this work has been asserted in accordance with the UK Copyright, Designs and Patents Act 1988.

The Library of Congress has cataloged the printed edition as follows:

Cobb, Andrew.
  Herbicides and plant physiology / Andrew H. Cobb, John P.H. Reade. – 2nd ed.
    p. cm.
  Includes bibliographical references and index.
  ISBN 978-1-4051-2935-0 (pbk. : alk. paper) – ISBN 978-1-4443-2780-9 (e-book)   1. Plants–Effect of herbicides on.   2. Plant physiology.   3. Herbicides–Physiological effect.   4. Weeds–Control.   I. Reade, John P. H.   II. Title.
  SB951.4.C63 2010
  632′.954–dc22

                                                                                      2010016815

A catalogue record for this book is available from the British Library.

Set in 10/12 pt Times by Toppan Best-set Premedia Limited

1   2010

Contents

Preface

Whilst recently acting in the capacity of External Examiner at a UK Higher Education Institution, AHC encountered the following question on an examination paper:

"Discuss why, after 50 years of using herbicides, weeds are still a major problem in agroecosystems."

The short answer, of course, is that the "weed problem" will never go away as long as crops are grown and harvested. The grower needs to be forever vigilant of the changing weed spectrum and of the need to use the appropriate methods available to ensure adequate control, including both crop and chemical rotation. The theory then is relatively simple, but in practice the grower is put under increasing pressure by variable returns, to repeatedly grow monocultures and rely on the cheapest methods of weed control and chemical inputs available. Too often a farmer will adopt a wait and see approach and then resort to a fire-fighting, last-minute strategy for weed control.

Variable returns and environmental considerations are indeed forcing the farmer to re-examine the need for all chemical inputs, including fertilisers and crop protection agents. This will reduce costs but with a consequence of lower yields and reduced crop quality. On the other hand, a more judicious, measured use of agrochemicals at an appropriate dose may provide the most cost effective management of a weed problem. To do this, the farmer must know the land, understand its previous history of cropping and have good records of the previous usage of crop protection chemistry. Even then, as we develop new cropping practices, such as minimum cultivation or even organic farming, we can expect the weed population and spectrum to change, population density to alter and different species to gain prominence. These all create new and demanding challenges to weed management and control strategies.

Our starting point then, in a study of herbicides, must be to understand weed biology and the competing demands of weeds and crops. Only then is reflection possible to choose the most appropriate herbicide for the weed issue in question.

Subsequent chapters include considerations of how herbicides have been discovered and developed (Chapter 2), how they gain entry into plants and move to their site of action (Chapter 3), and the basis for herbicide selectivity (Chapter 4). Detailed accounts are then given of how herbicides interact with the major physiological processes in plants leading to plant death. This begins with the inhibition of photosynthesis (Chapter 5), followed by pigment biosynthesis (Chapter 6), interactions with the plant growth regulator, auxin (Chapter 7), lipid biosynthesis (Chapter 8), amino acid biosynthesis (Chapter

9), cell division (Chapter 10) and cellulose biosynthesis (Chapter 11). Chapter 12 gives a detailed and up to date presentation on herbicide resistance, leading to an account of the development and underlying science of herbicide tolerant crops (Chapter 13). Finally, Chapter 14 highlights further targets for herbicide development that may yield new products in the following decade(s).

In reviewing literature for this second edition, it may be tempting at a superficial level to believe that there is little new to report in this discipline since the first edition was published in 1992. Indeed, superficially, there are few novel modes of action to report with little exploitation of new target sites. A more detailed scrutiny, however, reveals the development of new triazine inhibitors of PSII (p 98), of the HPPD inhibitors (p 128), more clarity on the mode of action of the chloroacetamides (p 167) and the discovery of a new class of microtubule inhibitors (p 205) and inhibitors of cellulose biosynthesis (p 213) to name but five. Why have there been fewer herbicide discoveries in the recent years? Two main contributory factors come to mind. The first is the increasing costs to develop a new product that satisfies all the regulatory requirements. The second lies in the use of molecular approaches to induce herbicide tolerance into crops, thereby extending the usage of existing products. These areas are considered further in Chapters 2 and 13, respectively. Our understanding of the metabolism, environmental impact, genetics and molecular biology of herbicide action, particularly in relation to resistance, has also greatly improved. Indeed, this understanding has led to lower doses of new agrochemicals that are more effective than previous versions and are more environmentally benign. Furthermore, new methods have been established for high-throughput screening (p 38) to discover new agrochemicals.

Since the first edition of this book our understanding of plant physiology and metabolism in general has improved, especially utilising our knowledge of the *Arabidopsis* genome. This has generated potential for new target sites that may yield the new herbicides of the coming decades.

The need for crop protection agents remains essential, as eloquently summarised by Len Copping (2001).

"In the time it takes to read this sentence, another 20 people will have been added to the world's population. And by this time next week, enough people will have been born to establish a new city about the size of Birmingham in the UK or Detroit in the US. This rapid expansion is forecast to stabilise by 2050 at 11 billion, a 100% increase from 1998. Enormous sociological and economic progress must occur to allow such increase without apocalyptic penalty, but of primary importance will be our ability to produce food in sufficient quantity and of appropriate quality to sustain an acceptable standard of living."

It is our view that only with the intelligent use of agrochemicals and especially crop protection agents will the food security of future generations be secured.

The Farm Ministers of the European Union have agreed to reduce the agrochemicals used in European agriculture, ostensibly to enhance the protection of human health and the environment. This could lead to the exclusion of any pesticide containing active ingredients considered hazardous, with the result that 15% of existing crop protection agents will be deregistered. As for herbicides, this could lead to the loss of active ingredients with an auxin-type mode of action, in addition to pendimethalin, linuron and ioxynil. All valuable components of our crop protection armoury. Ironically, even though these active ingredients could be banned for use by European farmers, we as consumers will still be eating perfectly safe food imported from non-European countries using the very same chemicals! This directive comes at a time when European politicians are, quite rightly, expressing concerns about food security and rising food prices. Yet, this decision could result in much lower crop yields, probably leading to further increases in food costs. It is hoped that the European Parliament will eventually take a more balanced view and that common sense and a pragmatic agricultural solution will prevail.

It is our view that more work is needed to communicate to the general public and EU legislators the distinction between hazard and risk. Thus, the caffeine in our morning coffee or the addition of salt to our food can kill (hazard), but the dose ingested will not (risk). Similar arguments can be presented for the natural but potentially toxic secondary metabolites that feature in our everyday diet, such as in potatoes or nuts. Or, applying the same logic to synthetic chemistry, to shampoos, aspirin and paracetamol. The latter is a most interesting and relevant example of a synthetic molecule that is widely used, since it is well established that paracetamol is hazardous, as it can cause fatal liver damage at high concentrations, but the actual risk of the dose needed to cure a headache is very low.

We hope that this volume will become a useful resource for those working in the plant protection industries and for advanced level undergraduates and postgraduate students of agricultural chemistry, plant physiology and biochemistry, and applied biology.

Andrew H. Cobb
John P.H. Reade

Reference

Copping, L. (2001) The Crop-Protection Industry. *Chemistry and Industry*, Issue 16, 500–501, 20 August 2001.

Chapter 1
An Introduction to Weed Biology

1.1 Introduction

The human race has been farming for over 10,000 years. Weeds would have been an unwelcome presence alongside crops ever since the first farmers saved and planted seeds in the region that is now present-day Turkey and the Middle East. Indeed, when these early farmers noticed a different plant growing, decided they did not want it and pulled it up, they were carrying out a form a weed control that is still used today: hand roguing.

But what are weeds? Weeds are all things to all people, depending on the viewpoint of the individual. To some they are plants growing where they are not wanted; to others they are plants growing in the wrong place, in the wrong quantity, at the wrong time; and to some they are regarded as plants whose virtues have yet to be fully discovered! The need to control weeds only arises when they interfere with the use of the land, and this is usually in the presence of a crop, such as in agriculture and horticulture. Weed control may also be necessary in other situations including amenity areas, such as parks and lawns, in water courses, or on paths and drives where the presence of plants may be regarded as unsightly. It should not be overlooked, however, that weeds contribute to the biodiversity of ecosystems and should only be removed when financial or practical implications make their presence unacceptable. With this in mind an appropriate definition of a weed is:

> Any plant adapted to man-made habitats and causing interference of the use of those habitats
>
> (Lampkin, 1990)

1.2 Distribution

On a global basis only about 250 species are sufficiently troublesome to be termed weeds, representing approximately 0.1% of the world's flora. Of these, 70% are found in 12 families, 40% alone being members of the Gramineae and Compositae. Interestingly, 12 crops from 5 families provide 75% of the world's food and the same 5 families provide many of the worst weeds (Table 1.1). This implies that our major crops and weeds share certain characteristics and perhaps common origins.

Herbicides and Plant Physiology, Second Edition By Andrew H. Cobb and John P.H. Reade
© 2010 A.H. Cobb and J.P.H. Reade

Table 1.1   Important plant families which contain both the major crops and the worst weeds of the world (from Radosevich and Holt, 1984).

Number of species classified in the world's worst weeds (%)	Family	Examples of major crops	Examples of major weeds	Common name
44	Gramineae	Barley, maize, millett, oats, rice, sorghum, sugar cane and wheat	*Elytrigia repens* (L.) *Alopecurus myosuroides* (L.) *Avena fatua* (L.) *Sorghum halepense* (L.) Pers *Echinochloa crus-galli* (L.)	Couch Black-grass Wild oat Johnson grass Barnyard grass
4	Solanaceae	White potato	*Solanum nigrum* (L.) *Datura stramonium* (L.) *Hyoscyamus niger* (L.)	Black nightshade Jimsonweed Henbane
5	Convolvulaceae	Sweet potato	*Convolvulus arvensis* (L.) *Cuscuta pentagona* (Engelm) *Ipomoea purpurea* (L.) Roth	Field bindweed Field dodder Tall morning glory
5	Euphorbiaceae	Cassava	*Euphorbia maculata* (L.) *Euphorbia helioscopia* (L.) *Mercurialis annua* (L.)	Spotted spurge Sun spurge Annual mercury
6	Leguminosae	Soybean	*Cassia obtusifolia* (L.) *Melilotus alba* (Desc) *Trifolium repens* (L.)	Sicklepod White sweetclover White clover

1.3   The importance of weeds

Most plants grow in communities consisting of many individuals. If the resources available (such as space, water, nutrients and light) become limiting then each species will be forced to compete. Weeds are often naturally adapted to a given environment and so may grow faster than the crop, especially since the crop species has been selected primarily for high yield rather than competitive ability. A unit of land may therefore be regarded as having a finite potential biomass to be shared between crop and weeds, the final proportion being determined by their relative competitive ability.

1.4   Problems caused by weeds

The most obvious problem caused by weeds is the reduction of yield through direct competition for light, space, nutrients and water. Weeds can have many further effects on the use of land, as illustrated in Table 1.2.

1.4.1   Yield losses

Crop losses approaching 100% are recorded in the literature (Table 1.3; Lacey, 1985). Such yield losses will, of course have a profound effect on a national economy both in terms of the need to import foodstuffs and the costs of weed control. Despite the many methods of weed management that are now available worldwide, it is estimated that

Table 1.2   Problems caused by weeds (from Naylor and Lutman, 2002).

Justification	Mechanism
Reduce crop yield	Interference with access to light, water and nutrients
Reduce crop quality	Admixture of contaminating seeds in arable crops Contamination of vegetable crops
Delay harvesting	Conservation of moisture may delay ripening and increase moisture level when harvested
Interfere with harvesting	Climbing plants can make combining more difficult Vigorous, late-growing weeds can interfere with harvesting potatoes and sugar beet
Interfere with animal feeding	Plants with spines or thorns inhibit animal foraging
Cause poisoning	Poisoning either through ingestion or contact
Taint animal products	Impart undesirable flavour, e.g. to milk
Act as a plant parasite	Competing for nutrients and water
Reduce crop health	Act as an alternative host for crop pests and diseases Increase vegetation at base of crop increasing moisture and disease
Reduce animal (and human) health	Act as intermediate host or a vehicle for ingestion of pests and parasites Photosensitivity Teratogens Carcinogens
Are a safety hazard	Reduce vision on roadsides Risk of fire under electricity lines, on garage forecourts
Reduce wool quality	Hooked seeds reduce value of fleece
Prevent water flow	Plant mass blocking ditches and irrigation channels
Exhibit allelopathy	Release of substances toxic to the growth of crop plants
Impact on crop establishment	Vegetation prevents establishment of young trees Competition for space with establishing crops

Table 1.3 Examples of yield losses due to weeds (from Lacey, 2001, by permission of Oxford University Press; *from Moss, 1987).

Crop	Yield loss (%)	Country
Cassava	92	Venezuela
Cotton	90	Sudan
Groundnuts	60–90	Sudan
Onions	99	UK
Rice	30–73	Colombia
Sorghum	50–70	Tanzania/Nigeria
Sugar beet	78–93	Texas, USA
Sweet potatoes	78	West Indies
Wheat*	66	UK
Yams	72	Nigeria

Table 1.4 Estimated percentage crop losses due to weeds, 1988–90 (from Oerke *et al.*, 1995).

	Estimated loss due to weeds (%)
Africa	16.5
North America	11.4
Latin America	13.4
Asia	14.2
Europe	8.3
Former Soviet Union	13.0
Oceania	9.6
Average	**13.1**

approximately 13% crop losses are still due to weeds alone (Table 1.4). Indeed, in 1974 the annual cost of weeds to agriculture in the USA was estimated at $US10 billion, with 50% due to yield reductions and 50% due to the cost of weed control (Rodgers, 1978).

In the tropics, parasitic weed species from the genera *Cuscuta* (dodders), *Orobranche* (broomrapes) and *Striga* (witchweeds) can have a profound effect on a range of crops. They absorb nutrients directly from the crop plant, which may not set seed at all in the case of cereals such as sorghum.

Weed control techniques are therefore aimed at the reduction in the competitive ability of weeds in a crop and the prevention of weed problems in a future crop. The former is increasingly based on chemical use, and the latter also requires suitable cultural and agronomic practices.

Yield loss may be usefully related to the number of weeds per unit area causing a defined yield loss in a defined crop, that is, as a Weed Threshold (Table 1.5) or as a Crop Equivalent (the amount of resource an individual weed uses expressed as the number of crop plants this resource would support; although in practice it is the biomass of the weed and the crop which is measured). Generally, these figures have only been determined for weed interaction with major crops, but they give a good indication of the ability of a particular species to compete with all crops.

Table 1.5 Relative competitive abilities of a number of common weeds found in winter cereals (from Lutman *et al.*, 2003).

Weed species	5% yield loss (plants m^{-2})	Weed species	5% yield loss (plants m^{-2})
Galium aparine	1.7	Poa annua	50.0
Anisantha sterilis	5.0	Epilobium spp.	50.0
Avena fatua	5.0	Polygonum aviculare	50.0
Lolium multiflorum	8.3	Sonchus spp.	50.0
Alopecurus myosuroides	12.5	Taraxacum officinale	50.0
Brassica napus	12.5	Fumaria officinalis	62.5
Sinapis arvensis	12.5	Geranium spp.	62.5
Tripleurospermum inodorum	12.5	Lamium purpureum	62.5
Cirsium spp.	16.7	Ranunculus spp.	62.5
Convolvulus arvensis	16.7	Veronica spp.	62.5
Fallopia convolvulus	16.7	Aethusa cynapium	83.3
Papaver spp.	16.7	Senecio vulgaris	83.3
Chenopodium album	25.0	Anagallis arvensis	100.0
Myosotis arvensis	25.0	Allium vineale	250.0
Persicaria maculosa	25.0	Aphanes arvensis	250.0
Silene vulgaris	25.0	Legousia hybrida	250.0
Stellaria media	25.0	Viola arvensis	250.0

Yield loss may also occur in addition to direct competition for resources. Allelopathy is the production of allelopathic chemicals by one plant species that may inhibit (or, in the case of positive allelopathy, stimulate) the growth of other species. Anecdotal evidence of negative allelopathic effects has been reported for a number of weed species, although supporting research is often lacking. Recent findings have been reviewed by Olofdotter and Mallik (2001) and others (see *Agronomy Journal* vol. 93). Given the ample evidence of allelopathy exhibited by crop species, it is highly likely that many weed species will also display these effects, and that it is only a matter of time before research demonstrating this becomes readily available.

Further examples of yield loss caused by weeds include the effects on non-plant organisms. One example of this is the presence of dandelion (*Taraxacum officinale*) in fruit orchards. Dandelion flowers are preferentially visited by insect pollinators and so pollination of fruit blossom (and therefore fruit yield) is reduced.

1.4.2 Interference with crop management and handling

Some weeds can make the operation of agricultural machinery more difficult, more costly, or even impossible. The presence of weeds within a crop may necessitate the need for extra cultivations to be introduced. This often leads to crop damage, reduced yields and increased pest and disease occurrence, although in sugar beet crops, where inter-row cultivation is often carried out and has previously been associated with yield loss, recent findings suggest that careful implementation can result in no loss of root yield or sucrose content (Dexter *et al.*, 1999; Wilson and Smith, 1999). This is possibly due to the development of tillage equipment that carries out more shallow cultivation and that is more carefully implemented, resulting in less seedling and root damage. Weeds can also affect the processes carried out prior to crop planting. For example, fat hen stems and leaves block the mesh of de-stoners, which are used prior to potato and other root crop planting. Species with rough, wiry stems that spread close to the ground (e.g. knotgrass, *Polygonum aviculare*) or are more erect in growth habit (e.g. fat hen, *Chenopodium album*) present major problems to the mechanical harvesting of many crops and can result in damage to machinery (e.g. pea viners) and subsequent harvesting delays. Other species can be troublesome when the crops are harvested by hand, such as the small nettle (*Urtica urens*) in strawberries and field bindweed (*Convolvulus arvensis*) in blackcurrants. The result of this is that fruit is not harvested and spoils on the plant.

1.4.3   Reduction in crop quality

Competition between crop and weed species can result in spindly leaf crops and deformed root crops which are less attractive to consumers and processors. A crop may have to be rejected if it contains weed seeds, especially when the crop is grown for seed, such as barley and wheat, and if the weed seeds are similar in size and shape to the crop, e.g. wild oats (*Avena fatua*) in cereal crops. Similar problems are encountered in the contamination of oilseed rape seed with seeds of weed species such as cleavers (*Galium aparine*). Where a proportion of the seed is saved for planting in subsequent seasons, this can cause a large increase in weed infestation. Contamination by poisonous seeds, such as darnel (*Lolium temulentum*) and corncockle (*Agrostemma githago*) in flour-forming cereals is also unacceptable and once led to vastly increased costs of crop cleaning. Such cleaning, however, has meant that these weeds are now probably extinct in agroecosystems in the UK. A further example that still causes major problems is black nightshade fruit (*Solanum nigrum*) in pea crops (Hill, 1977). In this case, the poisonous weed berry is of similar size and shape to the crop and so must be eradicated. Although grazing animals avoid poisonous species in pasture (e.g. common ragwort, *Senecio jacobea*), they may be difficult to avoid in hay and silage, and some species, notably the wild onion (*Allium vineale*), can cause unacceptable flavours in milk and meat.

1.4.4   Weeds as reservoirs for pests and diseases

Weeds, as examples of wild plants, form a part of a community of organisms in a given area. Consequently, they are food sources for some animals and are themselves susceptible to many pests and diseases. Because of their close association with crops, they may serve as important reservoirs or carriers of pests and pathogens, as exemplified in Table 1.6. Even where crop infestation does not occur, the presence of disease in weeds may cause problems, as is the case where grass weeds are infected with ergot (*Claviceps purpurea*), causing contamination of harvested grain with highly toxic ergot fragments.

Weeds may act as 'green bridges' for crop diseases, carrying the disease from one crop to another that is subsequently sown. Volunteer crops are particularly problematic in this case and can, in severe cases, negate the use of break crops as a cultural control measure for diseases. In addition, weeds can provide over-wintering habitats for crop pests, result-

Table 1.6 Some examples of weeds as hosts for crop pests and diseases (after Hill, 1977). Copyright © 1977. Reproduced by permission of Edward Arnold (Publishers) Ltd.

Pathogen or pest		Weed		Crop
1. Fungi				
Claviceps purpurea	(ergot)	Black-grass	(*Alopecurus myosuroides*)	Wheat
Gaeumannomyces graminis	(take-all)	Couch	(*Elytrigia repens*)	Cereals
Plasmodiophora brassicae	(clubroot)	Many crucifers		Brassicas
2. Viruses				
Tobacco ringspot		Dandelion	(*Taraxacum officinale*)	Tobacco
Cucumber mosaic		Chickweed	(*Stellaria media*)	Many crops
3. Nematodes				
Ditylenchus dipsaci	(eelworm)	Chickweed	(*Stellaria media*)	Many crops
		Spurrey	(*Spergula arvensis*)	
4. Insects				
Aphis fabae	(black bean aphid)	Fat hen	(*Chenopodium album*)	Broad and field beans

ing in quicker crop infestation in the spring. Ground cover provided by weeds can increase problems with slugs and with rodents, as the weeds provide greater cover and therefore reduced predation.

In 1994 and 1995 there were several severe outbreaks of the disease brown rot in potato in several European countries, especially in Holland, which was possibly exported to other countries via infected seed potatoes. This extremely virulent pathogen (*Pseudomonas solanacearum*, syn. *Burkholderia solanacearum*, syn. *Ralstonia solanacearum*) causes a vascular ring rot in the developing tuber and causes a major loss of yield. Although often considered a soil-borne organism, it was not found to persist for long periods in the soil following the harvest of infected crops. However, it was found to survive in the aquatic roots of infected woody nightshade (*Solanum dulcamara*) growing at the edge of irrigation channels. Thus, it may be the case that the pathogen overwinters in this wild host and is leaching into watercourses used to irrigate the crop spreads the disease. This perennial plant is now being eradicated from potato-growing areas. Several other species could also act as alternative hosts to the pathogen, including *Solanum nigrum* and *Tusilago farfara*, but further work is needed to confirm this.

1.5   Biology of weeds

Knowledge of the biology of a weed species is essential to the design of management strategies for that weed. An understanding of the life cycle of a species can be exploited in order to identify vulnerable times when weed management and control might prove more successful.

1.5.1   Growth strategies

According to Grime (1979), the amount of plant material in a given area is determined by two principal external factors, namely stress and disturbance. Stress phenomena include any factors that limit productivity, such as light, nutrient, or water availability; and disturbance implies a reduction in biomass by factors such as cultivation, mowing, or grazing. The intensity of both stress and disturbance can vary widely, with four possible combinations. However, only three growth strategies have evolved, as shown in Table 1.7. Although plants are unable to survive both highly stressed and disturbed environments, the other strategies have major significance to weed success.

Ruderals are the most successful agricultural weeds. These plants have typically rapid growth rates and devote most of their resources to reproduction. Because they inhabit recently disturbed environments there is little competition with other plants for resources, which therefore can be obtained without difficulty. They are generally short-lived ephemeral annuals that occupy the earliest phases of succession. Conversely, biennial and perennial weeds often employ a more competitive growth strategy in relatively undisturbed conditions. They use their resources perhaps less for seed production and more for support tissues, for example, to provide additional height for the interception of light, or more extensive root systems to obtain more water and minerals. Rapid growth rate may still be evident with high rates of leaf turnover. The third growth strategy, exhibited by the stress tolerators, is to reduce resource allocation to vegetative growth and seed production, so that the survival of relatively mature individuals is ensured in high stress conditions. Consequently, they have slow growth rates and are commonly found in unproductive environments.

Many arable weeds have characteristics common to both competitors and ruderals, and are referred to as competitive ruderals. Indeed, most of the annuals listed in *The World's Worst Weeds* (Radosevich and Holt, 1984) fit into this category, and are found in productive sites where occasional disturbance is expected. Examples include arable land that is cultivated, and meadows and grassland that are grazed or mowed. Interestingly, most crop plants also adopt a competitive ruderal strategy with their rapid growth rates and relatively large seed production. Competition between crop and weed is then related to their relative abilities to exploit the resources available.

The practice of growing crops in monoculture has exerted a considerable selection pressure in the evolution of weeds. Many characteristics have evolved that contribute to weed success and the main ones are listed in Table 1.8. Fortunately, not all of these

Table 1.7 Growth strategies of plants. Copyright © 1977. Reproduced by permission of Edward Arnold (Publishers) Ltd.

	Intensity of stress	
Intensity of disturbance	High	Low
High	Death	Ruderals
Low	Stress tolerators	Competitors

Table 1.8 The 'successful' weed (adapted from Baker and Stebbins, 1965).

Characteristic	Example species
1. Seed germination requirements fulfilled in many environments	*Senecio vulgaris*
2. Discontinuous germination (through internal dormancy mechanisms) and considerable longevity of seed	*Papaver* spp.
3. Rapid growth through the vegetative phase to flowering	*Cardamine hirsuta*
4. 'Seed' production in a wide variety of environmental conditions	*Poa annua*
5. Continuous seed production for as long as conditions for growth permit	*Urtica urens*
6. Very high 'seed' output in favourable environmental conditions	*Chenopodium album*
7. Self-compatible but not completely self-pollinating	*Alopecurus myosuroides*
8. Possession of traits for short and long distance seed dispersal	*Galium aparine*
9. When cross-pollinated, unspecialised pollinator visitors or wind pollinated	Grass weeds in general
10. If a clonal species, has vigorous vegetative growth and regeneration from fragments	*Cirsium arvense*
11. If a clonal species, has brittleness of leafy parts ensuring survival of main plant	*Taraxacum officinale*
12. Shows strong inter-specific competition by special mechanisms (e.g. allelopathic chemicals)	*Elytrigia repens*
13. Demonstrates resistance to herbicides through a number of resistance mechanisms	*Alopecurus myosuroides*

features are present in any one weed species, yet each character may give the weed a profound competitive advantage in a given situation. Some of these characteristics are discussed in more detail in the following sections of this chapter.

Invasive species have received far greater research focus in recent years (see, for instance, Shaw and Tanner, 2008 for a review), and DAISIE (Delivering Alien Invasive Species Inventories for Europe) currently reports 10,822 invasive species in Europe (this figure is for all invasive species, not just plants). Alien species present a real threat to biodiversity, and a number of political drivers have been put into place to combat their spread and reduce the occurrence of a number of alien species, including plants. These measures include the Convention of Biodiversity and the EU 2010 Halting Biodiversity Loss, both of which identify invasive weeds as being a key factor in biodiversity loss.

Alien plant species become a problem because they are growing in habitats away from their natural predators and so can spread with ease and do not need to invest valuable energy in biologically expensive chemical defence mechanisms. In addition to this, the Novel Weapons Hypothesis (NWH) proposes that in some cases allelopathic chemicals produced by alien species are more effective against native species in invaded areas than they are against species from the alien's natural habitat (Inderjit *et al.*, 2007). This gives further ecological advantage to some invasive species of plant.

Observations of the pathogens and predators that affect these invasive plants in their natural habitats may identify mycoherbicides and other biological controls that may prove useful in their management. Research into such controls for Japanese Knotweed (*Fallopia japonica*), Himalayan Balsam (*Impatiens glandulifera*) and Giant Hogweed (*Heracleum mantegazzianum*) have so far produced, at best, limited success (Tanner, 2008).

Moles *et al.* (2008) have recently proposed a framework for predicting plant species which may present a risk of becoming invasive weeds. This may prove more useful than Table 1.8 in risk assessment relating to alien species, prior to their becoming major weed problems.

1.5.2   Germination time

The success of some weeds is due to close similarity with a crop. If both the weed and the crop have evolved with the same agronomic or environmental pressures they may share identical life cycles and life styles. Thus, if weed seed maturation coincides with the crop harvest, the chances of weed seed spread are increased. This phenomenon is often apparent in the grasses such as barnyard grass (*Echinochloa crus-galli*) in rice, and wild oat (*Avena* spp.) in cereals. In Europe, spring wild oats (*Avena fatua*) germinate mainly between March and May, and the winter wild oat (*Avena ludoviciana*) shows maximum germination in November. Hence, the date of cereal sowing in relation to wild oat emergence is crucial to weed control. In general, weeds may be considered to occupy one of three categories; autumn germinators, spring germinators and those that germinate throughout the year. Figure 1.1 illustrates the germination patterns for a number of common arable weeds. Autumn-sown crops are more at risk from weeds that germinate during the autumn, when the crop is relatively small and uncompetitive. Conversely, weeds that cause problems in spring crops tend to be spring germinators (including the troublesome polygonums). Knowledge of the germination patterns of weeds plays a very important role in the designing of specific weed management strategies that will disrupt conditions conducive to the survival of the weed.

Further major problems are evident in sorghum, radish and sugar beet crops. Hybridisation of cultivated *Sorghum bicolor* (L.) Moench with the weed *S. halepense* (L.) Pers. results in an aggressive perennial weed that produces few seeds, but demonstrates vigorous vegetative growth. Similarly, hybridisation between the radish (*Raphanus sativus* L.) and the weed *R. raphanistrum* (L.) has produced a weedy form of *R. sativus* with dormant seeds and a root system that is more branched and penetrating than the crop. Lastly, hybridisation of sugar beet (*Beta vulgaris* (L.) subsp. *maritima*) has created an annual weed-beet that sets seed, but fails to produce the typically large storage root. In each of these examples of crop mimicry by weeds, chemical weed control is extremely difficult due to the morphological and physiological similarities between the weed and the crop.

1.5.3   Germination depth

Most arable weeds germinate in the top 5 cm of soil and it is this region that soil-acting (residual) herbicides aim to protect. Where minimum cultivation or direct drilling is carried out, the aim is to avoid disruption of this top region of soil in an attempt to minimise weed-seed germination. A small number weed species can germinate from

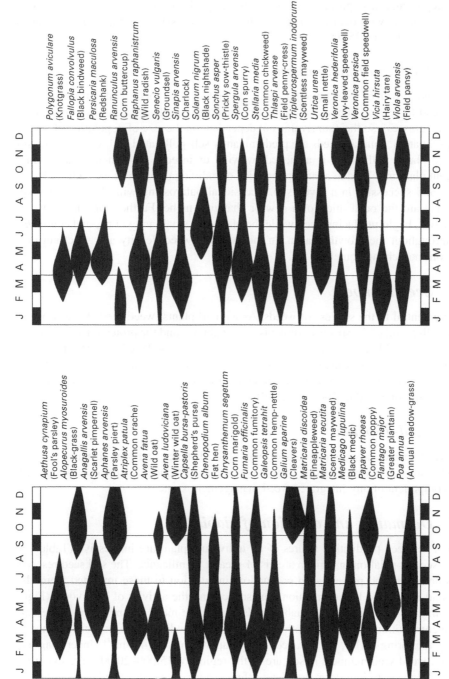

Figure 1.1 Germination periods of some common annual weeds. A greater width of the bar reflects greater germination. Reproduced from Hance, R.J. and Holly, K. (1990) *Weed Control Handbook*, 8th edn.

greater depths and this appears to be due to these species possessing larger seed. A good example of this is wild oat (*Avena* spp.), which can successfully germinate and establish from depths as low as 25 cm.

1.5.4   Method of pollination

Survival and growth of weed populations are dependent upon successful pollination. Weed species tend to rely on non-specific insect pollinators (e.g. dandelion) or are wind polli-nated (e.g. grasses), so their survival is not dependent on the size of population of specific insect pollinators. Annual weeds are predominantly self-pollinated and, when outcrossing does take place, pollination is achieved by wind or insects. This means that a single immigrant plant may lead to a large population of individuals, each as well adapted as the founder, and successful in a given site. Occasional outcrossing will alter the genotype, which may aid the occupation of a new or changing niche. Furthermore, many weeds, unlike crops, begin producing seed while the plants are small and young, and continue to do so throughout the growth season. In this way the weed density and spectrum in an arable soil may change quickly.

1.5.5   Seed numbers

Seeds are central to the success of weeds. As with all plants, weed seeds have two func-tions: the dispersal of the species to colonise new habitats, and the protection of the species against unfavourable environmental conditions via dormancy. Weeds commonly produce vast numbers of seeds, which may ensure a considerable advantage in a competitive environment, especially since the average number of seeds produced by a wheat plant is only in the region of 90 to 100 (Table 1.9).

1.5.6   Seed dispersal

Many weed species possess methods of both short distance and long distance seed dis-persal (Figure 1.2). By recognising these it is possible to reduce the spread of weed seeds, a vital component of any integrated weed management strategy.

1.5.7   Dormancy and duration of viability

Although the seed production figures of an individual plant are impressive (Table 1.9), the total seed population in a given area is of greater significance. The soil seed reservoir reflects both past and present seed production, in addition to those imported from else-where, and is reduced by germination, senescence and the activity of herbivores (Figure 1.3). Estimates of up to 100,000 viable seeds per square metre of arable soil represent a massive competition potential to both existing and succeeding crops, especially since the seed rate for spring barley, for instance, is only approximately $400\,m^{-2}$! Under long term grassland, weed seed numbers in soil are in the region of $15,000–20,000\,m^{-2}$, so conver-sion of arable land to long-term grassland offers growers a means of reducing soil weed-seed burden.

The length of time that seeds of individual species of weed remain viable in soil varies considerably. The nature of the research involved in collecting such data means that few

Table 1.9 Seed production by a number of common arable weeds and wheat (adapted from Radosevich and Holt, 1984 containing information from Hanf, 1983).

Weed	Common name	Seed production per plant
Veronica persica	Common field speedwell	50–100
Avena fatua	Wild oat	100–450
Galium aparine	Cleavers	300–400
Senecio vulgaris	Groundsel	1,100–1,200
Capsella bursa-pastoris	Shepherd's purse	3,500–4,000
Cirsium arvense	Creeping thistle	4,000–5,000
Taraxacum officinale	Dandelion	5,000 (200 per head)
Portulaca oleracea	Purslane	10,000
Stellaria media	Chickweed	15,000
Papaver rhoeas	Poppy	14,000–19,500
Tripleurospermum maritimum spp. inodorum	Scentless mayweed	15,000–19,000
Echinochloa crus-galli	Barnyard grass	2,000–40,000
Chamaenerion angustifolium	Rosebay willowherb	80,000
Eleusine indica	Goose grass	50,000–135,000
Digitaria sanguinalis	Large crabgrass	2,000–150,000
Chenopodium album	Fat hen	13,000–500,000
Triticum aestivum	Wheat	90–100

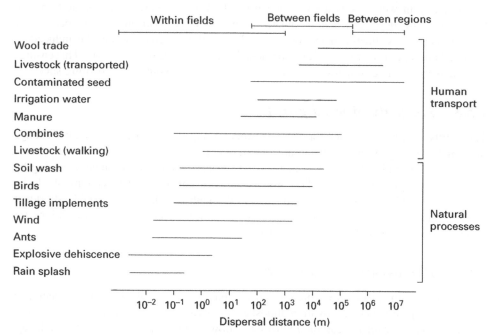

Figure 1.2 Some methods of weed seed dispersal with their estimated range in metres. Reproduced from Liebman M., Mohler C.L. and Staver C.P. (2001) *Ecological Management of Agricultural Weeds*. Cambridge University Press.

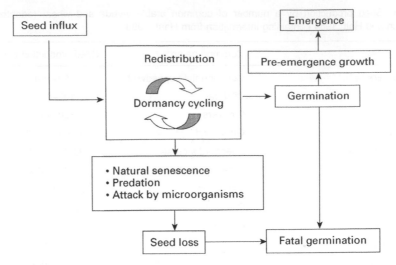

Figure 1.3 Factors affecting the soil seed population (from Grundy and Jones, 2002).

comprehensive studies have been carried out, but those that have (see Toole and Brown, 1946, for a 39 year study!) show that although seeds of many species are viable for less than a decade, some species can survive for in excess of 80 years (examples include poppy and fat hen). Evidence from soils collected during archaeological excavations reveals seeds of certain species germinating after burial for 100 to 600 (and maybe even up to 1700!) years (Ødum, 1965).

Dormancy in weed seeds allows for germination to be delayed until conditions are favourable. This dormancy may be innate and it is this that contributes to the periodicity of germination, as illustrated in Figure 1.1. In addition, dormancy may be induced or enforced in non-dormant seeds if environmental conditions are unfavourable. This ensures that the weed seed germinates when conditions are most conducive to seedling survival.

1.5.8   Plasticity of weed growth

The ability of a weed species for rapid phenotypic adjustment to environmental change (acclimation) may offer a considerable strategic advantage to the weed in an arable context. An example of the consequence of such plasticity is environmental sensing by fat hen (*Chenopodium album*). This important weed can respond to canopy shade by undergoing rapid stem (internode) elongation, although the plant is invariably shorter if growing in full sun. Similarly, many species can undergo sun–shade leaf transitions for maximum light interception (Patterson, 1985).

1.5.9   Photosynthetic pathways

Photosynthesis, the process by which plants are able to convert solar energy into chemical energy, is adapted for plant growth in almost every environment on earth. For most weeds and crops photosynthetic carbon reduction follows either the C_3 or the C_4 pathway, depending on the choice of primary carboxylator. In C_3 plants this is ribulose 1,5-bisphosphate carboxylase / oxygenase (RuBisCo) and the first stable product of carbon reduction is the three-carbon acid, 3-phosphoglycerate. Alternatively, in C_4 plants the primary carboxylator is phosphoenolpyruvate carboxylase (PEPC) and the initial detectable products are the four-carbon acids, oxaloacetate, malate and aspartate. These acids are transferred from the leaf mesophyll cells to the adjacent bundle sheath cells where they are decarboxylated and the CO_2 so generated is recaptured by RuBisCo. Since PEPC is a far more efficient carboxylator than RuBisCo, it serves to trap CO_2 from low ambient concentrations (micromolar in air) and to provide an effectively high CO_2 concentration (millimolar) in the vicinity of the less efficient carboxylase, RuBisCo. In this way, C_4 plants can reduce CO_2 at high rates and are often perceived as being more efficient than C_3 plants. In addition, because of their more effective reduction of CO_2, they can operate at much lower CO_2 concentrations, such that stomatal apertures may be reduced and so water is conserved.

The C_4 pathway is often regarded as an 'optional extra' to the C_3 system, and offers a clear photosynthetic advantage under conditions of relatively high photon flux density, temperature, and limited water availability, that is in tropical and mainly subtropical environments. Conversely, plants solely possessing the C_3 pathway are more advantaged in relatively temperate conditions of lower temperatures and photon flux density, and an assumed less limiting water supply (Figure 1.4).

Returning to the interaction between crop and weed, it is therefore apparent that, depending on climate, light to severe competition may be predicted. For example, a temperate C_3 crop may not compete well with a C_4 weed (e.g. sugar beet, *Beta vulgaris*, and redroot pigweed, *Amaranthus retroflexus*), and a C_4 crop might be predicted to outgrow some C_3 weeds (e.g. maize, *Zea mays*, and fat hen, *Chenopodium album*). Less competition is then predicted between C_3 crop and C_3 weeds in temperate conditions, with respect to photosynthesis alone.

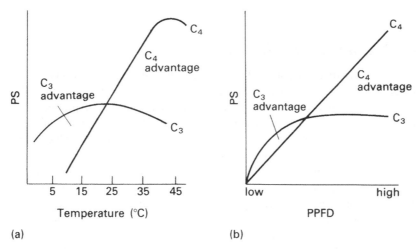

Figure 1.4 Expected rates of photosynthesis (PS) by C_3 and C_4 plants at (a) varying temperature and (b) varying photosynthetic photon flux density (PPFD). © Andy Cobb, 1992.

Table 1.10 Photosynthetic pathway of the world's ten worst weeds (from Holm *et al.*, 1977).

Latin name	Common name	Photosynthetic pathway	Number of countries where known as a weed
1. *Cyperus rotundus* (L.)	Purple nutsedge	C_4	91
2. *Cynodon dactylon* (L.) Pers.	Bermuda grass	C_4	90
3. *Echinochloa crus-galli* (L.) Beauv.	Barnyard grass	C_4	65
4. *Echinochloa colonum* (L.) Link.	Jungle rice	C_4	67
5. *Eleusine indica* (L.) Gaertn.	Goose grass	C_4	64
6. *Sorghum halepense* (L.) Pers.	Johnson grass	C_4	51
7. *Imperata cylindrica* (L.) Beauv.	Cogon grass	C_4	49
8. *Eichornia crassipes* (Mart.) Solms.	Water hyacinth	C_3	50
9. *Portulaca oleracea* (L.)	Purslane	C_4	78
10. *Chenopodium album* (L.)	Fat hen	C_3	58

In reality, C_4 weeds are absent in the UK but widespread in continental, especially Mediterranean, Europe. In the cereal belt of North America however, C_4 weeds pose a considerable problem and it is notable that eight of the world's top ten worst weeds are C_4 plants indigenous to warmer regions (Table 1.10). It will be of both interest and commercial significance if the C_4 weeds become more abundant in regions currently termed temperate (e.g. northern Europe), with the predictions of climate change.

1.5.10   Vegetative reproduction

Not all weeds classified as competitive ruderals are annuals. The exceptions are the herbaceous perennials, which have a high capacity for vegetative growth and include many of the most important weeds in the world. The vegetative production of new individuals can often be a very successful means of weed establishment. This is because the vegetative structures can rely on the parent plant for nutrients, which can confer a competitive advantage, especially at the start of the growth season. There are, however, disadvantages to vegetative reproduction. The principal ones being that since daughter plants are genetically identical to their parents they may not be well adapted to a changing environment and that widespread dispersal cannot occur by vegetative means alone, unlike with seeds. The vegetative structures themselves include stolons, rhizomes, tubers, bulbs, corms, roots and turions.

Stolons are long, slender stems that grow along the soil surface to produce adventitious roots and shoots; examples include the perennial bermuda grass (*Cynodon dactylon*), the annual crabgrass (*Digitaria sanguinalis*) and creeping buttercup (*Ranunculus repens*). Rhizomes are underground stems from which adventitious roots and shoots arise. Major examples include the perennials Johnson grass (*Sorghum halepense*), couch grass (*Elytrigia repens*), perennial sedges such as purple and yellow nutsedge (*Cyperus rotundus* and *C. esculentus*) and ground elder (*Aegopodium podagraria*). Purple nutsedge (*Cyperus rotundus*) has an extensive underground system of rhizomes and tubers. The rhizomes can penetrate and pass completely through vegetable root crops, and the tubers can remain dormant and carry the plant through very extreme conditions of drought, flooding or lack of aeration. *Cyperus rotundus* is a major weed of tropical and warm temperate regions of sugar cane, rice, cotton, maize and vegetables, groundnuts, soybeans and sorghum. Japanese knotweed (*Polygonum cuspidatum*), a highly invasive weed that has become a particular problem in many parts of the world, propagates largely by means of rhizomes (both locally and in moved soil) (Figueroa, 1989), as colonies rarely result from seed.

Tubers are enlarged terminal portions of rhizomes that possess storage tissues and axillary buds. Examples include the perennial sedges mentioned above, Jerusalem artichoke (*Helianthus tuberosus*) and the common white potato (*Solanum tuberosum*). Another particularly troublesome weed that produces tubers is the horsetail (*Equisetum arvense*). In this case, aerial shoots can be easily controlled, but deep-seated tubers will produce new shoots when conditions permit.

Bulbs are also underground organs that are modified buds surrounded by scale leaves, which contain the stored nutrients for growth, an example being wild onion (*Allium vineale*). Corms are swollen, vertical underground stems covered by leaf bases, for example, bulbous buttercup (*Ranunculus bulbosus*).

Many species produce long, creeping horizontal roots that give rise to new individuals, including perennial sow thistle (*Sonchus arvensis*), field bindweed (*Convolvulus arvensis*) and creeping or Canada thistle (*Cirsium arvense*). Some biennials and perennials form swollen, non-creeping taproots capable of regenerating whole plants. Common examples are dandelion (*Taraxacum officinale*) and curled and broad-leaved docks (*Rumex crispus* and *R. obtusifolius*). Several aquatic weeds produce vegetative buds or turions that have specialised nutrient-storing leaves or scales. These separate from the parent plant in unfavourable conditions, or are released after the decay of the parent, to remain dormant until favourable conditions return. Examples include Canadian pondweed (*Elodea canadensis*) and *Ceratophyllum demersum*.

Cultivation and soil disturbance will promote the fragmentation of all these vegetative structures. Propagation will then occur when the vegetative structure is separated from the parent plant. The brittleness of leafy parts ensures that although leaves may be removed manually or by grazing the means of vegetative reproduction remains in the soil. Only continuous cultivation will prevent the accumulation of stored nutrient reserves and so control these weeds.

1.6   A few examples of problem weeds

Black-grass (*Alopecurus myosuroides*) is of widespread distribution in Europe, temperate Asia, North America, and Australia, and has become a major problem weed where winter cereals are planted and reduced cultivation methods employed. In winter cereals about

80% of black-grass seedling emergence occurs from August to November, so crop and weed emergence coincide. The weed shows similar growth rates to the crop over the winter period, but is most aggressive from April to June with substantial grain losses being reported (Figure 1.5).

Black-grass flowers from May to August and is cross-pollinated. Seeds have short dormancy and viability (3% viable after 3 years), so ploughing, crop rotation, or spring sowing will remove the problem.

Bracken (*Pteridium aquilinum*) is a widespread, poisonous perennial weed of upland pastures and is found throughout the temperate regions of the world. This fern is thought to occupy between 3500 and 7000 km^2 in the UK alone, and may be spreading at 4% each year. It is difficult to control because of rhizomes that may grow up to 6 m away from the parent plant and are capable of rapid growth from underground apices and buds. The rhizome contains a starch reserve which acts as an energy reserve for the developing frond canopy. Translocated herbicides need to be repeatedly used if all rhizome buds are to be killed. Consequently, bracken is capable of rapid canopy establishment and is an aggressive coloniser of new areas. Indeed, some plants are estimated to be over 1000 years old. Other features that contribute to the success of this weed are its mycorrhizal roots, that ensure efficient nutrient uptake, especially in phosphate-deficient soils, and a potential production of 300 million spores per plant, which can remain viable for many years if kept dry.

Bracken creates a profound shading effect, suppressing underlying flora and gradually eliminating grass growth. Bracken also contains various carcinogens and mutagens, and is therefore poisonous to both humans and grazing animals. In addition, bracken may provide a haven for sheep ticks which can transfer numerous sheep and grouse diseases.

Cleavers (*Galium aparine*) is considered by many to be the most aggressive weed of winter cereals. It is of ubiquitous occurrence in hedgerows, and has become most invasive

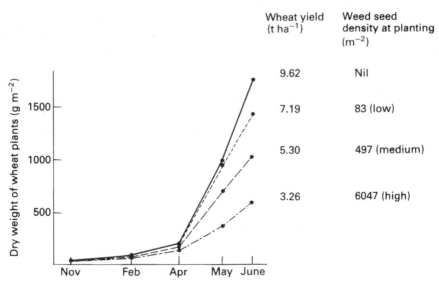

Figure 1.5   Effect of black-grass density on the growth and yield of winter wheat (from Moss, 1987).

in cereals and oilseed rape. Its climbing and scrambling habit allows it to rapidly outgrow the crop to form a dense weed canopy, eventually causing severe lodging, interference with harvesting procedures, large yield losses and severe crop contamination.

There is nowadays an increasingly widespread occurrence of crop species in succeeding crops when sown in rotation. These 'volunteer' crops include potatoes, cereals, oilseed rape, and sugar beet.

Potato 'ground keepers' are usually small tubers that are missed by the harvesters, although some are derived from true seeds. They can last several seasons, and pose a particular problem in subsequent pea and bean crops where they can only be eliminated by hand roguing. They also pose a considerable threat to the health and certification of subsequent potato crops since they can carry over pests and virus infections.

Volunteer cereals can also carry foliage diseases from one season to the next and to adjacent crops, examples being yellow rust (*Puccinia striiformis*), brown rust (*Puccinia hordei*), and powdery mildew (*Erysiphe graminis*). These cereals may become highly competitive weeds and can smother and kill young oilseed rape seedlings, for example. Even if the seedlings survive, growth is predictably stunted and winter kill more likely. Volunteer oilseed rape plants may also create an additional problem of oil purity. Since modern varieties are grown for low erucic acid and glucosinolate content, the presence of volunteer plants could severely contaminate the crop with unacceptably high levels of these compounds, which could cause the crop to be rejected.

Weed (sugar) beet has also become a serious problem in Europe, such that at least 45% of the UK crop is infested. These bolters also severely reduce crop yield.

In all cases harvesting techniques must be improved to avoid substantial seed return to the soil, and agronomic practices should be altered to reduce rapid germination of volunteers. It is also important that volunteers containing engineered resistance to herbicides are avoided at all costs. These plants would be difficult to control by chemical means, and could have serious consequences to the spread of herbicide-resistant genes in the population at large (Young, 1989).

The water hyacinth (*Eichornia crassipes* Mart. Solms) has been blamed for losing 10% of the water in the river Nile, equivalent to 7×10^9 m^3 annually, through increased transpiration, and this loss is in addition to its deleterious effects on irrigation systems, fishing activity, navigation and health (by harbouring vectors of human disease organisms) which, together with its pan-tropical spread, have earned it the name 'the million dollar weed' (Lacey, 1985).

1.7   Positive attributes of weeds

Although this section will outline the positive roles of weeds in agroecosystems, it might more properly be titled 'Positive Attributes of Non-sown Plant Species', to reflect the definition of a weed given at the beginning of this chapter.

Non-sown species of plant, whether native to a piece of land or accidentally introduced, can perform a number of very important roles. These need to be assessed prior to the implementation of any weed management practices as removal might cause more harm than good. Non-sown species have a valuable role in reducing soil erosion by water and wind. This is particularly important when a crop is small and after harvest, when erosion is likely to be more of a problem. An additional benefit is that plants growing in this situation will 'lift' and make available nutrients by absorbing them at depth through their roots and assimilating them into above-ground biomass. When the above-ground biomass dies, then nutrients are returned to the soil surface. If the non-sown species is a legume, then the added benefits of nitrogen fixation can be considered. In this way non-sown species are not only reducing erosion but also reducing nutrient leaching.

Although non-sown species may act as reservoirs and alternative hosts for pests and diseases they can also act as shelter for beneficial organisms that can contribute to biological control in crops. This shelter may be in the form of hedgerows or artificially created beetle banks, but the role of non-sown species must not be underestimated. In addition, complete removal of all non-crop species gives herbivorous organisms no choice but to eat the crop. Recent research has investigated whether organisms such as slugs might preferentially predate on non-crop species if they are present (Brooks *et al.*, 2003).

Non-sown species also contribute a major food source for birds and insects (Table 1.11), and therefore aid in the support of a biodiverse environment. This positive role must be considered alongside the negative effects of some species. Wild oat and black-grass have a negligible positive effect on biodiversity, while causing high yield losses. In such cases it is likely the weeds will be controlled. In other cases, where a positive role on biodiversity is identified, removal will depend to an extent on numbers present and financial implications. Even where a positive role has not been identified then certain species, such as corncockle and darnel, have now become very rare and conservation measures should be adopted where they are found, to reduce further decline. With an ever-growing focus on farming in an environmentally sensitive way, it is likely that there will become a greater emphasis on justifying why a non-crop species should be removed rather than justifying why it should remain.

Allelopathy, as mentioned in section 1.4.1, is usually used to describe the negative effect of one plant on another via the release of natural growth inhibitors. However, incidences of positive allelopathy have been reported, where allelopathic chemicals produced by one species have a positive effect on another species. An example is corncockle (now a rare arable plant) that grows alongside wheat. Corncockle produces agrostemmin which increases the yield and the gluten content of the wheat (Gajic and Nikocevic, 1973)

1.8   The ever-changing weed spectrum

Weed populations are never constant but are in a dynamic state of flux due to changes in climatic and environmental conditions, husbandry methods and the use of herbicides.

Such change is evident from a consideration of cereal production in the UK. Cereals were traditionally sown in the spring, but MAFF (Ministry of Agriculture, Fisheries and Food) statistics show that by 1985, 75% of the crop was sown in the autumn (Martin, 1987) with major changes in the weed flora. In the 1950s and 1960s the greater area of spring cereals encouraged spring-germinating broadleaf weeds such as knotgrass (*Polygonum aviculare*), redshank (*P. persicaria*), black bindweed (*Bilderdykia convolvulus*), poppy (*Papaver rhoeas*), charlock (*Sinapis arvensis*), chickweed (*Stellaria media*) and fat hen (*Chenopodium album*). The move to winter cereals has meant less competition from these spring weeds because the crop is already established. However, winter cereals have encouraged weeds that germinate and establish in the autumn and early spring,

Table 1.11 Ranking of the competitive effects of selected weed species and their value for birds and invertebrates[a] (Lutman et al., 2003).

Species	Competitive index	Value for birds	Value for insects
Alopecurus myosuroides	***		−
Avena fatua	****	−	−
Lolium multiflorum	****		
Poa annua	**	**	***
Aethusa cynapium	**		−
Anagallis arvensis	*		−
Aphanes arvensis	*		
Brassica napus	***	**	−
Chenopodium album	**	***	***
Cirsium spp.	***	*	***
Convolvulus arvensis	***		
Epilobium spp.	**		
Fallopia convolvulus	*	***	
Fumaria officinalis	**	*	−
Galium aparine	****	−	***
Geranium spp.	**		−
Lamium purpureum	**	−	**
Legousia hybrida	*		
Myosotis arvensis	**	−	−
Papaver spp.	***		*
Persicaria maculosa	**	***	**
Polygonum aviculare	**	***	***
Ranunculus spp.	**		
Senecio vulgaris	**	**	***
Sinapis arvensis	***	**	***
Sonchus spp.	**	*	***
Stellaria media	**	***	***
Tripleurospermum inodorum	***		***
Veronica spp.	**		−
Viola arvensis	*	**	−

[a] The number of asterisks refers to the species increased importance to birds/invertebrates or increasing competitive impact; '−' = no importance, blank = no information).

particularly cleavers (*Galium aparine*), speedwells (*Veronica* spp.), chickweed (*Stellaria media*) and field pansy (*Viola arvensis*). Similarly, traditional cereal cultivation techniques based on ploughing have given way to direct drilling and minimal cultivations using tines and discs which are less costly and energy intensive. Indeed, Chancellor and Froud-Williams (1986) have estimated that 40% of arable land in the UK is now cultivated without ploughing. One consequence is a reduction in the density of many annual broad-leaf weeds since they require deep soil disturbance to bring buried seeds to the surface; a lack of seed return due to herbicide use has also encouraged this decline. On the other hand, minimum cultivation has led to an increased abundance of annual grasses, particu-larly of species which readily establish near the soil surface and which have relatively short periods of dormancy. The main examples are black-grass (*Alopecurus myosuroides*),

Table 1.12 Main broadleaf weeds and grass weeds present in 2,359 winter cereal fields (from Whitehead and Wright, 1989).

Broadleaf weeds	Fields infested (%)	Grass weeds	Fields infested (%)
Stellaria media	94	Poa annua	79
Veronica persica	72	Avena spp.	42
Matricaria spp.	67	Alopecurus myosuroides	38
Galium aparine	58	Elytrigia repens	21
Lamium purpureum	47	Lolium spp.	14
Viola arvensis	45	Anisantha sterilis	13
Sinapis arvensis	36	Poa trivialis	7
Veronica hederifolia	30	Volunteer cereals	7
Capsella bursa-pastoris	23		
Volunteer oilseed rape	23		
Papaver rhoeas	18		
Fumaria officinalis	17		
Chenopodium album	13		
Aphanes arvensis	12		
Geranium spp.	11		

meadowgrass (*Poa* spp.) and sterile brome (*Anisantha sterilis*). Wild oats (*Avena fatua*) have also flourished in minimum cultivation, even though their seed requires moderate burial. These observations have been quantified in a major, large-scale survey of weeds present in winter cereals in the UK (Table 1.12).

The frequency of creeping perennials has also increased with minimal cultivation, for example field bindweed (*Convolvulus arvensis*) and Canada thistle (*Cirsium arvense*). Chancellor and Froud-Williams (1986) also point to the occurrence of unusual species in undisturbed arable land, particularly wind-dispersed seeds of the genera *Epilobium*, *Artemisia*, *Conyza* and *Lactuca*, and suggest that these may be the problem weeds of the future. Current increases in minimum and zero cultivation methods for establishment of cereal and oilseed crops, for environmental and financial reasons, will undoubtedly be mirrored by further changes in the weed spectrum of arable land.

Herbicide choice and use has also had a profound effect on the weed flora in cereals (Martin, 1987). The use of 2,4-D and MCPA since the late 1940s has caused a decline in many susceptible weeds, such as charlock (*Sinapis arvensis*) and poppy (*Papaver rhoeas*), although more tolerant species, including chickweed (*Stellaria media*), knotgrass (*Polygonum aviculare*) and the speedwells (*Veronica* spp.), have prospered. The introduction of herbicide mixtures in the 1960s with, for example, mecoprop and ioxynil, produced a much wider spectrum of weed control. Other herbicides have since been developed for an autumn-applied, residual action, including, for example, chlorsulfuron and chlorotoluron for improved control of grasses, and new molecules for the control of specific weeds, for example fluroxypyr for cleavers (*Galium aparine*). However, the same principles will always apply, namely that the selection pressure caused by sustained herbicide use will allow less-susceptible weed species to become dominant, and their continued use may encourage the selection of herbicide-resistant individuals within a species, as has now occurred to many herbicides (see Chapter 12).

Climate change is predicted to have significant effects on both the geographical distribution of weeds and on the severity of weed infestations. Evolutionary rate (for instance, in the development of herbicide resistance) has been demonstrated to vary dependent upon both temperature and moisture availability. This is probably a result of a combination of factors including generation time, population size and relative fitness of herbicide-resistant individuals. All of these factors may be affected by increased average global temperatures and subsequent differences in regional weather patterns (Anon., 2000). In addition, milder winters and warmer summers may allow for survival and population growth of species that were previously unsuited to a region's climate. Increases in the occurrence of *Phalaris* grasses in the UK in recent years may be a direct result of this (A.H. Cobb and J.P.H. Reade, personal observations).

1.9   Weed control

It is outside of the scope of this book to examine the finer details of weed control. Instead, a broad overview is presented and the reader is referred to other more comprehensive texts for further information such as Naylor (2002).

According to Lacey (1985), weed control encompasses:

(a)   the reduction of the competitive ability of an existing population of weeds in a crop;
(b)   the establishment of a barrier to the development of further significant weeds within that crop; and
(c)   the prevention of weed problems in future crops, either from the existing weed reservoir or from additions to that weed flora.

The first two objectives are met primarily by chemical means, and the third relies on agronomy and crop husbandry. Cultural practices are forever changing, along with the weed spectrum, and it is now increasingly recognised that an integrated approach utilising both cultural and chemical practices is necessary for optimal weed control.

1.9.1   Traditional methods

It was recognised in medieval times that the rotation of crops with fallow was the best means to conserve soil fertility and to prevent the build-up of pests, diseases and weeds. The later use of 'cleaning' crops (such as turnips and potatoes) allowed weed control by hand during active growth, and was balanced against 'fouling' crops (such as cereals) which could not be similarly weeded. By the mid-nineteenth century fertility was maintained from clover and livestock manure, and weed control by 'cleaning' crops, so that the unprofitable fallow period could be avoided. The advent of chemical fertilisers in the early twentieth century removed the need for clover, and profitability increased by the use of sugar beet as a combined cleaning and 'cash' crop. However, after the Second World War, increased urbanisation and industrialisation reduced the available workforce, and herbicides have gradually replaced the hoe. Similarly, farm practices have become increasingly mechanised, such that the continuous cultivation of one crop (monoculture) has become widespread, and reduced cultivation techniques are now in vogue.

1.9.2 Chemical methods

Chemical weed control is a twentieth-century technology. Copper sulphate was the first chemical used at the turn of the twentieth century to control charlock (*Sinapis arvensis*) in oats, and soon after came corrosive fertilisers (such as calcium cyanamide) and industrial chemicals (including sodium chlorate and sulphuric acid). Modern synthetic herbicides first appeared in France in 1932 following the patenting of DNOC (4,6-dinitro-*o*-cresol) for the selective control of annual weeds in cereals. Further dinitro-cresols and dinitro-phenols soon appeared, but these compounds had variable effectiveness and appeared to kill animals as well as plants. The discovery of the natural plant growth 'hormone' auxin in 1934 led to the further discovery of the synthetic growth regulators 2,4-D and MCPA based on phenoxyacetic acid chemistry. These compounds were the first truly selective herbicides that could reliably kill broad-leaved weeds in cereal crops, and they developed widespread popularity and use after the Second World War (Kirby, 1980). These compounds truly 'replaced the hoe' so that cereals could no longer be regarded as 'fouling' crops, and paved the way to the current practice of cereal monocultures.

Since the 1950s an increasing proportion of world cereal crops have become regularly treated with agrochemicals to achieve the control of an ever-widening variety of weeds. Nowadays chemical weed control has expanded to probably every crop situation in the world. Modern chemical weed control is not only more economic than traditional methods, but also has important technical advantages as weeds growing closest to the crop – and hence competing most for resources – can be controlled by selective herbicides. Furthermore, less crop-root disturbance is evident than with mechanical hoeing and fewer, if any, weed seeds are brought to the surface in the process. Finally, farmers now have chemical answers for most weed problems at a reasonable price.

1.9.3 An integrated approach

The development of Integrated Crop Management (ICM) practices means that Integrated Weed Management (IWM) systems have been developed that also embrace environmental and financial factors. IWM systems need to be effective enough for long-term maintenance of natural resources and agricultural productivity and also to have minimal adverse environmental impact combined with adequate economic returns to the farmer.

Key aspects of IWM systems include prevention of weed infestation, identification of weed species that are present, mapping and monitoring weed populations, prioritisation of management, management using a combination of mutually supportive techniques (manual, mechanical, cultural, biological and chemical methods, and evaluation of their success combined with documentation and perseverance.

Through the use of such systems it is hoped that weed management may be carried out in a sustainable manner, giving protection to both financial returns and to agroecosystems.

References

Anon. (2000) *Climate Change and Agriculture in the United Kingdom*. London: MAFF.
Baker, H.G and Stebbins, G.L. (1965) *The Genetics of Colonising Species*. New York: Academic Press.

Brooks, A., Crook, M.J., Wilcox, A. and Cook, R.T. (2003) A laboratory evaluation of the palatability of legumes to the field slug, *Deroceras reticulatum* Müller. *Pest Management Science* **59**(3), 245–251.

Chancellor, R.J. and Froud-Williams, R.J. (1986) Weed problems of the next decade in Britain. *Crop Protection* **5**, 66–72.

Dexter, A.G., Rothe, I and Luecke, J.L. (1999) Weed Control in Roundup Ready™ and Liberty Link™ Sugarbeet. In: *30th General Meeting of American Society of Sugar Beet Technologists* Abstracts, p. 9.

Figueroa, P.F. (1989) Japanese knotweed herbicide screening trial applied as a roadside spray. *Proceedings of the Western Society of Weed Science* **42**, 288–293.

Gajic, D and Nikocevic, G. (1973) Chemical allelopathic effects of *Agrostemma githago* upon wheat. *Fragm. Herb. Jugoslav* XXIII.

Grime, J.P. (1979) *Plant Strategies and Vegetation Processes*. London: Wiley.

Grundy A.C and Jones N.E. (2002) What is the Weed Seed Bank? In: Naylor, R.E.L. (ed.) *Weed Management Handbook*, 9th edn. Oxford: Blackwell Publishing / BCPC.

Hanf, M. (1983) *The Arable Weeds of Europe with Their Seedlings and Seeds*. Hadleigh, Suffolk: BASF United Kingdom Ltd.

Hance, R.J. and Holly, K. (1990) *Weed Control Handbook: Principles*, 8th edn. Oxford: Blackwell.

Hill, T.A. (1977) *The Biology of Weeds*. London: Edward Arnold.

Holm, L.G., Plucknett, D.L., Pancho, J.V and Herberger, J.B. (1977) *The World's Worst Weeds. Distribution and Biology*. Hawaii: University Press.

Inderjit, Callaway, R.M and Vivanco, J.M. (2007) Can plant biochemistry contribute to understanding of invasion ecology? *Trends in Plant Science* **11**(12), 574–580.

Kirby, C. (1980) *The Hormone Weed Killers: a Short History of Their Discovery and Development*. Croydon and Lavenham, UK: BCPC Publications / Lavenham Press.

Lacey, A.J. (1985) Weed control. In: Haskell, P.T. (ed.) *Pesticide Application: Principles and Practice*. Oxford: Oxford University Press, pp. 456–485.

Lampkin, N. (1990) *Organic Farming*. Ipswich, UK: Farming Press.

Lutman, P.J., Boatman, N.D, Brown V.K. and Marshall, E.J.P. (2003) Weeds: their impact and value in arable ecosystems. In: *The Proceedings of the BCPC International Congress: Crop Science and Technology 2003* **1**, 219–226.

Martin, T.J. (1987) Broad versus narrow-spectrum herbicides and the future of mixtures. *Pesticide Science* **20**, 289–299.

Moles, A.T., Gruber, M.A.M. and Bonser, S.P. (2008) A new framework for predicting invasive plant species. *Journal of Ecology* **96**(1), 13–17.

Moss, S.R. (1987) Competition between blackgrass (*Alopecurus myosuroides*) and winter wheat. *British Crop Protection Conference, Weeds* **2**, 367–374.

Naylor, R.E.L. (ed.) (2002) *Weed Management Handbook*, 9th edn. Oxford: Blackwell Publishing/ BCPC.

Naylor, R.E.L. and Lutman, P.J. (2002) What is a weed? In: Naylor, R.E.L. (ed.) *Weed Management Handbook*, 9th edn. Oxford: Blackwell Publishing / BCPC.

Ødum, S. (1965) Germination of ancient seeds – floristical observations and experiments with archaeologically dated soil samples. *Dansk Botanisk Arkiv* **24**(2), 1–70.

Oerke, E.C., Dehne, H.W., Schonbeck, F. and Weber, A. (eds) (1995) *Crop Production and Crop Protection: Estimated Losses in Major Food and Cash Crops*. Amsterdam: Elsevier.

Olofdotter, M. and Mallik, A.U. (2001) Allelopathy Symposium: introduction. *Agronomy Journal* **93**, 1–2.

Patterson, D.T. (1985) Comparative ecophysiology of weeds and crops. In: Duke, S.O. (ed.), *Weed Physiology*, Vol. I: *Reproduction and Ecophysiology*, Boca Raton, FL: CRC Press, pp. 101–130.

Radosevich, S.R. and Holt, J.S. (1984) *Weed Ecology: Implications for Vegetation Management.* New York: Wiley.

Rodgers, E.G. (1978) Weeds and their control. In: Roberts, D.A. (ed.) *Fundamentals of Plant Pest Control.* San Francisco, CA: W.H. Freeman. pp. 164–186.

Shaw, D.S. and Tanner, R. (2008) Invasive species: weed like to see the back of them. *Biologist* **55**(4) 208–214.

Tanner, R. (2008) A review on the potential for the biological control of the invasive weed, *Impatiens glandulifera* in Europe, In: Tokarska-Guzik, B., Brock, J.H., Brundu, G., Child, L., Daeler, C.C. and Pyšek, P. (eds) *Plant Invasions: Human Perception, Ecological Impacts and Management,* Leiden: Backhuys Publishers, pp. 343–354.

Toole, E.H. and Brown, E. (1946) Final results of the Duvel buried seed experiment. *Journal of Agricultural Research* **72**, 201–210.

Whitehead, R. and Wright, H.C. (1989) The incidence of weeds in winter cereals in Great Britain. *Brighton Crop Protection Conference, Weeds* **1**, 107–112.

Wilson R.G. and Smith J.A. (1999) Crop Production with glyphosate tolerant sugarbeet. In: *30th General Meeting of American Society of Sugar Beet Technologists,* Abstracts p. 50.

Young, S. (1989) Wayward genes play the field. *New Scientist* **123**, 49–53.

Chapter 2
Herbicide Discovery and Development

2.1 Introduction

Most crop protection problems can now be solved using chemicals, at a reasonable price to the farmer and grower. The use of herbicides in the developed countries has been particularly successful, with an estimated near-maximum market penetration in major crops (e.g. 85–100% in soybean, maize, and rice crops in the USA; Finney, 1988). The reason for this high usage is that it has proved financially rewarding to the farmer. Indeed, estimates suggest that each dollar spent in the USA on pesticides results in an additional income of four dollars to the farmer. Similarly, a pound on pesticides may generate an additional five pounds in the UK. However, reduced subsidies and lower farm incomes have led to lower chemical inputs by farmers, so that the growth rate of the herbicide market has declined in recent years. Also apparent has been an increase in price competition between companies, particularly with mixtures of chemicals whose patents have expired. Farmers, growers and consumers have all benefited from increased competition within the agrochemical industry, since chemical prices have dropped in real terms, while at the same time old or unsound products have been replaced by safer, more environmentally acceptable, lower dosage compounds. Furthermore, it should not be forgotten that expenditure on pesticides has not only proved financially rewarding to producers, but has also increased crop yield and quality, and contributed to self-sufficiency in many crops, particularly in Europe.

While the current armoury of agrochemicals is under threat by the 'green' lobby and the European Union (EU) legislature which has banned the use of many herbicides, the benefits of agrochemical use must not be forgotten. The judicious use of agrochemicals ensures the sustainable yields typical of conventional agriculture with the high quality and safety of agricultural products expected by the consumer. In this context, it is worth noting the study of Oerke and colleagues in 1994. They concluded that when no weed, disease or pest control was practised, 70% losses of potential yields of our major crops (rice, wheat, barley, maize, potato, soybean, cotton and coffee) could be predicted. With the use of agrochemicals this loss was estimated at 42%. Thus, the development of new crop protection agents and the more effective use of existing ones are important global priorities if we are to feed the world.

Herbicides and Plant Physiology, Second Edition By Andrew H. Cobb and John P.H. Reade
© 2010 A.H. Cobb and J.P.H. Reade

2.2   Markets

The agrochemical industry, based on the production of chemicals for crop protection, largely came into being after the Second World War with the commercialisation of the first truly selective broadleaf weed herbicides: 2,4-D (1945) and MCPA (1946). These non-toxic molecules were effective at low doses and were cheap to produce. Furthermore, they became available when maximum food production was essential and farm labour was scarce. Their success stimulated European and North American chemical companies to invest in research that led to the discovery of the wide range of herbicides now available. Early successes created a market value approaching US$3 billion in 1970 and an average of 6.3% real growth per annum was recorded over the following decade. By 1980 it had slowed to 4.5% per annum, averaged 2.2% growth during the 1980s and was predicted to average below 2% for the 1990s. By 1998 the market had become static, with only 0.1% real growth, but one still worth US$31 billion.

While global agrochemical sales rose by 3.8% in 2005, the market remained flat in real terms, after accounting for inflation and currency differences (Table 2.1) (Agrow, 2006). Comparison with 1988 data (cited by Cobb, 1992) shows that since 1998–2005, the fungicide share of the market has risen from 20.5% to 23.0%, herbicides from 43.6% to 45.8%, while the insecticide share has fallen from 29.7% to 26.3% (Table 2.1).

Real growth in recent times is noted in North America, Northern and Eastern Europe, and in China, although the market is truly global (Table 2.2).

Table 2.1   World agrochemical market (from World Agricultural Market, Agrow, 2006).

	Percentage of market (%)	Value (US$ billion)
Herbicides	45.8	15,389
Insecticides	26.3	8,387
Fungicides	23.0	7,728
Others (including plant growth regulators, nematicides, fumigants and biopesticides)	4.9	1,646
Totals	100.0	33,600

Table 2.2   World agrochemical sales (%), by region (from World Agricultural Market, Agrow, 2006).

Europe	26.0
North America	25.6
Asia	24.3
Latin America	16.8
Rest of the world	7.3
Total	100.0

What has caused this slowdown in market growth? Five contributing factors have emerged. First, past successes in chemical crop protection have led to a near-maximum market penetration in all the major crops, especially in Western Europe and the USA. Thus, there is an increasingly competitive market place for agrochemical companies to operate in. Second, new active ingredients are taking longer to discover, develop and register. Stricter legislative requirements have also led to delays in returns from investments, so that product profitability has declined. Indeed, estimates of at least US$250 million are often given for the cost of bringing a new product to the market (Figure 2.1). Third, the agrochemical arsenal is becoming increasingly mature as fewer examples of new chemistry acting at novel target sites are reported. Fourth, due to the successes of modern intensive agriculture in recent decades, there has been an overcapacity in farming and a marked decline in commodity prices. Consequently, sales of agrochemicals have declined. Fifth, the growth of the 'green' lobby and the more recent introduction of genetically modified (GM) crops have led to increasing consumer opposition to agrochemical use.

Consequently, the agrochemical industry itself has contracted significantly in recent years. The following chemical companies were involved in discovery research and development in the 1980s, but are no longer active: Celamerck, Chevron, Diamond Shamrock, Dr Maag, Duphar, Mobil, PPG Industries, Shell, Stauffer, 3M, Union Carbide and Velsicol. Mergers in the 1990s saw the agrochemical interests of Schering and Hoechst form AgrEvo, which in 1999 itself merged with Rhône-Poulenc to form Aventis, now a part of Bayer CropScience. In the USA, Dow and Eli Lilly formed DowElanco, now Dow AgroSciences, and the Swiss companies Ciba Geigy and Sandoz merged to become Novartis, which has now merged with Zeneca to form Syngenta, creating the biggest player in the crop protection industry with combined sales of over US$8 billion and a quarter of the global market.

The agrochemical industry is now dominated by six major multinational companies: Syngenta, Bayer, Monsanto, Du Pont, BASF and Dow, which invest between 8 and 11% of their sales in Research and Development including the search for new active ingredients. However, Monsanto has recently ceased this activity, preferring to focus

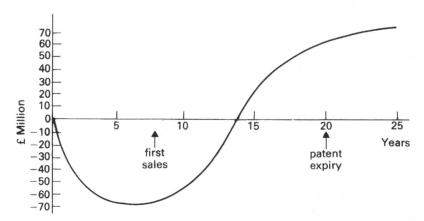

Figure 2.1 Typical cumulative cash flow for a successful new herbicide (updated, from Finney, 1988 and Rüegg et al., 2007).

on the use of glyphosate in genetically modified (GM) crops. Further consolidation is predicted as other global players consider offloading their agrochemical divisions to concentrate on what is perceived to be the more lucrative and buoyant pharmaceuticals market. Only time will tell how many agrochemical companies will survive this new wave of restructuring within the industry. What is clear is that the survivors will be those who have adapted best to change by embracing new technologies and innovative marketing, while at the same time providing competitively priced products that enable profitable farming.

All is not doom and gloom, however, and a significant market for crop protection products awaits the slimmed-down industry. There are expanding agricultural markets in Eastern Europe and Asia to be penetrated and there is much new technology to exploit. For example, the precise delivery of more optimally formulated agents will lead to lower chemical inputs into the environment, and recent developments in agricultural biotechnology are already changing both farming practices in general and agrochemical use in particular. Indeed, the ability to transfer genes for favourable traits to crops by recombinant DNA technology will undoubtedly transform agriculture and crop protection into totally new directions in the next decade.

While the first decade of GM was focused on the gene products controlling weeds (e.g. glyphosate resistance) and insects (via the BT toxin), the next decade is predicted to identify perhaps hundreds of genes that may secure increased yields in our key crops. Furthermore, it is interesting to note that in 2007 BASF and Monsanto agreed to collaborate to fund the search for stress-tolerant strains of maize, soybean, cotton and oilseed rape using GM technologies. The companies agreed a joint budget of €1.2 billion to find unique genes for stress-tolerant traits (Evans, 2010).

2.3 Prospects

Examination of the 2003 wheat crop shows the huge potential in expansion in eastern and central Europe and the former Soviet Union. These regions accounted for 57.8 Mha of wheat, 27% of the world area for that year, with yields of about 2.2 t ha^{-1}, producing 22.5% of the world's wheat. This is in comparison with average UK yields of about 8 t ha^{-1}. Access to more modern agrochemical products and fertilisers should profoundly increase these yields in years to come.

Equally, China, with a population of 1.3 billion and an ever-expanding agricultural product demand, is now probably the second largest producer of pesticides in the world, exporting almost $1.2 billion in agrochemical products in 2004.

Of the top ten global active ingredients, it is noteworthy that seven are herbicides, dominated by glyphosate, far and away the best selling agrochemical, while the most recently introduced product was fenoxaprop in 1984 (Table 2.3).

It is likely that sales of atrazine will decline, especially in Europe, to be replaced by, for example, mesotrione in maize. The increasing use of the more recently introduced strobilurin fungicides and the neonicotinoid insecticides will surely be recognised in any future version of Table 2.3.

In recent years the rate of adoption of transgenic crops has been particularly rapid, with the effect, for example, of displacing the soybean herbicide market to a reliance on one

Table 2.3 Top ten post-patent active ingredients ranked by sales in 2003.

	Active ingredient	Type	Sales in 2003 (US$ million)	Debut year
1	Glyphosate	(H)	2,933	1974
2	Paraquat	(H)	375	1962
3	Chlorpyrifos	(I)	350	1965
4	Metolachlor	(H)	295	1975
5=	2,4-D	(H)	290	1945
5=	Mancozeb	(F)	290	1943
7	Atrazine	(H)	260	1957
8	Fenoxaprop	(H)	255	1984
9	Pendimethalin	(H)	225	1976
10	Chlorothalonil	(F)	210	1963

Key: H, herbicide; I, insecticide; F, fungicide.

chemical alone, glyphosate. Indeed, it is estimated that about half of the 72 million acres of soybeans sown in the USA in 1999 were tolerant to glyphosate. Sales of glyphosate reportedly rose by 25% in 1998 and other companies have responded by cutting costs of their soybean herbicides to compete. In addition, as the patent for glyphosate, owned by Monsanto in the USA, expired in 2000, many other companies have developed their own formulations to compete with existing products. Further developments in the global glyphosate market are likely. Furthermore, in 1999 transgenic versions of all major crops for tolerance to herbicides were sown in 70 million acres (28.5 million hectares) in the USA. In 2006, 102 million hectares of land was used to grow GM crops in the world, over 50% in the USA. According to Evans (2010), at the end of February 2010, 14 million farmers in 25 countries planted 134 million ha of GM crops in 2009, of which 13 million farmers were in developing countries. Thus, the major agrochemical companies have recently expanded their life science businesses or acquired new seed companies to exploit these major technological developments, for example with herbicides.

Transgenic versions of monocotyledon crops, especially wheat and rice, are reportedly in development, although the technological challenges are greater. The next generation of transgenic crops will feature aspects of crop quality as well as crop protection, with a projected market value of US$5 billion in 2020 (Thayer, 1999). Many improved agronomic traits have been developed in the laboratory and only time will tell which ones are adopted commercially. The list is almost endless (Evans, 2010). Protection of crops from insects and diseases, in addition to tolerance to herbicides, is well under way, and major advances have been reported in our understanding of drought, cold and salt tolerance at the molecular level. Enhanced antioxidant, fibre and lignin production, altered amino acid and lipid content and elevated starch synthesis have all been achieved in various crops in recent years. Such traits, coupled with, for example, herbicide tolerance, are seen as ways of increasing agricultural productivity in an environmentally benign fashion.

Acceptance of this technology to date by farmers has been widespread in the USA, Argentina and Canada, and markets are increasing in Australia, China, Mexico and South Africa. Consumer concerns are evident in Europe, however, and moratoria on commercial sales of transformed crops are likely to delay their introduction.

Without doubt, there are major reservations in Europe about the developments of GM crops. Consumers are especially concerned about food safety issues, labelling of products as containing ingredients from GM crops, the perceived problems of transferring genes from one organism to another, the concept of escape of resistant genes to other organisms, perceived threats to biodiversity, the risk of widespread antibiotic resistance and simply the fear of the unknown. Thus, the role of the scientific community must be not to simply list the benefits of this brave new botany, but also to give a balanced view of all the above concerns with impartial statements based on facts and sensitivity to opposing views.

An 'all or nothing' debate on GM products, as promulgated by some pressure groups is sterile. Polarised views only serve to confuse the public when they are attempting to take an informed position on this issue. We must surely accept that there are major ethical and environmental concerns about GM foods, and that there are varying degrees of acceptability to all parties. Thus, the true test of consumer commitment to GM and GM-free foods is yet to come. Will the consumer pay for a premium for a chemical-free and GM-free product, verified by detailed analytical testing, or in 2020 will we look back to the debate and wonder what the fuss was all about?

Agrochemical companies have invested heavily to acquire seed companies specifically to produce and market transgenic crops, and have developed new biotechnology ventures to keep up with new developments in this rapidly expanding field. Strong sales growth has been predicted for herbicide-tolerant crops. Will these forecasts be realised? Only time will tell. What is certain is that agricultural biotechnology is here to stay and will transform the crop protection industry in ways that we are only now beginning to comprehend.

2.4   Environmental impact and relative toxicology

The environmental impact and toxicological considerations of new herbicide development are coming increasingly to the fore. Nowadays, before its commercial release, a new product has to satisfy an extensive battery of tests to determine potential toxicity to a wide range of organisms, effects on key environmental functions, and its movement and fate within the environment. It is of interest to compare some current criteria for environmental acceptability with those tests considered necessary in previous decades (Table 2.4).

The registration and approval of new agrochemicals in Europe is both a time-consuming and complex process. It is governed by the European Commission (EC) Directive 91/414/EEC, which harmonises the national product approval requirements throughout Europe and consists of two main phases. In the first, the active ingredient must be approved at EU level and in the second, formulations appropriate to national markets have to be registered by European member states. The European public should be reassured that only active ingredients demonstrated to be without risk to human or animal health or to the environment will be allowed to be used and listed in the Directive.

The whole process of product registration now has three stages, namely:

1   generation of a data package for the active ingredient,
2   preparation and submission of a dossier to the member state, and
3   assessment and approval by the member state and the European Food Safety Authority.

Table 2.4 Typical tests required for the environmental acceptability of pesticides (from Graham-Bryce, 1989).

1950s	1980s
Acute toxicity to bees, fish and birds	Acute toxicity to 3 species of fish; effects on fish growth and bioaccumulation. Acute toxicity to algae, aquatic invertebrates (4 species), bees and birds (2 species.) Reproduction study in *Daphnia*. Dietary toxicity to birds. Effects on soil fauna including earthworms and rates of leaf litter decomposition. Effects on nitrification and carbon dioxide evolution from soils. Rates of biodegradation. Physicochemical properties.

Acquiring the data package can take four to five years and cost about €10 million. Approximately 250 tests need to be conducted on a new active ingredient in the areas of quality, safety and efficacy. **Quality** involves testing the behaviour of the active ingredient (ai) and formulated products, including physicochemical properties, storage stability and methods for analysis. **Safety** involves all aspects of toxicity studies including mammalian toxicity (details of metabolism, acute and chronic toxicity, genotoxicity, developmental and reproductive effects), residues (plant and livestock metabolism), crop and livestock trials performed according to Good Agricultural Practice, environmental fate (behaviour in soil, water and air especially regarding metabolites) and ecotoxicity (acute, short-term and reproductive effects in birds, aquatic organisms and non-target organisms, such as bees, beneficial insects and earthworms). **Efficacy** entails many field trials using the proposed formulated product on many crops in defined climatic areas. These studies determine the proposed label recommendations and the good agricultural practice for the product.

The preparation and submission of the dossier can take a year for a new active ingredient and includes study summaries, risk assessments and efficacy of formulations and co-formulations, that is, mixtures with other components. The dossier needs to be prepared to a set format (Guideline Document 1663 / VI / 94 Rev 8) and in accord with the Organisation for Economic Co-operation and Development (OECD) format (Rev 1, 2001), which means that the dossier should be acceptable by the EU, the USA, Canada, Japan and Australia. The full dossier can account to 60,000 pages.

The dossier then undergoes a complete check by the Member State – that may last a further year, although provisional approval product registration may be granted for a period of 3 years. A draft assessment report by the Member State is then reviewed by the European Food Safety Authority, which can take another year. A positive recommendation is then reviewed by the EC Standing Committee on the Food Chain and Animal Health and either endorsed, changed or rejected. This process can take a further six months. Next, the active ingredient needs to be authorised in every Member State where the product will be sold, taking into account local climate, cropping patterns and diet. This National Approval is only applicable to the formulated product and can take a further year. Finally, registration lasts for a maximum of 10 years, although active ingredients can be reviewed at any time.

New EU legislation (EU No. 1907/2006) concerning the registration, evaluation and restriction of chemicals (REACH) looks likely to further reform the way agrochemicals are managed in Europe. REACH regulations (2007) require that approximately 30,000 chemicals produced and marketed in the EU in amounts above 1 tonne per annum will require a full technical dossier. Chemicals used above 10 tonnes per annum will also require a chemical safety assessment. Thus 'old' chemistry will become subject to increasingly rigorous and demanding testing prior to use.

This legislation has been very controversial. On the one hand, some pressure groups would prefer a total ban on untested chemicals, while Trades Unions are positive that better Health and Safety procedures will result to protect their members. Then again, REACH is perceived by some as a threat to the European chemicals industry, ensuring an exodus of business to the less well-regulated nations, such as the USA, China and India. Clearly, the issue of substituting 'old' chemistry with 'new' products having a fully approved dossier will tax the agrochemicals industry for some time to come.

The selective toxicity of new herbicidal molecules has also become increasingly important in recent years. Successful products are expected to potentially inhibit target weed processes with negligible risk to the crop or other organisms. Hence, most new herbicides have LD_{50} values for acute toxicity to rats in excess of 2,000–3,000 mg of active ingredient per kilogram of body weight (Table 2.5) and are therefore less toxic than aspirin, caffeine, nicotine, or even table salt.

It is a further requirement for pesticide registration that residues are routinely measured both in foodstuffs and in the general environment. This includes when a crop is harvested and after storage, and in animal products such as milk and meat. Results of

Table 2.5   Toxicity of some herbicides and common chemicals to rats (updated after Graham-Bryce, 1989).

	Acute oral LD_{50} to rats (mg kg^{-1})
Herbicides	
Chlorotoluron	>10,000
Asulam	>5,000
Imazapyr	>5,000
Sulfometuron-methyl	>5,000
Glyphosate	4,320
Other chemicals	
Table salt	3,000
Aspirin	1,750
Bleaching powder (hypochlorite)	850
fluoride toothpastes	52–570
Shampoo (zinc pyrithione)	200
Caffeine	200
Nicotine	50

such residue analyses have lately created some debate, and much confusion, both in the popular press and in the minds of many consumer groups. Indeed, the proliferation of references to 'organically grown' or 'pesticide-free' produce is evidence of an increasingly held belief that all pesticides are harmful to life even at 10^{-15} g! Pesticide registration authorities must therefore attempt to regularly define safe limits for each product and so convince the consumer of product safety or otherwise. The definition of a reasonable versus an unreasonable risk to the consumer, however, has become a legal and scientific minefield. The maximum permitted amount of residual pesticide is determined from three factors, namely:

(a)   the smallest dose in parts per million (ppm) to produce detectable harmful effects in laboratory animals (i.e. Acute Reference Dose, ARfD);
(b)   a safety factor, which should be large enough to compensate between humans and test animals, usually 100; and
(c)   a food factor, based on the proportion of the particular food in an average diet, (i.e. Acceptable Daily Intake, ADI).

In this way, a minimum harmful dose of, say, 10 ppm, a safety factor of 100 and a food factor of 0.2 will result in a maximum permitted value or Maximum Residue Level (MRL) of 0.5 ppm (i.e. 10 ÷ 100 × 2). Extensive and regular monitoring of foodstuffs suggests that if manufacturers' recommendations for application are followed, then risks to the consumer are minimal or negligible. Constant vigilance is required by both governments and producers to ensure that pesticide use is regularly scrutinised for both real and imaginary hazards, and so allay unnecessary public anxiety.

Media coverage in recent years has suggested that pesticide residues in the human diet constitute an unacceptable risk and, simply, that any detectable residue is too high and potentially carcinogenic. This perceived hazard of pesticide use, however, seldom takes into account the presence of natural toxins in our food. Ames and colleagues (1990) have compared the abundance and toxicity of natural toxins with synthetic pesticides and have concluded that 99.99% of our dietary toxin intake is from natural foods. They have estimated that Americans consume 1.5 g of natural toxins each day in roasted coffee, potatoes, tomatoes, whole wheat, brown rice and maize, which is about 10,000 times more than the amount of pesticide residues consumed. Furthermore, surprisingly few of these natural toxins have been tested for carcinogenicity. Thus, Ames *et al.* (1990) concluded that the health hazards of synthetic pesticide residues are insignificant when compared to human exposure to natural toxins in our diet.

2.5   The search for novel active ingredients

The ideal herbicide should:

- be highly selective to plants and non-toxic to other organisms,
- act quickly and effectively at low doses,
- rapidly degrade in the environment, and
- be cheap to produce and purchase.

This is a difficult list of criteria to fulfil and seldom are all of these properties shown in one active ingredient. However, the search to widen our herbicide portfolio continues.

High plant selectivity is achieved by targeting processes unique to plants, and the chloroplast is a unique organelle where these processes are located. Thus, the inhibition of photosynthesis and the biosyntheses of pigments, cofactors, amino acids and lipids is invariably lethal. Other major targets located outside the chloroplast but still unique to plants include the biosyntheses of cell wall materials, microtubules for cell division and the receptors for plant hormones (Cobb, 1992).

The number of target sites that have been exploited, however, is remarkably few, at between 15 and 20. The consequence of a limited number of target sites is that weed resistance to existing herbicides is becoming increasingly prevalent (see Chapter 12). The problem is so serious that scientists from all of the major agrochemical companies, academic institutions and national organisations have formed the Herbicide Resistance Action Committee (HRAC) to standardise herbicide classification according to mode of action and to highlight management strategies for the control of resistant weeds. An important cornerstone to the prevention of herbicide resistance is the use of herbicides with different target sites in mixtures, sequences and rotations. Scientists are becoming increasingly aware, however, of the problems of cross-resistance to herbicides of different chemical groups, and evidence is emerging that resistance due to enhanced metabolic detoxification of herbicides is becoming increasingly widespread. Thus, the combination of herbicide mixtures that exploit different target sites and are metabolised by different routes is proposed to keep resistance under control. If resistance is allowed to accumulate within our major weeds, then our current armoury of herbicides will be rendered useless in the foreseeable future, with drastic consequences to crop yields and quality.

How can this fate be averted? The scientific challenge is to discover new chemistry with novel modes of action and commercial potential as herbicides. The literature implies limited success in this regard with two target sites, namely acetolactate synthase (ALS) and protoporphyrinogen oxidase (Protox) dominating herbicide discovery in recent years. Indeed, the only novel target site to have emerged and been commercially exploited in 15 years is the enzyme 4-hydroxyphenylpyruvate dioxygenase (HPPD), EC 1.13.11.276 (see Chapter 6 for further details).

The discovery of new herbicide activity may involve three lines of approach, namely:

1   the rational design of specific inhibitors of key metabolic processes,
2   the use of known herbicides or phytotoxic natural products as lead compounds for further synthesis, or
3   the random screening of new chemicals.

To date there have been no publications to suggest that the first approach is feasible. Although we can identify target processes or enzymes that may have potential for herbicide design and discovery, we have not yet come to terms with the complexities of plant metabolism to exploit them. The literature contains many published attempts to design herbicides as enzyme inhibitors that were potent *in vitro* but commercially unsuccessful. Only an increased understanding of plant physiology and biochemistry will allow us to proceed beyond this theoretical phase and so turn rational design into a distinct possibility.

The second approach is essentially imitative, and is often referred to as 'analogue synthesis' or 'me-too' chemistry. It does provide new targets and standards for synthetic chemists in particular, although the search for an analogue with new activity is seldom predictable. Lead compounds may have shown biological activity either commercially or have known activity, for example as secondary metabolites, allelochemicals, or other 'natural' plant products (i.e. biorational design). Certainly, the academic and patent literature are closely scrutinised for indications of new activity.

The random screening of novel chemicals against target weeds is the approach most likely to lead to the discovery of a new class of herbicide. In this way, the agrochemical and chemical industries, sometimes in collaboration with academic institutions, employ chemists to synthesise novel molecules and biologists to screen for their activity. The outcome of the random screening process is not simply left to chance but nowadays involves a stepwise assessment of the potential of new compounds in primary, secondary, and tertiary or field screens.

The primary screen aims to establish lead structures, namely those with sufficient activity against target species at a suitably low dose to warrant further study. Since most agrochemical companies screen thousands of chemicals each year, this process is both costly and crucial (Table 2.6). Thus, if selection criteria are low, many compounds of marginal activity will be selected and screening costs may become astronomical, but if standards are too high, a potential lead may be missed. Since it is practically impossible to screen all compounds in the field, companies attempt to simulate these conditions in glasshouses or controlled environments. In this way, new chemicals are commonly applied to glasshouse-grown weeds and crops, and any growth regulatory or phytotoxic symptoms are scored visually at regular intervals. Interestingly, new compounds are routinely tested in other primary screens, for example molecules synthesised as herbicides will be tested for fungicidal or insecticidal activity. Indeed, quite surprising results have been reported from such screens with the generation of new leads (e.g. Giles, 1989).

Table 2.6 Costs for the development of an agricultural chemical (modified from Giles, 1989).

Development phase	Length(years)	Compounds tested per eventual registered product	Total costs (£ million)
First synthesis and glasshouse screens	1	22500[a]	45
Re-synthesis and first field experiments	2	150	1
Optimisation of synthesis, large-scale field trials and product safety	2	7.5	7
Further optimisation of synthesis, full field development, product safety and registration	3	1.5	8
Totals	8	1	61

[a] Note that Berg et al.(1999) considered this 'hit rate' to be in the region of 1 in 46,000 for 1995, as a result of increasing competition, environmental issues and toxicological considerations.

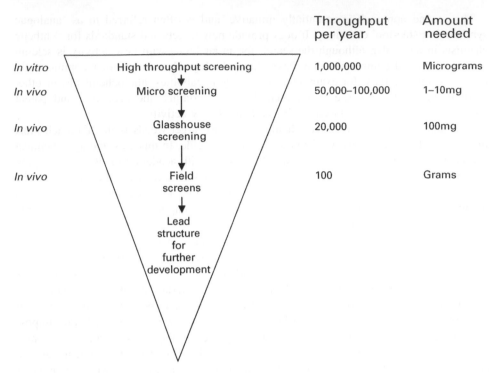

		Throughput per year	Amount needed
In vitro	High throughput screening	1,000,000	Micrograms
In vivo	Micro screening	50,000–100,000	1–10mg
In vivo	Glasshouse screening	20,000	100mg
In vivo	Field screens	100	Grams
	Lead structure for further development		

Figure 2.2 Modern herbicide screening.

Most companies have now introduced high throughput technologies (HTTs) into their discovery processes (Figure 2.2). HTT refers to a range of tools and techniques to enable rapid and parallel experiments, such as herbicide screening, to increase productivity and the development of new leads. The benefits of high throughput screening can include:

1 faster discovery and optimisation of new lead compounds,
2 greater efficiency and productivity,
3 faster and improved target innovation, and
4 faster optimisation of the formulation process.

Examples in use include *Lemna* plants, green algae, cell suspensions from target weeds and germinating cress (*Lepidum sativum* L.) seeds arranged, for example, in 96 well plates. Robotics and automated assays are used to note changes in control populations. ***In vitro*** high throughput screens may reveal new classes of herbicide chemistry. Processes such as combinatorial chemistry have been developed for this purpose (e.g. see Ridley *et al.*, 1998 for a detailed account). The modern screening process therefore starts with a large chemical library and assumes a maximum hit-rate of 0.1%, at best.

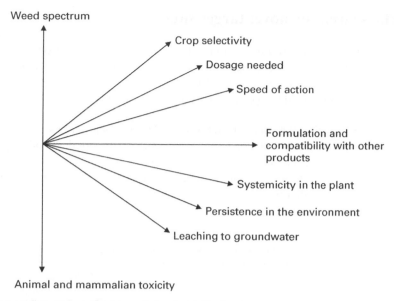

Figure 2.3   Information needed to optimise herbicide development.

It is the aim of the secondary screen to optimise these initial observations by further chemistry to yield compounds with the desired characteristics for commercial potential. To be successful, a logical sequence of further study is necessary to establish whether the new structures are sufficiently active or novel to warrant further development. Thus, the characteristics shown in Figure 2.3 need to be precisely tested, with a clear list of priorities. Furthermore, biologists and chemists need to work closely if such optimisation is to be achieved.

Field Screens are used to test hypotheses formulated in laboratory and glasshouse testing, and so confirm that molecules are as active in the field as initially predicted. These screens form the basis of many decisions that may yet seal the fate of a new herbicide class. Certainly, promise must be confirmed before very expensive and thorough toxicological and environmental studies are pursued. Compounds showing reliable and reproducible activity in the glasshouse may prove unpredictable in the field when exposed to a wide range of environmental and climatic constraints. Thus, extensive field evaluation is needed to ensure that, for example, the compound is adequately formulated (e.g. for rainfastness, cuticle penetration, and compatibility with other pesticides) and used at the correct growth stages to give optimal weed control with minimal crop damage.

To satisfy the aforementioned criteria takes many years and the investment of huge sums of money (Table 2.6). However, as stated by Giles (1989), 'those companies which are able to optimise these processes to reduce their risks but to take advantage of opportunities arising from new types of chemistry ... have a good chance of discovering products with which to ensure their future; those that do not will have less chance of finding products and a reduced chance of survival'.

2.6 The search for novel target sites

In addition to the challenges of product innovation – reducing the time from discovery to market, global competitiveness and the impact of new environmental legislation – a further issue facing the industry is the discovery of new target sites.

The list of known herbicide target sites is surprisingly short and currently stands at 19 (Table 2.7).

However, Berg *et al.* (1999) have listed at least 33 additional targets that have been demonstrated experimentally but not commercialised, and reported 26 other patented herbicide targets (Table 2.8 a and b).

This implies a huge investment that may yield novel, commercially important target sites and new chemistry that may generate lead synthesis and ultimately new herbicides.

Table 2.7 Existing targets for herbicide action (from Berg *et al.*, 1999).

Photosystem II	Tubulin organisation
Photosystem I	Protoporphyrinogen oxidase
Acetolactate synthase	Phytoene desaturase
EPSP synthase	Zeta-carotene desaturase
Glutamine synthase	Dihydropteroate synthase
'Auxin receptor'	HPP Dioxygenase
'Auxin transporter'	'Gibberellin biosynthesis'
ACC carboxylase	Uncouplers
Tubulin assembly	Cellulose biosynthesis
plus very long-chain fatty acid biosynthesis inhibitors (Wakabayashi and Böger, 2002)	

Table 2.8(a) Thirty-three additional herbicide targets demonstrated *in vitro*.

Adenylosuccinate synthase	Deoxy-xylulosephosphate synthase
Squalene cyclase	Deoxy-xylulosephosphate reductoisomerase
Obtusifoliol demethylase	HMG-CoA reductase
Acetohydroxyacid reductiosomerase	Farnesyldiphosphate synthase
Imidazolglycerol-P dehydratase	Squalene synthase
Isopropylmalate dehydrogenase	'Sphingolipid biosynthesis'
Histidinol dehydrogenase	Panthotenate synthase
Anthranilate synthase	Oxopantoyllactone reductase
Homoserine dehydrogenase	AMP/Adenosine deaminase
Ornithine carbamoyl transferase	Dihydrofolate reductase
Aspartate aminotransferase	Aminoacyl-tRNA synthase
Pyruvate dehydrogenase	Protein phosphatase 2A
Glycine decarboxylase	Tyrosine kinase
Acyl-CoA synthase	RNA polymerase III
ß-Ketoacyl synthase	
Aminolevulinic acid dehydratase	
Hydroxymethylbilane synthase	
Glutamate-semi-aldehyde amino transferase	

Table 2.8(b)   Twenty-six additional patented herbicide targets.

Homoserine kinase	Sterol Δ14 reductase
Threonine synthase	Adenylosuccinate lyase
Dihydrodipicolinate synthase	Ribose-5P isomerase
Desoxyarabinoheptulosonate-7P synthase	Transketolase
Dihydroxyacid dehydratase	Fructose-6P 1-P transferase
Branched-chain amino transferase	Isocitrate dehydrogenase
Isopropylmalate dehydratase	Galactose dehydrogenase
Anthranilate phosphoribosyl transferase	D1 protease
ATP phosphoribosyl transferase	Rubisco methylase
Cysteine synthase	Cytokinin receptor
Deoxyxylulosephosphate synthase	Glutamate receptors
Isopentenyldiphosphate isomerase	Sodium channel cyclin-dependent protein kinase
	'Cytoskeletal components'

Genome analysis (or functional genomics) is likely to be a powerful tool in future herbicide discovery and the model plant being used for this approach is *Arabidopsis thaliana* L. Heynh (thale cress). The publication of the complete *Arabidopsis thaliana* genome sequence in December 2000 was a key-step in our understanding of the higher plant genome. This cruciferous weed has been used as an increasingly important model system for the study of plant molecular biology. Scientists have been able to exploit its short generation time coupled with a small genome that is amenable to molecular techniques. Hence, of the estimated 23,000 protein-coding genes it is thought that nearly 40% have currently unknown cellular roles, which in theory are all potential herbicide targets. The challenge is to identify which of these genes code for proteins, the inhibition of which would lead to the cessation of plant growth, or the identification of a rate-limiting metabolic step, or the accumulation of toxic intermediates – or a mixture of all three.

How this may be achieved in practice has been described by Berg *et al.* (1999). In essence, lethal mutants are generated by various means, and the gene or genes involved are identified. This requires a large database and sequence analysis of each for up to 30,000 genes and powerful computing backup. Once the target has been identified its importance requires validation, for example by antisense technology. Thereafter it can be used to design an assay for high throughput screening to identify new leads for herbicide development.

A comprehensive knowledge of gene function will surely provide innovative targets in the future and may lead to new active ingredients, justifying the major investment required to pursue this approach. Indeed, Berg *et al.* (1999) estimate that up to 3,000 new possible targets may be identified by this method.

2.7   Mode of action studies

The term 'mode of action' may be defined as the complete sequence of events leading to plant injury, and therefore includes all areas of interaction between a herbicide and a crop or test species. A great many features contribute to successful weed control and may be categorised into areas of herbicide uptake, movement and metabolism (Figure 2.4).

The term 'primary or target site' is commonly used to describe the biochemical location at which a herbicide may potently inhibit an important process. This site is expected to show the fastest response to the herbicide and should also be the most sensitive site to yield a commercially useful effect. This precise site of action is therefore the location of a molecular interaction which triggers a series of secondary events leading to the death of a target weed. An example of such a target site is the D1 protein in the thylakoid membranes of the chloroplast. Many herbicides bind to this protein and so inhibit photosynthetic electron flow through the thylakoid. Consequently, active oxygen species are generated which cause plant death by photo-oxidative damage (Chapter 5). The D1 protein is a well-known example of a target site with a specific and strategic membrane location. Other examples of known target sites include natural receptors to plant growth substances (e.g. the auxin-type herbicides, Chapter 7), and specific enzymes in the biosyntheses of lipids and amino acids (Chapters 8 and 9, respectively).

There is no proven or simple route to discover the target site of a new herbicide. On the contrary, the investigator requires an appreciation of the vagaries of plant metabolism, an enjoyment of detective work and a large measure of good fortune! A sequential investigation is often followed to establish a target site, as described below. However, this may give the false impression of standardised procedures. Although such an approach is generally sound and valid, it should be noted that the investigator may uncover uncharted areas of plant metabolism, in which the novel herbicide may be regarded as

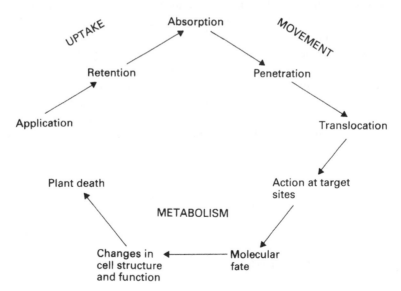

Figure 2.4 Herbicide mode of action.

a new and powerful probe. In this way photosynthetic inhibitors have allowed the molecular dissection and characterisation of Photosystem II (Chapter 5), and, as a further example, our understanding of aromatic amino acid biosynthesis was greatly aided from the study of glyphosate action (Chapter 9).

A sequential investigation to discover the target site of a herbicide is as follows:

1 Symptom development in target species.
2 Structural analysis of sensitive tissues.
3 Monitoring of key physiological processes.
4 Biochemical studies.

Detailed observations of symptom development provide the starting point for herbicidal detective work. If these observations are recorded daily the speed of response is quickly established and the most sensitive tissues identified. In addition, the development of secondary effects may have useful diagnostic value, as will the addition of 'standard' herbicides to the study. Thus, if symptoms are similar to those produced by existing herbicides, a shared target site may be implied. For example, photosynthetic inhibitors typically induce leaf chlorosis before necrosis, and auxin-type herbicides cause characteristic twisting and bending effects in young, elongating tissues of susceptible plants. In the case of soil-applied herbicides, the emergence of 'white' cotyledons is a good indication of the inhibition of carotenoid biosynthesis by the so-called 'bleaching' herbicides. These observations may be followed by ultrastructural comparisons of sensitive tissues from treated and untreated plants. The progressive development of abnormalities in both extracellular and intracellular organisation is a powerful indicator of which metabolic compartments are involved in herbicide action. Isotopic tracers and autoradiography may be used at this stage to pinpoint this cellular site with precision.

The clues generated from the visual and microscopic assessment of treated tissues may then be used to monitor key physiological processes in the presence or absence of the herbicide. Such studies point to the most likely areas of inhibition in the discovery of the target site. For example, photosynthesis and respiration may be measured in intact tissues, or isolated chloroplasts or mitochondria using oxygen electrode techniques. In recent years, many groups and companies have developed model systems for such studies, including the use of isolated cells and protoplasts from target species, cell cultures and green algae. In general, various macromolecular syntheses can be examined in these systems, especially by monitoring the incorporation of radiolabelled precursors into stable products, including leucine into protein, thymidine into DNA, uracil into RNA and acetate or mevalonate into lipids. In addition, physiological viability may be assessed using fluorescent stains, and membrane integrity and function examined by measuring potassium flux. In each case, comparison with herbicide standards should be encouraged and both time-courses and dose-responses should be established to determine the precise sensitivity of the process to herbicidal inhibition.

Once a likely target pathway is indicated, a sequence of reactions is examined in more detail. Initially, concentrations of pathway intermediates should be investigated, both in the presence and in the absence of the herbicide. If one compound accumulates in the presence of the inhibitor, then this is good evidence that the target site of an enzyme has been found. For example, glyphosate treatment results in accumulation of shikimate 3-phosphate, implying the inhibition of 5-enoyl-pyruvyl shikimic acid 3-phosphate (EPSP) synthase, and the inhibition of protoporphyrin oxidase by acifluorfen is pinpointed by the accumulation of protoprophyrin IX in treated plants. Such an accumulation of a metabolic intermediate implies that further key molecules are not being formed. This can be confirmed by studying whether the addition of the key metabolite is able to overcome the inhibition. In this way the inhibition of branched-chain amino acid biosynthesis by the sulphonylureas and imidazalinones can be overcome by the addition of leucine, isoleucine and valine. Additional proof that a true target site has been discovered may come from further studies with mutants. Probably the best-documented example of such mutagenesis is the herbicide site of the D1 protein in the chloroplast thylakoid. That a single substitution at position 264 from serine to glycine is sufficient to cause total resistance to atrazine in some weeds is surely powerful evidence of a target site! Further studies may then focus on the chemical and biochemical properties of the target site. For instance, the kinetics of the inhibition and the precise mechanism of inhibition may be investigated. Such studies are not only of academic value, but are also central to our understanding of herbicide selectivity at the molecular level and may establish precise structure–activity relationships for the development of molecules with optimal activity. Additionally, site-directed mutagenesis may then be employed to create crop tolerance.

In a recent review, Grossmann (2005) has described a functional array of bioassays conducted at BASF, Limburgerhof, aimed at diagnosing the mode of action of a new herbicide molecule. These assays are designed to differentiate between the distinct responses of complex structures (plant, tissue, meristematic cell, organelle), developmental stages, types of metabolism and physiological processes. He has coined the term 'physionomics' to describe this physiological profiling as providing the first clues to mode of action. This term follows the use of other '-omics' technologies in studying herbicide discovery and mode of action:

Functional Genomics: Generating mutants and screening them for functional gene identification.
Transcriptomics: Profiling gene expression utilising DNA microarrays and RNA extraction.
Proteomics: Protein profiling by gel electrophoresis of extracted proteins.
Metabolomics: Metabolic profiling using metabolite extraction and separation by gas chromatography / liquid chromatography–mass spectrometry.
Physionomics: Physiological profiling following functional bioassays.

Biochemical studies are often routinely performed in cell-free systems on enzymes of fungal or bacterial origin. Several dangers are implicit in this *in vitro* route, and particular care is needed not to extrapolate results obtained *in vitro* to predict *in vivo* activity in the whole organism. Thus, the perfect herbicide *in vitro* may have negligible practical value if it cannot reach its active site. Indeed, many a promising, lipophilic candidate has not been developed further owing to its being confined to the cuticle and not entering the weed! To this end, increasing use is being made of the physicochemical properties of a molecule that will allow a prediction of its systemicity and stability in both target species and the environment. These include the octan-1-ol / water partition coefficient (K_{ow}), and the dissociation constant (pK_a). No single measure of organic phase/water distribution can

Table 2.9 Herbicide systemicity and log K_{ow}.

| Mobility | log K_{ow} | | |
	-3 to 0 (hydrophilic)	0 to 3 (intermediate)	3 to >6 (lipophilic)
Non-systemic			trifluralin
Xylem mobile		triazines, phenylureas	diflufenican, diphenylethers
Both xylem and phloem mobile	glyphosate, aminotriazole, glufosinate	auxin-type herbicides, sulphonylureas, imidazolinones, sethoxydim	

be predicted for herbicides because of the wide variety of organic phases within a plant, such as hydrocarbon waxes, triglycerides, proteins, lignins or even carbohydrates. However, Briggs and colleagues (Bromilow *et al.*, 1986) have found that the partition coefficient can give a good prediction of systemicity. For acids and bases the dissociation constant of the chemical and the natural pH of the various plant compartments will also determine the proportion of ionised and non-ionised forms present. Ionisation decreases log K_{ow} and so can have a dramatic effect on the movement of a compound in a plant. Most phloem-mobile, systemic compounds are therefore weak acids with log K_{ow} in the range of -1 to 3, but the immobile soil-applied herbicide diflufenican has a log K_{ow} value of 4.9 (Table 2.9).

Metabolism of the herbicide within a plant may also have a profound effect on movement. Metabolism in general attempts to reduce lipid solubility and prevent toxic action. For example, aryl hydroxylation may be expected to cause increased mobility from the parent molecule (pK_a change from neutral to 10), and its subsequent conjugation to glucose will create a less mobile glucoside (pK_a change from 10 to neutral) that becomes subject to further compartmentation or inactivation within the cell (Bromilow *et al.*, 1986). Figure 2.5 presents an overview of how mobility may be related to log K_{ow} and pK_a.

Mode of action studies greatly aid, and in many cases totally explain, the selectivity shown by many herbicides. Indeed, susceptibility, tolerance, and resistance are being increasingly defined at the metabolic level. Selectivity may of course be due to differential leaf interception, retention, or uptake, but once inside the plant several metabolic criteria are now evident. Generally, selectivity may be achieved at several steps as suggested in Figure 2.6.

Many herbicides are themselves inactive (Figure 2.6A) and need to be metabolically activated before phytotoxicity is observed. Thus, paraquat and diquat are activated by light in the thylakoid to generate toxic active oxygen species (Chapter 5); the butyl esters of MCPA and 2,4-D are converted to active acids in susceptible species by β-oxidation (Chapter 7), and some graminicides, such as the aryloxyphenoxypropionates, require conversion from ester to acid for optimal activity (Chapter 8). Active herbicides may be metabolised before they can reach their target site (Figure 2.6B).

Many enzymes, often mixed-function oxygenases, have been implicated in herbicide metabolism studies. Maize and sorghum contain a high concentration of glutathione *S*-transferase, so that atrazine is conjugated and detoxified before it reaches its thylakoid

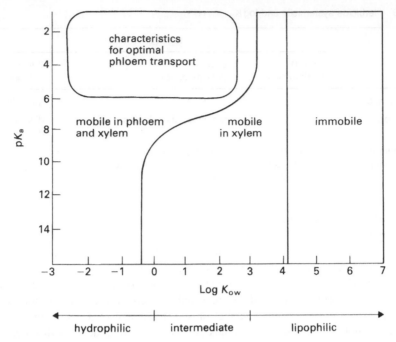

Figure 2.5 Relationship between herbicide mobility, log K_{ow} and pK_a (after Bromilow *et al.*, 1986 with permission).

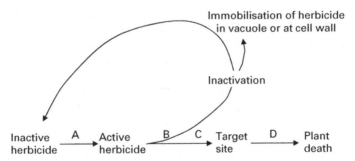

Figure 2.6 Herbicide metabolism in relation to selectivity. Points A–D are stages where selectivity may be achieved.

site of action, wheat readily hydroxylates chlorsulfuron and the metabolites are inactivated by conjugation with glucose. Similarly, soybean selectivity hydroxylates and glycosylates bentazone; metamitron is selectively deaminated in sugar beet, and diuron is initially demethylated in some species prior to inactivation, again by hydroxylation and glycosylation.

Herbicide metabolites may not always be inactive. On the contrary, evidence is accumulating that glycosyl ester formation may be reversible in many instances. Thus, conjugates stored in the vacuole or bound to cell wall components may be regarded as herbicide reservoirs, which may be called into play at a later stage.

Selectivity may also be achieved at the target site either by differential binding or inactivity (Fig. 2.6C). For example, the spectacular selectivity shown by the aryloxyphe-noxypropionate and cyclohexanedione graminicides appears to be primarily due to the inability of these compounds to inhibit acetyl coenzyme A carboxylase in dicotyledonous crops (Chapter 8). Monocot–dicot differences are also implied in the mode of action of auxin-type herbicides. In this case, it is speculated that the monocotyledonous auxin recep-tor is not accessible to the phenoxyalkanoic acids (Chapter 7).

Finally, toxic species may be inactivated before phytotoxicity is observed (Figure 2.6D). An example is the enhanced activity of superoxide dismutase in some grasses that enables relative tolerance to paraquat.

A final example of the importance of mode of action studies lies in an appreciation of the increasing problem of herbicide resistance. The prolonged use of persistent and potent herbicides with a common target site is a certain recipe for the eventual development of resistant weeds. Only by herbicide, and preferably crop, rotation can this problem be overcome. Thus, the use of herbicides with different target sites will lessen the selection pressures that favour resistance.

The most successful herbicides with negligible mammalian toxicity are those which inhibit, often selectively, metabolic processes that are unique to plants. These are princi-pally photosynthesis, the action of plant growth regulators and the biosyntheses of pig-ments, lipids, and amino acids. These target processes will be considered in separate chapters of this book.

2.8   A lower limit for rates of herbicide application?

One of the great successes of the agrochemical industry in the last six decades has been the vast reduction in the amounts of chemicals applied to obtain weed control, from kilo-grams to grams of active ingredients per hectare. Indeed, the most recently introduced PROTOX and ALS inhibitors have rates quoted as low as $2–5\,g\,ai\,ha^{-1}$ (Wakabayashi and Böger, 2002). This raises the question of whether there may be a lower limit of application that still maintains high efficacy. Clearly, the environmental benefits of lower doses would be enormous, but how low can we go? The following calculations may be of use in this discussion.

Assumptions

1 A weed leaf area of $10\,cm^2$.
2 The leaf has $300,000\,cells\,cm^{-2}$.
3 The leaf therefore has 3×10^6 cells.
4 One herbicide molecule needs to bind to one specific site per cell.
5 3×10^6 molecules need to gain entry into the leaf.
6 If the herbicide has a molecular weight of 250 daltons, then $250\,g$ will contain 6.023×10^{23} (Avogadro's number) of molecules.
7 3×10^6 molecules are contained in $3 \times 10^6 \times 250 \div 6.023 \times 10^{23} = 1.25 \times 10^{-15}\,g$ per plant!
8 This suggests that a lower theoretical limit of 1.25 femtograms of herbicide are needed to gain entry into the plant for herbicidal action.

However

9   Perhaps only 10% of the dose gains entry into the plant, and only 10% of that reaches the target site in an active form. Thus, 1.25×10^{-13} g must be applied to each plant, equivalent to approximately 10^{-6} g ha^{-1}.

10 These calculations suggest the possibility of a lowest field rate in the range of micrograms per hectare, some six orders of magnitude lower than rates currently used. Can these theoretical values be achieved in the field? The above model contains some major assumptions and extrapolations that may not be experimentally sound. A few important issues are listed below.

11 Leaves are seldom arranged perpendicular to the spray, so issues of leaf interception, sprayer technique, formulation and retention, will strongly influence the penetration of the herbicide.

12 Environmental variables have a profound impact on herbicide efficacy and are beyond the control of the most efficient operator and agrochemical company! These will influence entry into the plant, the movement of the active ingredient and the metabolic state of the plant to tolerate chemical attack.

13 The assumption that one herbicide molecule per cell may be phytotoxic is likely to be invalid, taking into account the metabolic defences mechanisms that we know are present in each leaf cell and the probability of many sites per cell.

Given the above caveats, it may still be expected that at least another order of magnitude may be achieved in reducing herbicide dosage to less than 1 gram per hectare in the decades ahead. This target is surely not beyond the innovative scientists in the modern agrochemical industry.

References

Agrow (2006). Global agrochemical market flat in 2005. Issue 490, p. 15, February 24th, 2006.

Ames, B.N., Profet M. and Gold L.S. (1990) Dietary pesticides (99.99% all natural). *Proceedings of the National Academy of Sciences, USA* **87**, 777–781.

Berg, D., Tietjen K., Wollweber D. and Hain, R. (1999) From genes to targets: Impact of functional genomics on herbicide discovery. *Proceedings of the British Crop Protection Conference, Weeds* **1**, 491–500.

Bromilow, R.H., Chamberlain K. and Briggs G.G. (1986) Techniques for studying the uptake and translocation of pesticides in plants. *Aspects of Applied Biology* **11**, 29–44.

Cobb, A.H. (1992) *Herbicides and Plant Physiology*, 1st edn. London: Chapman and Hall.

Evans, J. (2010) Food security is all in the genes. *Chemistry and Industry*, **9**, 14–15.

Finney, J.R. (1988) World crop protection prospects: demisting the crystal ball. *Brighton Crop Protection Conference, Pests and Diseases* **1**, 3–14.

Giles, D.P. (1989) Principles in the design of screens in the process of agrochemical discovery. *Aspects of Applied Biology* **21**, 39–50.

Graham-Bryce, I.J. (1989) Environmental impact: putting pesticides into perspective. *Brighton Crop Protection Conference, Weeds* **1**, 3–20.

Grossmann, K. (2005) What it takes to get a herbicide's mode of action. Physionomics, a classical approach in a new complexion. *Pest Management Science* **61**, 423–431.

Oerke E.C., Weber A., Dehne H.-W., Schonbeck F. (1994) Conclusion and Perspectives. In: Oerke, E.C., Dehne, H.W., Schonbeck, F. and Weber, A. (eds) *Crop Production and Crop Protection: Estimated Losses in Food and Cash Crops*. Amsterdam: Elsevier, pp. 742–770.

Ridley, S.M, Elliot, A.C., Yeung, M. and Youle D. (1998) High-throughput screening as a tool for agrochemical discovery: automated synthesis, compound input, assay design and process management. *Pesticide Science* **54**, 327–337.

Rüegg, W.T, Quadranti, M. and Zoschke, A. (2007) Herbicide research and development: challenges and opportunities. *Weed Research* **47**, 271–275.

Wakabayashi, K.O. and Böger, P. (2002) Target sites for herbicides: entering the 21st Century. *Pest Management Science* **58**, 1149–1154.

Chapter 3
Herbicide Uptake and Movement

3.1 Introduction

The effectiveness of a herbicide treatment ultimately depends on the amount of active ingredient that reaches the target site. There are, however, many barriers that prevent the herbicide molecule passing from the outside of the leaf through the cuticle and underlying cells, utilising transport systems, to reach the site of action. Indeed, the physical and chemical nature of the cuticle, the contents of the herbicide formulation and the environmental and physiological history of the plants in question will all influence herbicide efficacy. Many herbicides are also applied to the soil and their effectiveness, again, is determined by factors influencing uptake into the root and transport to the target site. This chapter will consider the passage of herbicides from outside the plant to the target tissues.

3.2 The cuticle as a barrier to foliar uptake

The outer leaf surface is covered with a waxy cuticle that waterproofs the leaf and provides the first line of defence between the plant and the environment. Its structure and chemical content are both varied and complex, but the successful passage across it is a vital aspect of herbicide efficacy. Generally, the cuticle is 0.1–13 µm thick and contains three components: an insoluble cutin matrix, cuticular waxes, and epicuticular waxes (Figures 3.1 and 3.2). It is not a homogeneous layer and varies greatly from species to species.

Waxes found on the surface of the cutin matrix are termed the epicuticular waxes and have a very diverse structure and composition. They can easily be removed by brief immersion of the leaf in organic solvents and analysis reveals a complex mixture of very long-chain fatty acids (VLCFAs), hydrocarbons, alcohols, aldehydes, ketones, esters, triterpenes, sterols and flavonoids (Table 3.1 and see Holloway, 1993; Post-Beittenmiller, 1996). Alkanes and ketones predominate in leek and brassica leaf epicuticular waxes, but are seldom observed in barley or maize. Similarly, peanuts are enriched in alkanes compared to maize where primary alcohols are abundant (Table 3.2).

Herbicides and Plant Physiology, Second Edition By Andrew H. Cobb and John P.H. Reade
© 2010 A.H. Cobb and J.P.H. Reade

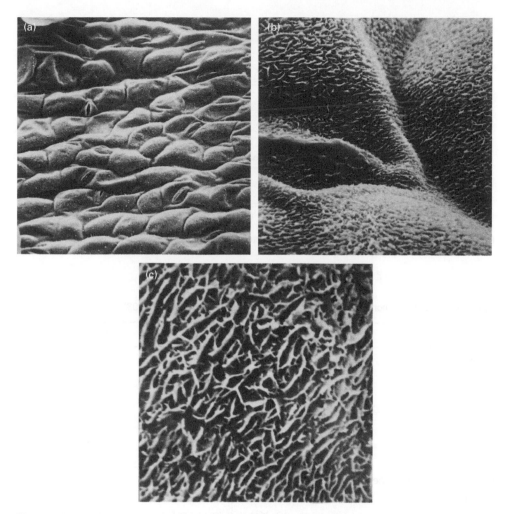

Figure 3.1 The upper leaf surface of fat hen (*Chenopodium album* L.) as shown by scanning electron microscopy at different magnifications: (a) ×540, (b) ×5,450 and (c) ×11,000 (after Taylor *et al.*, 1981).

When one class of homologue predominates, characteristic crystals of epicuticular wax form, which are very distinctive as rods, granules, crusts or aggregates. These structures may not be uniformly distributed over the whole leaf surface and differences may exist between upper (adaxial) and lower (abaxial) surfaces (e.g. Figure 3.1), and are often less evident on stomatal guard cells. Their presence often gives the leaf a dull or transparent appearance, while leaves with no epicuticular wax projections appear shiny or glossy.

Cutin is a polyester based on a series of hydroxylated fatty acids, commonly with 16 or 18 carbon atoms, the relative proportion of which varies according to species. Although found in all plants, it is one of the least understood of the major plant

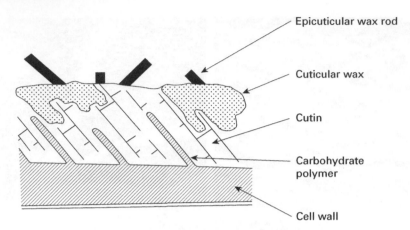

Figure 3.2   A generalised scheme of cuticle structure (after Price, 1982).

Table 3.1   Most common epicuticular wax components (after Holloway, 1993).

Class	Formula	Range of n
n-alkanes	CH_3-$(CH_2)n$-CH_3	C_{17}–C_{35} (often C_{29} or C_{31})
n-alkyl-monoesters	CH_3-$(CH_2)n$-COO-$(CH_2)n$-CH_3	C_{32}–C_{72}
n-aldehydes	CH_3-$(CH_2)n$-CHO	C_{16}–C_{34} (often C_{26} or C_{28})
n-1-alkanols	CH_3-$(CH_2)n$-OH	C_{18}–C_{36} (often C_{26} or C_{28})
n-alkanoic acids	CH_3-$(CH_2)n$-COOH	C_{14}–C_{36} (often C_{26} or C_{28})
Less common components include:		
n-ketones	$$CH_3\text{-}(CH_2)n\text{-}\overset{\displaystyle O}{\overset{\displaystyle \|}{C}}\text{-}(CH_2)n\text{-}CH_3$$	C_{23}–C_{38} (often C_{29} or C_{31})
n-sec-alcohols	$$CH_3(CH_2)n\text{-}\overset{\displaystyle OH}{\overset{\displaystyle \|}{CH}}\text{-}(CH_2)n\text{-}CH_3$$	C_{21}–C_{33} (often C_{29} or C_{31})
β-diketones	$$CH_3\text{-}(CH_2)n\text{-}\overset{\displaystyle O}{\overset{\displaystyle \|}{C}}\text{-}CH_2\text{-}\overset{\displaystyle O}{\overset{\displaystyle \|}{C}}\text{-}(CH_2)n\text{-}CH_3$$	C_{29}–C_{33}

polymers. In most cutins the dominant monomer is an ω-hydroxy fatty acid, the self-polymerisation of which will produce a linear polyester chain (Table 3.3). The mid-chain oxygen-containing functional groups (such as hydroxyl and epoxy) may be esterified to other ω-hydroxy fatty acids by polyester synthases, creating a branched structure.

Table 3.2 Variations in epicuticular lipid classes (from Post-Beittenmiller, 1996 and Taylor *et al.*, 1981). Values are percentage of total. Reproduced with permission of Annual Reviews, Inc, via Copyright Clearance Center.

Class	Leek	Barley	Maize	Brassica	Peanut	*C. album*
Fatty acids	6.4	10.3	0	1.9	38.1	0
Aldehydes	18.0	1.7	20.0	3.9	2.4	30.3
Alkanes	31.0	0	1.0	40.3	35.7	6.6
Sec-alcohols	0	0	0	11.9	0	0
Ketones	51.8	0	0	36.1	0	0
Primary-alcohols	0	83.0	63.0	1.9	23.8	44.7
Esters	0	4.7	16.0	3.9	0	17.7

Table 3.3 Common cutin monomers, normally C16 and C18, if fatty acids (from Pollard *et al.*, 2008).

Monomer type	Abundance (%)
Unsubstituted fatty acids	1–25
ω-hydroxy fatty acids	1–32
α,ω-dicarboxylic acids	<5
Epoxy fatty acids	0–34
Polyhydroxy fatty acids	16–92
Polyhydroxy dicarboxylic acids	Trace
Fatty alcohols	0–8
Glycerol	1–14
Phenolics (ferulic acid)	0–1

We still do not know, however, how these components are precisely arranged or how they contribute to cutin function. The outer surface of the cuticle is most lipophilic and becomes more hydrophilic towards the underlying epidermal cells. Various workers have suggested that polar pathways may exist through the cuticle where herbicide movement can take place. Miller (1985) has reported the occurrence of such channels in many plant families but their significance in herbicide uptake remains to be demonstrated. Further transcuticular pathways may be provided by carbohydrate polymers extending into the cuticle from the cell wall (Figure 3.2), but their role in herbicide movement is again obscure.

The biosynthesis of the cuticular waxes occurs almost exclusively in the epidermal cell cytoplasm. As detailed in Table 3.1, the majority of epicuticular wax components are derived from VLCFAs, primarily 20–32 carbons in length. They are synthesised from C_{16}–C_{18} plastidic fatty acid precursors by cytoplasmic membrane-bound elongases, using malonyl-CoA as the two-carbon donor. The wide diversity of wax components arises from the operation of three parallel pathways (Figure 3.3).

The recent use of forward and reverse genetic approaches in *Arabidopsis* has led to the identification of oxidoreductase and acyltransferase genes involved in cutin biosynthesis (as reviewed by Pollard *et al.*, 2008).

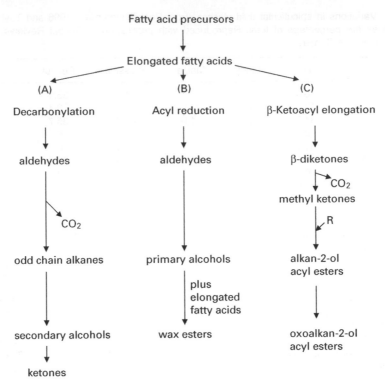

Figure 3.3 An overview of the three primary pathways of epicuticular wax biosynthesis (from Post-Beittenmiller, 1996). Reproduced with permission of American Institute of Biological Sciences via Copyright Clearance Center.

There appear to be three enzyme families involved:

- Fatty acid oxidases of the CYP86A sub-family
- An acyl-activating enzyme of the long-chain acyl-CoA synthase (LACS) family
- Acyltransferases of the glycerol-3-P acyl-CoA *sn*-1 acyltransferase family (GPAT)

These reactions of acyl activation, ω-oxidation of acyl chains and acyl transfer to glycerol could take place in several sequences and pathways, although the ω-oxidised-acylglycerols are considered to be the polyester building block.

These enzymes are thought to be located in the endoplasmic reticulum, although the cellular site of the polyester synthases is not known. How the waxes find their way to the cuticle remains uncertain. Passage along the polar pores has been suggested, or simple diffusion may occur through spaces in the cell wall. Lipid transfer proteins (LTPs) have now been demonstrated in the epidermal cells of several species and located in the cell wall. It is speculated that these LTPs transport lipids through the endoplasmic reticulum and deposit them outside the cell, although this pathway remains to be proven. LTPs are small (9–10kDa), basic proteins that are widespread and abundant in plants, constituting as much as 40% of the soluble protein pool in maize seedlings. They consist of 91–95 amino acids differing widely in sequence but always containing 4 disulphide bridges, with a three-dimensional structure that contains an internal hydrophobic cavity. They are able to bind acyl chains and transport them from the endoplasmic reticulum to the cell wall for cutin biosynthesis (Kader, 1997).

Cuticular lipid biosynthesis is very sensitive to environmental conditions and signals such as light intensity, photoperiod, humidity, chilling, soil moisture content and season, which all have an effect on cuticular development and hence herbicide efficacy. In particular, the change from high to low humidity can trigger wax production by more than an order of magnitude, an important factor to consider when extrapolating data on herbicide trials from the glasshouse to the field environment.

Generally, the cuticle will thicken during conditions that are unfavourable to plant growth, including low temperatures, photon flux density and water availability, and so herbicide absorption is maximised when opposite conditions prevail.

3.3   Physicochemical aspects of foliar uptake

Most chemicals penetrate most plants poorly when applied alone and so require an adjuvant for uptake to occur. The adjuvant increases the amount of uptake into the lipophilic environment of the cuticle. This uptake can be predicted from the octanol / water partition coefficient (K_{ow}), the dissociation constant pK_a and the parameter $\Delta \log P$. This latter term, derived from work on the penetration of the blood–brain barrier, viz.

$$\Delta \log P = \log P_{ow} - \log P_{alk}$$

subtracts the alkane/partition coefficient (generally determined using hexane or cyclohexane) from the octanol/water value and is thought to have considerable relevance to cuticular penetration, since epicuticular wax is essentially hydrocarbon in character. Thus, Briggs and Bromilow (1994) consider that permeability through the cuticular wax varies inversely with $\Delta \log P$, while permeability along the aqueous, polar cuticular route is inversely related to $\log K_{ow}$. Since $\Delta \log P$ is positive for most compounds, those entering the wax then move readily into the cuticle, as they are more strongly absorbed by the octanol-like cuticle. On the other hand, uptake via the aqueous route occurs for compounds with high water solubility. This relationship is illustrated in Figure 3.4.

Surfactants are amphipathic molecules (i.e. possess a hydrophilic head and a hydrophobic tail) and are often added to aqueous formulations for many purposes, including cuticle retention on leaf surfaces, absorption and penetration to the target site. Four classes of these surface-active agents have been identified:

1   *Anionic.*   Here, surface-active properties are provided by a negatively charged ion. For example, a hydrophobic group is balanced by a negatively charged hydrophilic group, such as a carboxyl ($-COO^-$).
2   *Cationic.*   In this case, the surface-active properties are provided by a positively charged ion. Thus, a hydrophobic group is balanced by a positively charged hydrophilic group, such as a quaternary ammonium ($- \overset{|}{\underset{|}{N}}{}^{+}$).

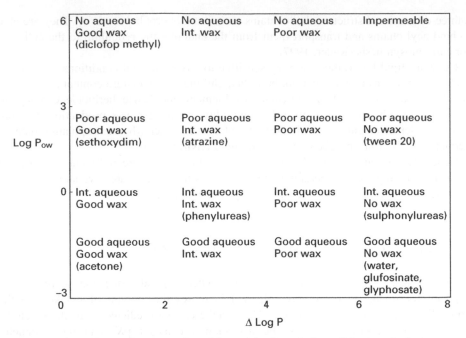

Figure 3.4 Effect of log P_{ow} and Δ log P on foliar uptake through the cuticle via epicuticular wax or aqueous routes. Int., intermediate; names in parentheses refer to estimates for some active ingredients and adjuvants (after Briggs and Bromilow, 1994). © Ernst Schering Research Foundation.

3 *Non-ionic*. No electrical charge is evident. Thus, a hydrophobic group consists of alkylphenols, alcohols or fatty acids and is balanced by non-ionisable hydrophilic groups, such as ethylene oxide (–CH_2CH_2O–).

4 *Amphoteric*. These molecules have hydrophilic groups with the potential to become cationic in an acid medium or anionic in alkaline conditions.

The ratio between the hydrophilic and lipophilic groups is termed the hydrophilic–lipophilic balance (HLB). Thus, compounds of low HLB are relatively water-soluble and so surfactants can be selected for specific purposes. They can also be further categorised as spray modifiers or activators. Spray modifiers reduce the surface tension of the spray droplets and so improve the wetting and spreading properties of the formulation, resulting in a greater degree of retention on the leaf. Activators are added to improve foliar absorption, an example being a range of polyoxyethylenes with HLB values of 10–13.

How surfactants aid cuticular penetration still remains uncertain and large differences are observed between plant species. The most obvious effect of surfactants is to reduce leaf surface tensions and contact angles, thus increasing the spread of the chemical to cover a greater proportion of the leaf surface. A high contact angle (Figure 3.5A) between the droplet and the surface also means that the droplet is easy to dislodge and run off the leaf. The lowest values for contact angle and surface tension are reached at the critical

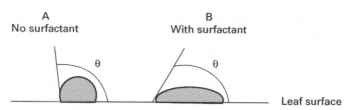

Figure 3.5 Schematic representation of droplets on a leaf surface in the absence (A) and presence (B) of a surfactant. θ, contact angle.

micelle concentration of the surfactant, usually about 0.1% to 0.5%, on a volume to volume proportion of the formulation. At such conditions contact angles may be lowered, for example, from 110 to 60 degrees (Figure 3.5B) and surface tensions from 70 to 35 mNm^{-2}.

3.4   Herbicide formulation

Herbicides will not kill weeds as active ingredients unless they are formulated to reach the target site. In the broadest sense, formulated means combined with a liquid or solid carrier so that they can be applied uniformly, transported and perform effectively. The formulation, then, is a physical mixture of the active ingredient and formulants, and the formulation chemist has a very broad brief, namely to ensure that the product retains activity with time, is easy and safe to use, and is as cheap as possible. Thus, the chemical may need to be stored below freezing-point in the winter and yet be exposed to 30 °C in the summer, and so freezing and combustion with loss of activity and safety are unacceptable. Similarly, the control of foaming when the formulation is mixed is necessary, since excessive foam leads to lack of accuracy of measurement and possible environmental contamination. Thus, the formulation chemist should be included early on in the development of a new product to ensure optimum delivery and biological efficacy.

Formulations are either liquid or solid, and many variations are possible. *Aqueous concentrates* contain a soluble active ingredient in a carrier, usually water. The concentration is dependent on the solubility of the active ingredient. An *emulsifiable concentrate* (EC) consists of a herbicide dissolved in an organic solvent with an emulsifier added to create an oil/water emulsion when water is added. The EC is commonly 60–65% (by weight) of herbicide dissolved in 30–35% solvent with 3–7% emulsifier, and typically forms an opaque or milky emulsion when added to water. The solvent itself can also be phytotoxic and aid herbicide activity. *Dusts* are finely powdered dry materials that provide good surface coverage. In *microcapsules*, the herbicide is enclosed in small, porous polymer particles that release the herbicide slowly over the growing season. *Suspension concentrates* are concentrated aqueous dispersions of herbicides and are virtually insoluble in water. They contain no organic solvent but require other agents to ensure suspension. *Ultra-low volume* formulations, also referred to as controlled droplet application, entails the use of specialist sprayers depositing very low volumes on leaf surfaces, often by electrostatic attraction. *Water dispersible granules* are solid formulations as fine particles containing 2–10% active ingredient. They are applied directly and are valued for soil-applied herbicides. *Wettable powders* are finely divided solids that are easily suspended in water. The active ingredient is added to an inert material, such as a clay, and a wetting agent. The relative strengths and weaknesses of the main herbicide formulations are listed in Table 3.4.

Concern over the use of solvents and other ingredients has created pressure for the chemical industry to change or re-examine existing formulations. Formulators are now challenged with investing in solvent-free formulations, which will force a shift away from some traditional solvent-based emulsifiable concentrates to water-dispersible granules and aqueous concentrates (e.g. Morgan, 1993). For example, Syngenta changed their formulation of the graminicide fluazifop-butyl from a solvent-based emulsifiable concentrate to a water-based, emulsion-in-water (EW) formulation. The advantage of

Table 3.4 Herbicide formulations (from Briggs and Bromilow, 1994). © Ernst Schering Research Foundation.

Name	Ingredients	Advantages	Disadvantages
Aqueous concentrate (SL)	• Active ingredient • Wetter • Surfactant • Water	• Cheap and easy to produce • Low volatility • Low phytotoxicity • Easy to mix	• Expensive to pack and transport • Frost sensitive • May corrode metal • Cannot contain high ai concentrations
Dust (DP)	• Active ingredient • Carrier	• Cheap and easy to produce • Easy to apply • Safe to handle	• Risk of user contamination and drift • Bulky to store and transport • Flowability affected by damp
Emulsifiable concentrate (EC)	• Active ingredient • Solvent • Emulsifier • Activator	• Easy to produce • Easy to handle and mix • Good when ai is insoluble in water • High efficacy	• Expensive to pack and transport • Fost sensitive • Risk of thickening • May corrode containers • Can be phytotoxic • Volatile
Microcapsule (CS)	• Active ingredient • Solvent • Emulsifier • Thickener • Anti-foaming agent • Preservative	• Cheap to produce • Low dusting • Easy to handle • Low solvent	• Expensive production equipment • Frost sensitive • May thicken at high temperatures • Expensive to package

(Continued)

Table 3.4 (*Continued*)

Name	Ingredients	Advantages	Disadvantages
Suspension concentrate (SC)	• Active ingredient • Diluent • Wetting agent • Dispersant • Anti-freezing agent • Anti-foaming agent • Preservative • Water	• No solvent • High active ingredient concentration • Easy to mix and store • Compatible with aqueous concentrates	• May settle out in storage • Frost sensitive • Can cause phytotoxicity
Ultra-low volume (ULV/CAD)	• Active ingredient • Oil • Viscosity modifier	• Applied in very small quantities • Ready to use • Cheap to store, pack and transport	• Application can be labour intensive • Risk of patchy application • High toxicity • Needs specialised equipment
Water-dispersable granule (WG)	• Active ingredient • Carrier • Wetting agent	• Low dusting • Cheap to pack • Easy to handle • Frost-tolerant • No solvent	• Expensive production equipment
Wettable powder (WP)	• Active ingredient • Carrier • Wetting agent	• Cheap and easy to produce and pack • Easy to handle • Frost-tolerant • No solvent	• Produces dust • Difficult to measure and mix

the new EW formulation is that the herbicide is now dissolved in much less solvent and surrounded by emulsifying agents in water, and this leads to improved handling, transport and storage characteristics.

The possible contamination of drinking water is also a highly emotive issue in some European countries, such as Denmark, where groundwater provides a large proportion of the drinking water. In this case, considerable effort is needed to reduce herbicide leaching, and formulations such as slow-release microcapsules may show promise. Solid formulations are also possibly safer than liquid ones and so may be expected to gain an increased market share.

The availability and cost of formulation technology is a further consideration. Microencapsulation, for example, requires relatively expensive technology that may not be easily accessible outside of Europe and the USA. The trend towards safer, low solvent and solid formulations, however, should be applauded, and perhaps further encouraged by stricter controls being imposed at the product registration stage.

Further concern has recently been expressed about possible adverse consequences arising from the release into the environment of molecules with oestrogenic properties. Implicated agrochemicals include the alkylphenyl polyethoxylates, in particular the nonylphenols, affecting humans (lower sperm quality and count, testicular and male breast cancer), fish (including gender changes) and reptiles (developmental abnormalities). The risks need to be clearly defined and many controlled experiments are needed, so that both the industry and public at large are aware of the real risks involved. Clearly, the agrochemical industry is very concerned with this issue and the replacement of the nonylphenols with safer products is awaited.

The convention used to name the non-ionisable alcohol ethoxylates is to give the alcohol chain followed by the number of $-CH_2CH_2O-$ units. Ethoxyethanol, $CH_3CH_2OCH_2CH_2OH$, is therefore C_2E_1.

Volatility is expressed as the equivalent hydrocarbon (EH), which is the number of carbon atoms in the benzene or alkane groups with the same expected boiling point and vapour pressure. Thus, chlorine has a carbon equivalent of 2 and hence the EH of chlorobenzene is 8. The boiling point of chlorobenzene is 132 °C, similar to that of the C8 hydrocarbons p-xylene (135 °C), octane (125 °C), dimethylcyclohexane isomers (120–125 °C) and allylcyclopentane (125 °C). Oxygen in an ether link and silicon are both equivalent to one carbon. So, for the ethoxylated trisiloxanes contained in Silwet L-77 (mean ethylene oxide content 8, oligomer range 2–14) (Figure 3.6).

It is predicted that a compound EH20 will only have a half-life of 30 minutes on a leaf, while EH25 will undergo considerable vapour loss over 1 day, and EH30 is equivalent to the loss of a few $g\,ha^{-1}\,d^{-1}$, which may be acceptable in the field.

These values explain why many common solvents are all highly volatile with a very short-lived influence on penetration (e.g. xylene and chlorobenzene (EH8), acetone and isopropylamine (EH5), heptylacetate, acetophenone and dimethyl sulfoxide (EH11). Fatty acids, esters and alcohols are less volatile and possess higher lipophilicity (e.g. butyl oleate has an EH = 24 and $\log P_{ow}$ = 9.5). However, small chain fatty acids (e.g. C_8–C_{10}) appear phytotoxic in their own right.

Perhaps the most common surfactant adjuvants are the ethoxylated alcohols and phenols (Table 3.5). Other than the short-chain alcohols with E2 to E15, volatility does not present a problem. The $\log P_{ow}$ values suggest that the C_8–C_{10} alcohol ethoxylates are potentially mobile in plants following penetration, but the C_{12} and higher alcohol ethoxylates and the alkylphenol ethoxylates with E-values <15 would not be expected to be mobile.

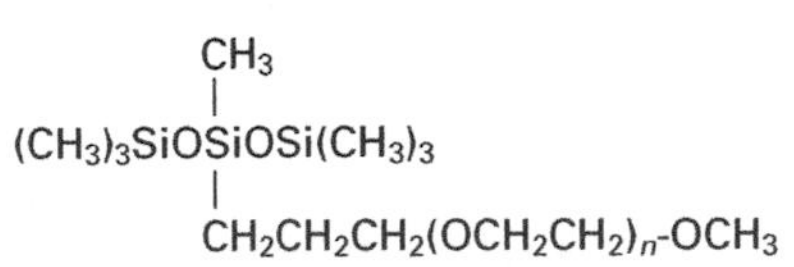

$$CH_3$$
$$|$$
$$(CH_3)_3SiOSiOSi(CH_3)_3$$
$$|$$
$$CH_2CH_2CH_2(OCH_2CH_2)_n\text{-}OCH_3$$

Figure 3.6 Structure of ethoxylated trisiloxanes in Silwet L-77. When $n = 2$, the equivalent hydrocarbon is C15 + 3 (Si) + 5 (ether O), i.e. EH = 23.

Table 3.5 Estimated physical properties of ethoxylated alcohols and allyl phenols (from Briggs and Bromilow, 1994).

Adjuvant	EH	$\log P_{ow}$	Water solubility mol 1^{-1}	$\log P_{alk}$
C8 alcohol				
E2	19	2.7	2×10^{-3}	1.1
E15	28	2.4	4×10^{-3}	−0.1
E10	>30	1.9	1×10^{-2}	−2.1
E15	>30	1.4	4×10^{-2}	−4.1
E20	>30	0.9	1×10^{-1}	−6.1
C10 alcohol				
E5	30	3.4	4×10^{-4}	1.1
E10	>30	2.9	1×10^{-3}	−0.9
E15	>30	2.4	4×10^{-3}	−2.9
E20	>30	1.9	1×10^{-2}	−4.9
C12 alcohol				
E5	>30	4.4	4×10^{-5}	2.3
E10	>30	3.9	1×10^{-4}	0.3
E15	>30	3.4	4×10^{-4}	−1.7
E20	>30	2.9	1×10^{-3}	−3.7
C18 alcohol				
E5	>30	7.4	4×10^{-8}	5.9
E10	>30	6.9	1×10^{-7}	3.9
E15	>30	6.4	4×10^{-7}	1.9
E20	>30	5.9	1×10^{-6}	−0.1
Octylphenol				
E5	>30	4.1	7×10^{-5}	1.6
E10	>30	3.6	2×10^{-4}	−0.4
E15	>30	3.1	8×10^{-4}	−2.4
E20	>30	2.6	2×10^{-3}	−4.4
Nonylphenol				
E5	>30	4.4	4×10^{-5}	2.5
E10	>30	3.9	4×10^{-4}	0.5
E15	>30	3.4	4×10^{-4}	−1.5
E20	>30	2.9	1×10^{-3}	−3.5

Increasing ethoxylation shows only a small decrease in $\log P_{ow}$, equivalent to a small increase in water solubility. However, $\log P_{alk}$ decreases sharply with ethoxylation by about 0.4 per E unit, indicating decreased lipophilicity, and so $\Delta \log P$ increases. Consequently, penetration rate should be lower for the long-chain higher ethoxylates, which are too polar to penetrate *via* the epicuticular wax and at the same time too water-insoluble to penetrate by an aqueous route.

Also abundant on leaf surfaces are trichomes (leaf hairs) and stomata (pores for gaseous exchange). The involvement of the former in foliar penetration remains unknown. Trichomes may form a dense, inpenetrable mat on the leaves of some species and so may represent a further physical barrier to herbicide uptake, since they can prevent or slow down the passage of the herbicide formulation to the cuticle. On the other hand, stomata have frequently been implicated in foliar uptake and the general view held in the 1970s was that stomata could provide a direct route for herbicide penetration. In order to do so, a surface tension of less than $30\,mN\,m^{-2}$ would be necessary, which is seldom the case with most herbicide formulations. An exception to this is the organosilicones, which includes the ethoxylated trisiloxane, Silwet L-77. These molecules cause excellent surface spreading and so can enhance or maximise the deposition of the active ingredient. Their concentration is crucial, however, since high doses may lead to an excessive run-off. These 'super-spreaders' do allow sufficient reductions in surface tension to permit stomatal infiltration. This response can only take place immediately following application, while the spray deposit remains in a liquid form, after which cuticular uptake is the sole pathway (Stephens *et al.*, 1992). This may be a useful property when rainfall is expected within a few hours after application, imparting rainfastness to the formulation, rather than rain washing off the herbicide from the leaf.

A further example of developments with uptake and formulation is with glyphosate, used for total vegetation control. This important herbicide is traditionally formulated as an isopropylamine salt with a cationic surfactant and exhibits relatively slow uptake into perennial grass weeds. A new trimesium (i.e. trimethyl sulphonium) salt with a novel polyalkylglucoside adjuvant has been claimed to show higher activity due to enhanced uptake, and to be rainfast within an hour after application.

Since most adjuvants evolved from empirical screening, it is important for the user and the public to be informed how the additive works and why it is included in the formulation. Unfortunately, this information is seldom, if ever, available and there is a current lack of regulatory harmonisation, especially in Europe, where different laws operate in the different Member States. Forthcoming legislation will mean more stringent registration requirements for adjuvants in the future. Holloway (1994), in anticipation of these changes, has proposed a wide-ranging list of possible criteria for the safe and efficacious use of existing and new products. He proposes several important questions: does the adjuvant affect spray atomisation; droplet evaporation during flight; droplet drift, deposition retention or spreading; overall target coverage; rate of droplet evaporation; physical form and moisture content of the spray deposit; rate and amount of uptake, translocation and metabolism? He also suggests the need to know: on what plant species is the adjuvant most beneficial; at what concentration is it most effective; are there any problems with intrinsic phytotoxicity; is the product compatible with other formulants; how safe is the adjuvant to use; what is the toxicity to non-target organisms; how rapidly is the product biodegraded in soil and plants; what are the potential cost benefits? This information would ensure higher standards for adjuvant efficacy and provide a more rational and scientific footing for future adjuvant use.

The hypothesis that surfactants alone might be phytotoxic to plant tissues was recently tested at the molecular level by Madhou and colleagues (2006), who studied gene expression in *Arabidopsis thaliana* in response to a foliar application of the etheramine surfactant NUL1026 at 0.2% (w/v). They found that the expression of 196 genes was significantly altered 1 hour after treatment. A number of genes were upregulated, coding for enzymes involved in both detoxification and signalling pathways. This is consistent with a stress response of overlapping gene expression to wounding, pathogen, abiotic stress and hormone treatment. It may therefore be that surfactant use prepares the plant defences for external attack. Follow-up studies are awaited to further understand the plant response to xenobiotic attack.

3.5   Uptake by roots from soil

The uptake of herbicides by plant roots and their movement in the xylem or phloem is becoming increasingly well understood on the basis of their physicochemical properties. Uptake can take place from the soil either via the air or water phase. Knowing that diffusion constants are about 10,000 times greater in air than in water, and the ratio of concentration in air and water, it is possible to predict which phase is most important in root uptake. Hence, herbicides such as trifluralin, EPTC, triallate and bifenox are likely to move as vapour in moist soils, while movement will be by diffusion in the aqueous phase by monuron and simazine. Lipophilic herbicides with a high vapour pressure will be strongly absorbed by roots from a moist soil. Volatility should also be considered and soil incorporation is often necessary (e.g. with trifluralin) to prevent rapid loss by volatilisation.

Uptake *via* the soil aqueous route has been studied in detail by Briggs and colleagues at Rothamsted Experimental Station, UK, using barley plants grown in nutrient solution and radiolabelled test compounds (e.g. Briggs et al., 1982, 1987, 1994). Distribution of a non-ionised compound between the roots and the bathing solution was defined by the root concentration factor (RCF), where

$$RCF = \frac{\text{Concentration in roots}}{\text{Concentration in nutrient solution}}$$

It was found that the RCF was directly related to the log of the octanol/water coefficient ($\log P_{ow}$), thus uptake increased with increasing lipophilicity.

The uptake of acidic compounds by roots is very different from that of the non-ionised herbicides above, but is dependent on the pH of the soil solution. Thus, the uptake of 2,4-D into barley roots over a 24-hour period from nutrient solution was 36 times greater at pH 4.0 than at pH 7.0 (Briggs *et al.*, 1987). This may be explained as an ion-trap effect, whereby weak acids are accumulated in compartments of higher pH by virtue of the greater permeation rates across membranes of the undissociated form compared to the anion (Figure 3.7).

Once inside the root hair, the herbicide has to be transported to the vascular system for long-distance transport to the target site. Two routes are possible via the apoplastic or symplastic pathways. The former entails movement down a concentration gradient along the cell walls, while the latter entails the cytoplasmic continuity of the root cortical cells via plasmodesmata. Water and solutes cannot enter the xylem by an entirely apoplastic route, but must move through the symplast of the endodermis. This is because the tangential walls of the endodermal cells are thickened by the deposition of suberin to form the water-impermeable Casparian strip (Figure 3.8). Efficiency of transport from the root cortex to the xylem is low, at less than 10%, which further illustrates the difficulty that weak acids have in crossing biological membranes when they are largely ionised at a physiological pH.

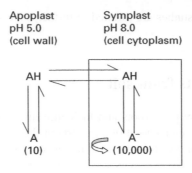

Apoplast
pH 5.0
(cell wall)

Symplast
pH 8.0
(cell cytoplasm)

Figure 3.7 Accumulation of a weak acid (pK_a 4.0) within root cells by the ion-trap effect.

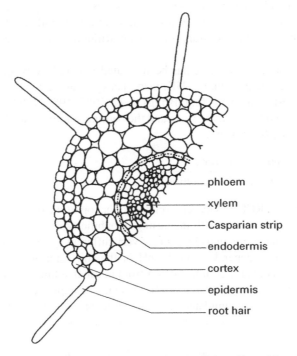

Figure 3.8 Cross section of a dicotyledonous root, showing the position of the endodermis and the Casparian strip (from Bromilow and Chamberlain, 1991, with kind permission of Springer Science and Business Media).

3.6 Herbicide translocation from roots to shoots

Movement in plants can be predicted from pK_a and logP_{ow} values and is therefore largely controlled by the physicochemical properties of the herbicide (Figure 3.4). As a general rule, lipophilic compounds (i.e. logP_{ow} >4) are non-systemic, while compounds of inter-

Table 3.6 Metabolism of 2,4-D butyl ester and systemicity of its metabolites (from Bromilow *et al.*, 1986 with permission).

Compound[a]	log K_{ow}	pK_a	Mobility
2,4-D butyl ester	4.4	Non-ionised	Immobile
2,4-D acid	2.9	3.0	Xylem/phloem
2,4-D glucose	0.6	Non-ionised	Xylem
Ring-OH 2,4-D	2.2	3.0	Xylem/phloem
Ring *O*-glucosyl 2,4-D	0.0	3.0	Xylem/phloem
Diglucosyl conjugate of -COOH and ring-OH	−2.3	Non-ionised	Immobile

[a] Structures of the metabolites of 2,4-D are on page 153.

mediate lipophilicity (log P_{ow} −0.5 to +3.5) move in the xylem, and weak acids are also mobile in the phloem to the physiological sinks, or growth points, in the plant. Most compounds enter the phloem quite freely – in contrast to sucrose, which requires an active transport system – and leave it equally readily. Movement is therefore according to sink strength. Movement in the xylem is at least 50 to 100 times faster than in the phloem and is therefore dependent on the transpiration stream at any particular time. These rates can reach huge values, especially in trees where, for example, it has been calculated that an oak tree can transpire a ton of water a day, or a maple tree growing in the open was measured to transpire 48 gallons of water per hour! Even in young seedlings (2–3 leaf stage) of black-grass rates of 0.25 g m^{-2} leaf h^{-1} have been recorded from plants growing in greenhouse conditions (Sharples *et al.*, 1997). Thus, water can move through the xylem at velocities of greater than 10 m h^{-1}, while movement through the phloem may be 50 to 100 times less.

The above patterns of systemicity have been confirmed for many herbicides by autoradiography using radiolabelled molecules, and further details for individual herbicides can be found in Bromilow and Chamberlain (1991).

Metabolism of a herbicide within the plant will also influence its movement, since metabolism generally reduces lipophilicity. Whereas aryl hydroxylation may be expected to increase the mobility of the parent molecule, its subsequent conjugation to glucose, for example, may create a less mobile glucoside that becomes compartmentalised or inactivated within the cell vacuole. An example of the consequences of metabolism on mobility is illustrated in Table 3.6.

3.7 A case study: the formulation of acids

Most commercial formulations of the phenoxyalkanoic acids or 'auxin-type' herbicides (Chapter 7) contain the active ingredient (e.g. Figure 3.9) in the salt or ester form. Of the salts, the amine forms are commonly used, although the sodium, potassium and ammonium salts are also found. Some examples of the cationic amine salts are shown in Table 3.7. These salts are highly soluble in water and formulated as aqueous concentrates. This ensures that they can be applied at relatively high rates of active ingredient but at relatively low volume.

Phenoxyalkanoic acid esters may also be applied as emulsifiable concentrates and as such exhibit greater herbicidal efficacy than the parent acids. This is because the ester is more lipophilic and more rapidly absorbed by the target weed. The esters of low-molecular-weight alcohols (such as methyl, ethyl and propyl), however, are volatile and may cause unwanted phytotoxicity to non-target plants, including crops. This problem has been overcome with the use of low-volatile esters (Figure 3.10).

A further example has been the introduction of the butoxy-methyl ethyl ester of fluroxypyr (Figure 3.11) (Snel *et al*, 1995), which can be formulated as an emulsion-in-water or a wettable powder without the aromatic solvents required by its EC predecessor, the methylheptyl ester.

The aryloxyphenoxypropanoates, such as diclofop (Figure 3.9), are also formulated as esters, traditionally as emulsifiable concentrates that partition rapidly into the cuticle. In this case, the esters of the lower molecular weight alcohols are relatively stable, as

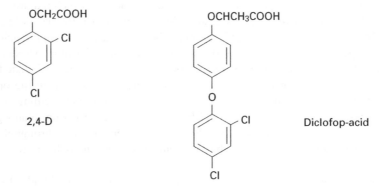

Figure 3.9 Structures of a typical phenoxyalkanoic acid (2,4-D) and an aryloxyphenoxypropanoate (diclofop-acid).

Table 3.7 Amine cations of phenoxyalkanoic acid amine salts (Loos, 1975).

Organic Group (R)	Cation structure and name		
	$\left[\begin{array}{c} R \quad H \\ \diagdown \diagup \\ N \\ \diagup \diagdown \\ H \quad H \end{array}\right]^{+}$	$\left[\begin{array}{c} R \quad H \\ \diagdown \diagup \\ N \\ \diagup \diagdown \\ R \quad H \end{array}\right]^{+}$	$\left[\begin{array}{c} R \quad R \\ \diagdown \diagup \\ N \\ \diagup \diagdown \\ R \quad H \end{array}\right]^{+}$
CH_3-	Methylamine	Dimethylamine	Trimethylamine
CH_3CH_2-	Ethylamine	Diethylamine	Triethylamine
$HOCH_2CH_2-$	Ethanolamine	Diethanolamine	Triethanolamine
$\overset{\displaystyle OH}{\underset{\displaystyle CH_3CHCH_2-}{\|}}$	Isopropanolamine	Diisopropanolamine	Tri-isopropanolamine

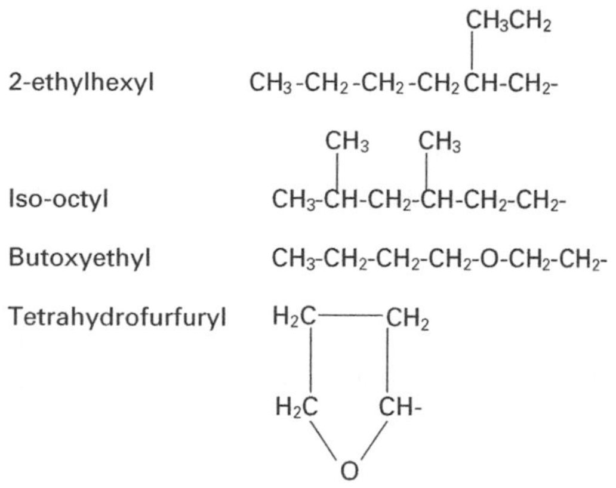

2-ethylhexyl

$$CH_3CH_2$$
$$|$$
$$CH_3-CH_2-CH_2-CH_2CH-CH_2-$$

Iso-octyl

$$CH_3 \quad CH_3$$
$$| \quad\quad |$$
$$CH_3-CH-CH_2-CH-CH_2-CH_2-$$

Butoxyethyl

$$CH_3-CH_2-CH_2-CH_2-O-CH_2-CH_2-$$

Tetrahydrofurfuryl

$$H_2C\text{———}CH_2$$
$$| \quad\quad |$$
$$H_2C \quad CH-$$
$$\diagdown \quad \diagup$$
$$O$$

Figure 3.10 Examples of some low volatility esters used in herbicide formulation.

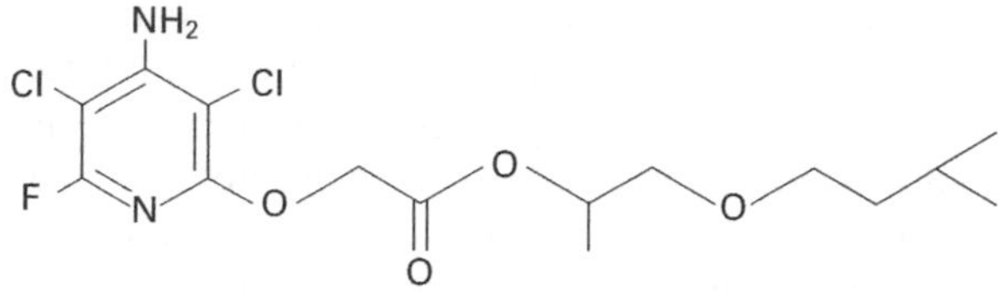

Figure 3.11 Butoxy-methyl ethyl ester of fluroxpyr.

evidenced by the commercialised examples diclofop-methyl, quizalofop-ethyl and flu-azifop-butyl. The ester itself is not the active moiety but requires conversion to the potent acid as a prerequisite for herbicidal activity.

De-esterification or hydrolysis *in vivo* was first demonstrated by the pioneering work of Crafts (1960) and an enzyme responsible was first partially purified and characterised from wild oat by Hill *et al.* in 1978. These authors found that this enzyme was non-specific and could de-esterify several ester types. Hence, the esters may be regarded as 'pro-her-bicides', requiring enzymatic activation in the plant tissues.

The phenoxyacetic acids (including 2,4-D, MCPA and mecoprop) have $\log K_{ow}$ values in the range of 2.9 to 3.6 and pK_a values close to 3.0. Thus, they move effectively in the phloem but are strongly held in adjacent tissues by ion-trapping. On the other hand, the aryloxyphenoxypropanoates are more lipophilic ($\log K_{ow}$ 3.4–4.4) and have pK_a values in the region of 3.5. As may be predicted from such lipophilic molecules, their phloem mobility is very poor. However, the potency of these herbicides at the grass meristem is such that even though less than 1% of applied dose is mobile, it is sufficient to control grass weeds (see Chapter 8, and Carr *et al.*, 1986).

3.8   Recent developments

A new class of adjuvant producing increased efficacy and an improved environmental profile is the amido propyl amines. *N,N*-dimethyl-1,3-propanediamine is condensed to

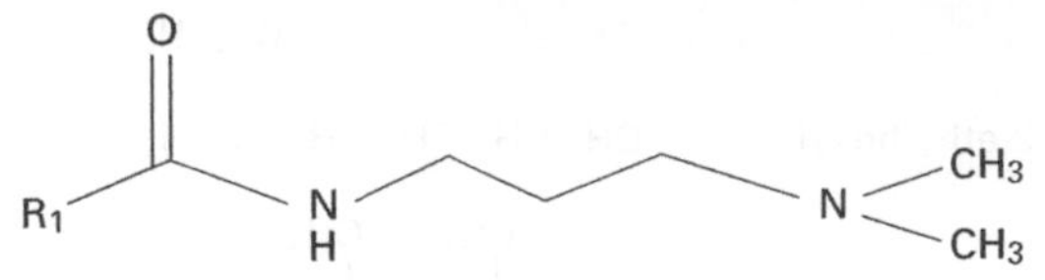

Figure 3.12 Structure of amido propylamines, where R_1 can be C8 / C10, COCO or SOY fatty acids.

fatty acid molecules (Figure 3.12). Surface properties are dependent on the pH environment. At high pH, these molecules are non-ionic, but at pH values below 8 the amine will protonate and the molecule will become more cationic. This may lead to more absorption to the negatively charged leaf surface.

Van der Pol and colleagues (2005) have developed lactate esters (*n*-propyl lactate, *n*-butyl lactate and 2-ethylhexyl lactate) that enhance the efficacy of the dimethylamine salt of 2,4-D and the 2-ethylhexyl ester of 2,4-D on fat hen (*Chenopodium album*). The lactate esters may act as solvents or penetrate the leaf cuticle itself and enhance the mobility of the herbicide through the cuticle. No phytotoxic effects were observed on tomato seedlings, suggesting promising potential for these 'green' solvents.

In recent years increasing attention has been paid to the problems associated with spray drift and how they might be overcome. Polymers of polyacrylamide, hydroxypropyl guar and ethyl hydroxyethyl cellulose (EHC) have been incorporated into the tank mix to reduce spray drift. Hazen (2005) found increased glyphosate retention in *Echinochloa crus-galli* when EHC was added to the formulation leading to enhanced bioefficacy.

References

Briggs, G.G., Bromilow, R.H. and Evans, A.A. (1982) Relationship between lipophilicity and root uptake and translocation of non-ionised chemicals by barley. *Pesticide Science* **13**, 495–504.

Briggs, G.G. and Bromilow, R.H. (1994) Influence of physicochemical properties on uptake and loss of pesticides and adjuvants from the leaf surface. In: Holloway, P.J., Rees, R.T. and Stock, D. (eds) *Interactions Between Adjuvants, Agrochemicals and Target Organisms*. Berlin: Springer-Verlag, pp. 1–27.

Briggs, G.G, Rigitano, R.L.O. and Bromilow, R.H. (1987) Physicochemical factors affecting the uptake by roots and translocation to shoots of weak acids in barley. *Pesticide Science* **19**, 101–112.

Bromilow, R.H. and Chamberlain, K. (1991) Pathways and mechanisms of transport of herbicides in plants. In: Kirkwood, R.C. (ed.) *Target Sites for Herbicide Action*. New York: Plenum, pp. 245–284.

Bromilow, R.H., Chamberlain, K. and Briggs, G.G. (1986) Techniques for studying the uptake and translocation of pesticides in plants. *Aspects of Applied Biology* **11**, 29–44.

Carr, J.E., Davies, L.G., Cobb, A.H. and Pallett, K.E. (1986) Uptake, translocation and metabolism of fluazifop-butyl in *Setaria viridis*. *Annals of Applied Biology* **108**, 115–123.

Crafts, A.S. (1960) Evidence for hydrolysis of esters of 2,4-D during absorption by plants. *Weeds* **8**, 19–25.

Hazen, J.L. (2005) Retention and bioefficacy with ethyl hydroxyethyl cellulose (EHC) as a tank mix adjuvant to reduce spray drift. In: Proceedings of the British Crop Protection Council Congress. *Crop Protection and Technology* **2**, 891–896.

Hill, B.D., Stobbe, E.H. and Jones, B.L. (1978) Hydrolysis of the herbicide benzoylprop-ethyl by wild oat esterase. *Weed Research* **18**, 149–154.

Holloway, P.J. (1994) Evaluation of Adjuvant Modes of Action. In: Holloway, P.J., Rees, R.T. and Stock, D. (eds) *Interactions between Adjuvants, Agrochemicals and Target Organisms*. Berlin: Springer-Verlag, pp. 143–148.

Holloway, P.J. (1993) Structure and chemistry of plant cuticles. *Pesticide Science* **37**, 203–232.

Kader, J.C. (1997). Lipid-transfer proteins: a puzzling family of plant proteins. *Trends in Plant Science* **2**, 66–70.

Loos, M.A. (1975) Phenoxyalkanoic acids. In: Kearney, P.C. and Kaufman, D.D. (eds) *Herbicides; Chemistry, Degradation and Mode of Action*. New York: Marcel Dekker, pp. 1–128.

Madhou, P., Raghavan, C., Wells, A. and Stevenson, T.W. (2006) Genome-wide microarray analysis of the effect of a surfactant application in *Arabidopsis*. *Weed Research* **46**, 275–283.

Miller, R.H. (1985) The prevalence of pores and canals in leaf cuticular membranes. *Annals of Botany* **55**, 459–471.

Morgan, L.J. (1993) Formulants and additives and their impact on product performance. *Brighton Crop Protection Conference, Weeds* **3**, 1311–1318.

Pollard, M., Beisson, F., Li, Y. and Ohlrogge, J.B. (2008) Building lipid barriers: biosynthesis of cutin and suberin. *Trends in Plant Science* **13**, 236–246.

Post-Beittenmiller, D. (1996) Biochemistry and molecular biology of wax production in plants. *Annual Review of Plant Physiology and Plant Molecular Biology* **47**, 405–430.

Price, C.E. (1982) A review of the factors influencing the penetration of pesticides through plant leaves. In: Cutler, D.F., Alvin, K.L. and Price, C.E. (eds) *The Plant Cuticle*. New York: Academic Press, pp. 237–252.

Sharples, C.A, Hull, M.R and Cobb, A.H. (1997) Growth and photosynthetic characteristics of two biotypes of the weed black-grass (*Alopecurus myosuroides* Huds.) resistant and susceptible to the herbicide chlorotoluron. *Annals of Botany* **79**, 455–461.

Snel, M., Banks, G., Mulqueen, P.J., Davies, J. and Paterson, E.A. 1995. Fluroxypyr butoxy-1-methylethyl ester; new formulation opportunities. *Brighton Crop Protection Conference, Weeds*, **1** 27–34.

Stephens, P.J.G., Gaskin, R.E. and Zabkiewicz, J.A. (1992) Pathways and mechanisms of foliar uptake as influenced by surfactants. In: Foy, C.L. (ed.) *Adjuvants for Agrochemicals*. Boca Raton, FL: CRC Press, pp. 385–398.

Taylor, F.E., Davies, L.G. and Cobb, A H. (1981) An analysis of the epicuticular wax of *Chenopodium album* leaves in relation to environmental change, leaf wettability and the penetration of the herbicide bentazone. *Annals of Applied Biology* **98**, 471–478.

Van der Pol, J.F., van der Linden, J.T. and de Ruiter, H. (2005) Phytotoxicity and adjuvancy of lactate esters in 2,4-D based agrochemical formulations. In: Proceedings of the British Crop Protection Council Congress, *Crop Science and Technology* **1**, 447–452.

Chapter 4
Herbicide Selectivity and Metabolism

4.1   Introduction

The main reason for the considerable success of modern herbicides is their selective phytotoxic action. Thus, only the weed is killed and its competitive effects are overcome by the crop. However, selectivity is a very relative term and totally dependent on dose, such that an application rate may alone determine crop tolerance and weed susceptibility. Since the concentration that reaches a target site is crucial to phytotoxicity, it follows that any factor that alters the concentration of active herbicide *in vivo* will then contribute to selectivity. How selectivity is achieved is therefore a complex subject, usually based on exploiting differences between the crop and the weed. These differences are many and varied, some of which are briefly considered below.

At the level of the whole plant, differences in morphology can be a major contributor to herbicide selectivity, especially in the case of a broadleaf weed in a cereal crop. Foliar arrangement ensures that more surface area of the weed leaf is exposed for greater spray interception than the more upright, narrower leaf of the crop. The extent of spray retention may therefore ensure selectivity when the same dose may be equally phytotoxic to both crop and weed. Furthermore, the growth areas (meristems) of a broadleaf weed tend to be more accessible to the spray compared with the 'protective sheath' of coleoptile or leaves that surround the cereal meristem.

In the case of soil-applied herbicides, the relative positions and growth rates of the weed and the crop are the most important factors. Depth protection is valuable in the pre-emergent control of many weeds in large-seeded, slow growing crops. Sugar beet, for example, may be sown deeper in the soil to gain protection from surface-incorporated herbicides. Indeed, many weeds that are shallow-rooted may rapidly establish before the sugar beet seedlings, and may therefore be controlled by a fast-acting contact treatment before the crop emerges. In this way the timing of herbicide treatment is an important contributor to selectivity.

Differences in absorption and uptake may be exploited by choice of herbicide formulation, as may be rates of both short- and long-distance transport within the plant. Translocation patterns will also alter sites of herbicide accumulation. Selectivity

---

Herbicides and Plant Physiology, Second Edition By Andrew H. Cobb and John P.H. Reade
© 2010 A.H. Cobb and J.P.H. Reade

is commonly achieved by the presence of detoxifying enzymes and their relative rates of activity. Furthermore, it is now established that differences in herbicide sensitivity at a target site may provide the basis of selectivity in some cases. Usually, however, selectivity results from a complex interaction of several factors, although differential metabolism is nowadays regarded as the major contributor. Plants show large differences in their ability to metabolise herbicides, and a significant difference in metabolism may often be closely correlated with tolerance or susceptibility.

Plants have evolved a wide range of metabolic systems to detoxify the chemicals which they encounter in the environment. This is well illustrated in the differential metabolism of selective herbicides between resistant crops and susceptible weeds. Indeed, it is now clear that in many cases, the basis of herbicide selectivity is the ability of the crop to metabolically detoxify the herbicides, whereas the weeds are less able to do this or may even activate the herbicide. It should also be noted, although outside the scope of this text, that plant metabolism is also able to modify the action of systemic fungicides, insecticides and industrial pollutants that are encountered.

Plants rely on their biochemical mechanisms of defence against foreign chemicals (xenobiotics). As a general rule, the enzymes involved have a broad substrate specificity. Thus, the foreign compounds are often structurally related to the intermediates of plant secondary metabolism, especially flavonoids. These enzymes are normally present and functional throughout the life of the plant (i.e. they are constitutive). In some cases, however, their activity is induced by the xenobiotics.

For detoxification to be successful, the phytotoxic chemical must be quickly metabolised to a less or non-toxic product. The rate of this process can decide whether a plant survives or succumbs to chemical attack. Often both crop and weed may have the same biochemical pathways, but metabolic rates are higher in the crop. Also, differences in herbicide metabolism and compartmentation are observed in crops and weeds.

4.2   General principles

In order to penetrate the waxy surfaces of leaves, pesticides are invariably highly lipophilic molecules. As a general rule, oxidative metabolic attack serves to both enhance the reactivity and the polarity of the molecule. Herbicides modified in this way become less phytotoxic, although there are some notable examples in which bioactivation can take place. The products of these primary reactions are then conjugated, often with sugars or amino acids, rendering them highly water-soluble, biologically inert and readily stored in the plant cell vacuole away from any potential sites of action. These storage forms may become susceptible to hydrolysis and, in some cases, may constitute a pool of potentially active herbicide. Immobilisation by binding to lignin or other insoluble cellular constituents may become the final fate of a herbicide residue or, in some instances, water-soluble metabolites may be excreted from the roots.

It is now clear that herbicides can be metabolised in many ways and this can have a direct bearing on crop tolerance or susceptibility. An understanding of how herbicides are metabolised is therefore central to enhancing herbicide design and selectivity.

4.2.1   *Phases of herbicide metabolism*

Herbicide metabolism has been well documented in the last two decades and much progress has been made in our understanding of the enzyme systems involved and where they act in the plant cell. The four phases commonly observed are as follows:

Bioactivation
Phase I   Metabolic attack
Phase II   Conjugation
Phase III   Sequestration

Some herbicides have been shown to undergo bioactivation within plant cells, where a pro-herbicide is converted to a phytotoxic agent by the action of plant enzymes. Before this activation they may be less or non-phytotoxic, so the plant can be instrumental in manufacturing the substance that will eventually kill it. Bioactivation can involve removal of chemical groups that have aided in herbicide uptake and this can have the added benefit of trapping the herbicide within the cell. Herbicide detoxification is achieved by key enzymes that carry out two major functions. First, they alter the chemical structure of the herbicide to render it biologically inert. Second, these reactions serve to increase both the reactibility and the polarity of the herbicide, so that is can be removed from the cytoplasm and either stored in the vacuole or bound to the cell wall. Metabolically, this is achieved by introducing or uncovering polar groups (Phase I metabolism). In some cases polar groups may already be in place in the original herbicide structure, in which case Phase II metabolism can be carried out without the need for Phase I reactions. In other cases, conjugate formation may be reversible and this may serve to store phytotoxic molecules.

4.2.1.1   *Bioactivation*

Bioactivation is the process whereby 'pro-herbicides', often herbicidally inactive molecules, are enzymatically converted to phytotoxic compounds in the plant cell. Many herbicides are therefore formulated as inactive hydrophobic esters to enable them to penetrate the waxy leaf cuticle. Ester hydrolysis then reveals the biologically active acid or alcohol groups. Examples include the conversion of bromoxynil octanoate to bromoxynil, and diclofop-methyl to diclofop acid (Figure 4.1). Both the aryloxyphenoxypropionate graminicides, such as fenoxaprop-ethyl, and the phenoxy-carboxylic acids, such as 2,4-DB, are rapidly de-esterified in both crops and weeds. De-esterification may, in some cases, inactivate herbicides, as in the metabolism of sulphonylurea esters, such as chlorimuron-ethyl, in which the de-esterified product is herbicidally inactive. Another example is the conversion of EPTC to the herbicidally active sulphoxide derivative. Imazamethabenz-methyl may also be regarded as a pro-herbicide. In susceptible weeds, hydrolysis results in a potent inhibition of branched chain amino-acid biosynthesis, whereas hydroxylation of the intact ester occurs in resistant maize and wheat. Bioactivation of DPX-L8747 by *N*-dealkylation in susceptible species leads to an active herbicide, whereas in resistant crops hydroxylation following the formation of a glutathione conjugation of the intact pro-herbicide leads to non-toxic metabolites. These examples demonstrate how bioactivation may be a mechanism of selectivity between crop and weed. Opening of the oxadiozolidine ring of methazole to form 1-(3,4-dichlorophenyl) urea and *N*-demethylation of

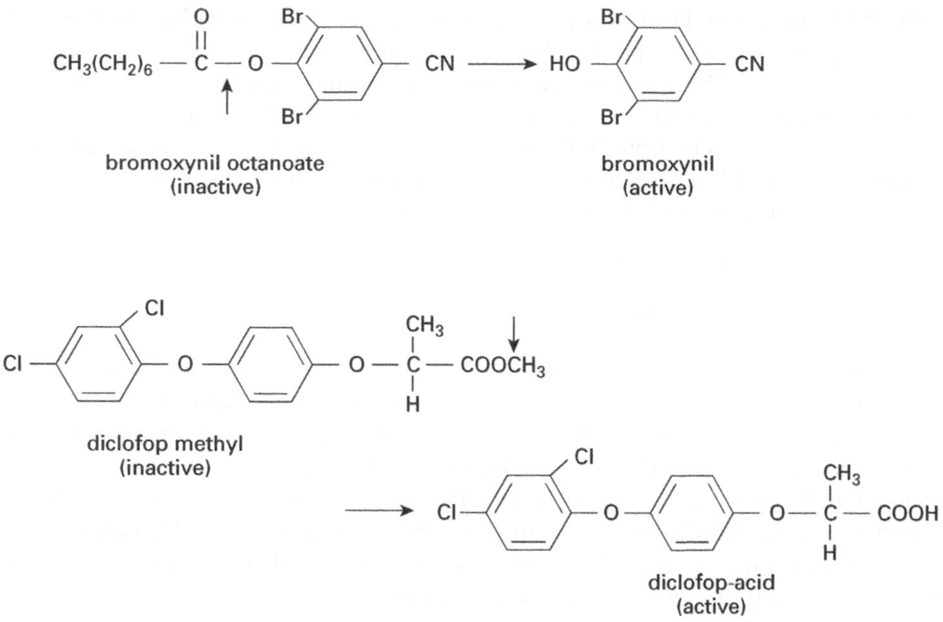

Figure 4.1 Bioactivation of inactive pro-herbicides to active molecules.

pyridazinone to form a potent phytoene desaturase inhibitor, further demonstrates the wide range of chemical reactions that can lead to bioactivation. In the case of the bioherbicide bialaphos, a tripeptide obtained from *Streptomyces*, metabolic cleavage results in the release of glufosinate, which is an important herbicide in its own right. Interestingly, resistance to the herbicide triallate in a biotype of *Avena fatua* has been demonstrated to be due to a reduced ability to convert triallate to the phytotoxic product triallate sulphoxide. This is the only reported instance of resistance being due to an inability of a weed to bioactivate a herbicide.

Cummins *et al.* (2001) found a large number of diverse proteins in wheat capable of hydrolysing herbicide esters and these activities differed from those in competing grass weeds. Indeed, crude extracts from black-grass plants were more active in ester hydrolysis than wheat. They purified a 45 kDa esterase from wheat that could bioactivate bromoxynil octanoate, but showed no activity towards diclofop-methyl. Conversely, esterase activity towards diclofop-methyl was higher in the weed than the crop. Their observations support the rapid bioactivation by ester hydrolysis of graminicides in grass weeds, contributing to their selectivity. The location of this esterase activity is thought to be the cell wall.

4.2.1.2   Metabolic attack

This phase of metabolism aims to introduce or reveal chemically active groups, such as –OH or –COOH, which can undergo further reactions. The most common way in which plants attack herbicides is by hydroxylation of aromatic rings or of alkyl groups by a family of enzymes known as the cytochrome P450 mono- or mixed function oxidases (P450s).

The P450s are a very large family of enzymes now thought to be the largest family of enzymatic proteins in higher plants. They all have a haem porphyrin ring containing iron at a catalytic centre. These enzymes are responsible for the oxygenation of hydrophobic molecules, including herbicides, to produce a more reactive and hydrophilic product. The reaction utilises electrons from NADPH to activate oxygen by an associated enzyme, cytochrome P450 reductase. One atom from molecular oxygen is incorporated into the substrate (R), while the other is reduced to form water:

$$R\text{-}H + O_2 \longrightarrow R\text{-}OH + H_2O$$

$$NADPH + H^+ \qquad NADP^+$$

The enzymes are located on the cytoplasmic side of the endoplasmic reticulum and are anchored by their N-terminus (Figure 4.2). They are found in all plant cells but in very low abundance. This, coupled with their lability *in vitro*, has meant that they are difficult to study biochemically. All P450s have a highly conserved region of 10 amino acids surrounding the haem group and it is this region that is responsible for the binding of O_2, its activation and the transfer of protons to form water. The rest of the P450 amino acid sequences are highly variable and this probably explains the wide variety of reactions and substrate specificity shown by this enzyme superfamily.

Their name is derived from the maximum absorbance at 450 nm when the reduced enzyme is bound to the inhibitor carbon monoxide, and the 'P' signifies protein. This inhibition by carbon monoxide is typically overcome by light. These features are all used in setting criteria for the involvement of P450 enzymes in herbicide metabolism. These criteria are:

- requirement of O_2,
- requirement for NADPH,
- association of enzyme activity with the microsomal fraction produced by centrifugation at $100,000 \times g$, enriched in the endoplasmic reticulum,
- inhibition by CO, which is reversible by light,
- inhibition by anti-reductase antibodies, and
- inhibition of *in vitro* activity by known P450 inhibitors, including aminobenzotriazole, paclobutrazole, piperonyl butoxide and tetcyclasis.

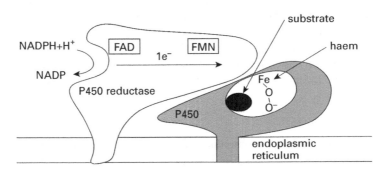

Figure 4.2 A diagrammatic representation of plant P450 and P450 reductase enzymes on the endoplasmic reticulum (after Werck-Reichhart *et al.*, 2000).

The P450 proteins are between 45 and 62 kDa in size. While their amino acid sequences may vary considerably, their three-dimensional structure is highly conserved, especially in the haem-binding region. The haem binds to the protein at a cysteine residue and the flanking sequence (Figure 4.3) is a characteristic of all P450s.

The conserved oxygen-binding sequence is about 150 residues upstream from the haem and consists of Ala or Gly–Gly–X–Asp or Glu–Thr–Thr or Ser. In both haem- and oxygen-binding sequences, X denotes any other amino acid.

When these conserved sequences were used to study P450s in plants, a surprisingly large number and diversity of P450s was found. Indeed, more than 500 plant P450 genes are now known in over 50 families, indicating that the P450s are the largest group of plant proteins. The precise roles of the proteins encoded by these genes is, however, largely unknown.

A nomenclature has been designed for P450 genes based on the identity of the amino acid sequences of the proteins they encode (Figure 4.4). The genes have been numbered in chronological order depending on their date of submission to the P450 nomenclature committee (http://drnelson.utmem.edu/CytochromeP450.html).

Typical families are numbered from CYP71 to CYP99. For example, CYP71 C6v1 in wheat is inhibited by glyphosate and CYP76B1 catalyses the dealkylation of the phenylurea herbicides (Wen-Sheng *et al.*, 2005).

The discovery of new P450 genes continues in plants. In contrast, only about 50 P450 genes in 17 families have been described in humans. So why are there so many P450s in plants? The answer seems to be that they play a very wide role in plant secondary metabolism. They have been shown to be involved in the biosynthesis and metabolism of a wide variety of compounds, including terpenes, flavonoids, sterols, hormones, lignins, suberin, alkaloids and phytoalexins. They are also induced by pathogen attack, xenobiotics and by light-induced stress, unfavourable osomotic conditions, wounding and infection.

It is currently believed that herbicide molecules also fit the active sites of these P450s involved in biosynthesis, suggesting a broad diversity of substrate selectivity.

Regarding their roles in herbicide metabolism, much remains to be done to establish substrate specificity and both the molecular and the metabolic regulation of these

$$Fe$$
$$|$$
$$Gly-Cys-X-Arg-X-Gly-X-X-Phe$$

Figure 4.3   The haem-binding sequence that is characteristic of all P450s.

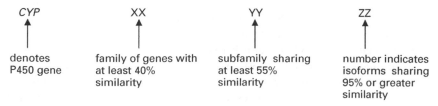

CYP	XX	YY	ZZ
denotes P450 gene	family of genes with at least 40% similarity	subfamily sharing at least 55% similarity	number indicates isoforms sharing 95% or greater similarity

Figure 4.4   Nomenclature for naming P450 genes.

important enzymes. Such understanding will be invaluable in predicting and elucidating herbicide selectivity, as well as in the discovery and design of new selective herbicides.

The main reactions catalysed by P450s are shown in Figure 4.5. In herbicide metabolism these are hydroxylation and dealkylation, which progresses via a hydroxylation step. Examples of herbicides metabolised by P450s in plant systems include sulfonylureas (including primisulfuron, nicosulfuron, prosulfuron, triasulfuron and chlorimuron), substituted ureas (chlorotoluron, linuron), chloroacetanilides (metolachlor, acetochlor), triazolopyrimidines (flumetsulam), aryloxyphenoxypropionates (diclofop), benzothiadiazoles (bentazon) and imidazolinones (imazethapyr). Selectivity to herbicides can be due to ability of the crop to metabolise herbicides via P450s, an ability that may not be possessed by susceptible weeds. In some cases, however, this metabolism is not enough to prevent crop damage, either because of low rates of P450 metabolism or phytotoxicity of products produced by these reactions. Crop damage may only be prevented if reactions from Phase II (conjugation) are successful in carrying out further detoxification.

Some Phase I reactions may be catalysed by peroxidases (E.C. 1.11.1.7) which are commonly found in leaves at high concentrations, being able to catalyse oxidations using hydrogen peroxide. They are currently thought to be involved in proline hydroxylation, indole acetic acid (IAA) oxidation and lignification, and have been implicated in the metabolism of aniline compounds produced in the degradation of phenylcarbamate, phenylurea and acylaniline herbicides.

4.2.1.3 Conjugation

In this phase of herbicide metabolism, the molecule becomes conjugated to natural cell metabolites, such as amino acids, sugars, organic acids or the tripeptide glutathione (γ-glutamyl–cysteinyl–glycine). This has the effect of both further reducing phytotoxicity and increasing the solubility of the herbicide or its metabolite, which may also serve to target the conjugate to the vacuole.

The most widely studied conjugation reaction in relation to herbicide detoxification is that of glutathione conjugation, carried out by the enzyme family glutathione S-transferases (GSTs), E.C. 2.5.1.18. Glutathione is abundant in plants, often exceeding 1 mM concentration in the leaf cell cytoplasm, where it functions as a scavenger of free radicals, protecting photosynthetic cells from oxidative damage. As GSTs are a large group of similar enzymes found in all eukaryotes, differences in the spectrum of GSTs present plays an important role in selectivity of herbicides. GSTs have a range of endogenous functions involving their abilities to detoxify and act as redox buffers.

The GSTs are abundant, soluble enzymes of about 50 kDa, each composed of two subunits of equal size, containing an active site located in the N-terminus that binds glutathione and is highly conserved in all GSTs. Herbicides and other xenobiotics are bound at the hydrophobic C-terminal half of the subunit. This site varies considerably and accounts for the differing specificity of GSTs towards herbicides.

Arabidopsis thaliana has at least 30 distinct GST genes and their nomenclature is complex. The subunits have been classified on the basis of their amino acid sequence and similarities in gene structure. The classes are termed phi (F) zeta (Z), theta (T) and tau (U) and are used in conjugation with the species name, for example:

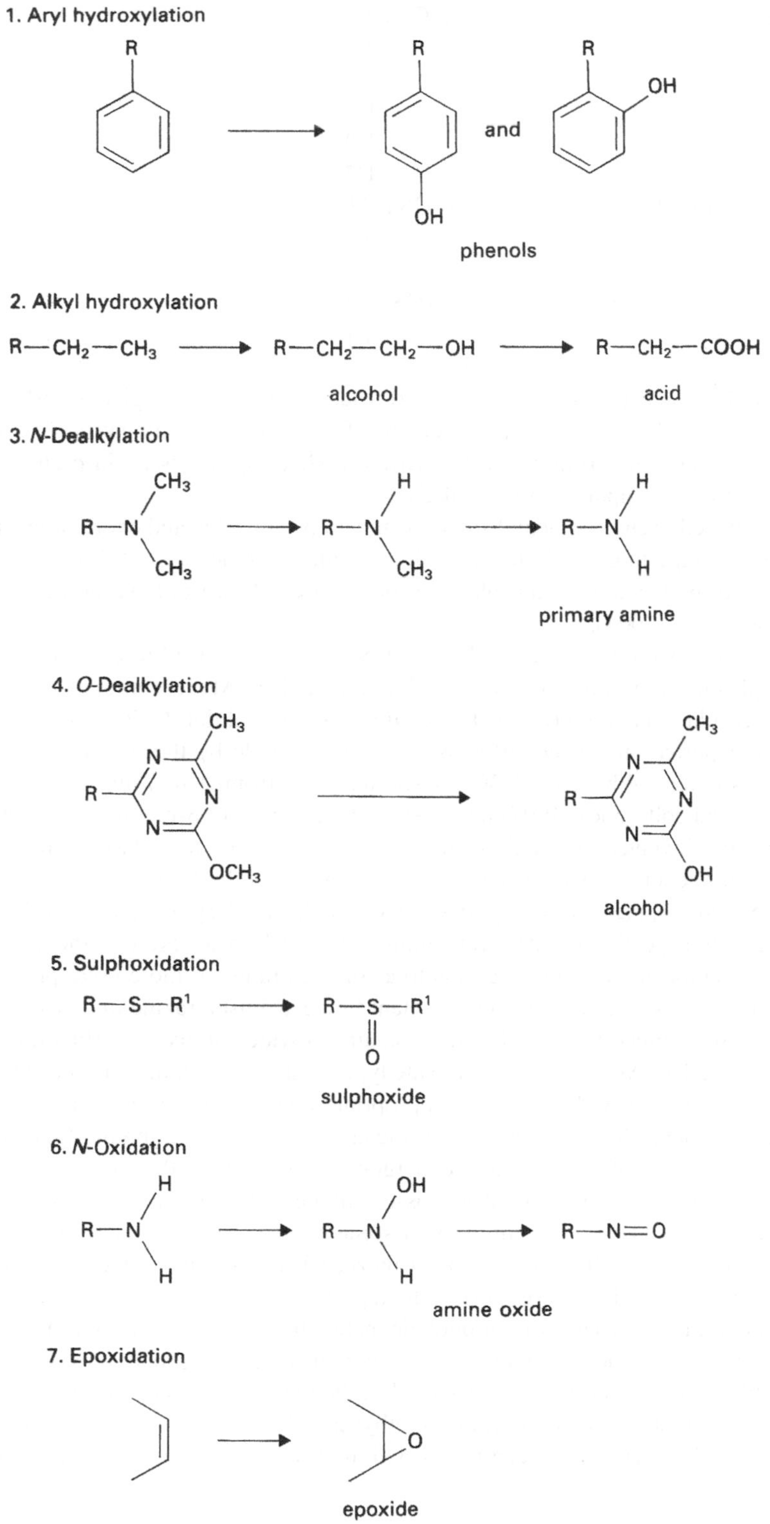
1. Aryl hydroxylation
R
R
R
OH
and
OH
phenols
2. Alkyl hydroxylation
R—CH2—CH3
R—CH2—CH2—OH
R—CH2—COOH
alcohol
acid
3. N-Dealkylation
CH3
R—N
CH3
H
R—N
CH3
H
R—N
H
primary amine
4. O-Dealkylation
CH3
N
R
N
N
OCH3
CH3
N
R
N
N
OH
alcohol
5. Sulphoxidation
R—S—R1
R—S—R1
O
sulphoxide
6. N-Oxidation
H
R—N
H
OH
R—N
H
R—N=O
amine oxide
7. Epoxidation
O
epoxide

Figure 4.5 The main reactions carried out by cytochrome P450 monooxygenases (P450s).

In *Zea mays*:	*Zm* GST F1
	F2
	F3
	U1
	U2
	U3
In *Arabidopsis thaliana*:	*At* GST T1
	F1
	U1
In *Alopecurus myosuroides*:	*Am* GSTU1
	F1

Why are there so many GSTs in plants and what are their physiological roles? Although often cited in the literature, the involvement of GSTs in the conjugation of secondary metabolites remains far from clear. A function in shuttling anthocyanin pigments to the tonoplast for vacuole uptake appears likely.

Plants exposed to environmental stress or infection show elevated expression and activity of GSTs, suggesting a role in maintaining cellular homeostasis following oxidative stress. Indeed, herbicide resistant black-grass has been shown to have higher activity of GST (Reade *et al.*, 2004).

Edwards and Dixon (2000) consider that GSTs are well placed to use physiologically high cytoplasmic concentrations of glutathione (0.2–1.0 mM) to conjugate electrophilic herbicide residues effectively. As the conjugates can inhibit GST activity, they are actively transported out of the cytoplasm into the vacuole by the ATP-binding cassette (ABC)-transporters. When the ABC-transporter gene from *Arabidopsis* was expressed in yeast, Lu and colleagues (1997) demonstrated the uptake of *S*-metolachlor-glutathione, confirming the involvement of these transporters across the tonoplast membrane.

Once in the vacuole, a specific vacuolar carboxypeptidase hydrolyses the glycine residue and the dipeptide conjugate is re-exported to the cytoplasm. The glutamate group is removed by a glutamyl transpeptidase and the remaining cysteinyl derivatives are further transformed by *N*-malonylation and further oxidation to a complex range of more polar products.

GST activity was first demonstrated in plant tissue against atrazine in maize extracts in 1970. Since this observation, GST activity against a wide variety of herbicides has been reported (Table 4.1). As with other conjugate types, glutathione conjugation can be carried out against the parent herbicide if an appropriate conjugating group is present, or can follow Phase I metabolism. An example of the latter is the conjugation of glutathione with thiocarbamates, only after they have undergone conversion to their corresponding sulphoxides. Crops are often reported to possess higher GST activities against herbicides than susceptible weeds and this might offer some degree of selectivity between crop and weed. Activity against chloroacetamides (maize, wheat, sorghum, rice), oxyacetamides (maize), atrazine (maize, sorghum), fenoxaprop, fluorodifen, flupyrsulfuron-methyl, dimethenamid (all wheat) and the sulphoxide metabolite of EPTC (sorghum) have all been reported. In soybean, homoglutathione is found in place of glutathione. Conjugations utilising this against several chloroacetanilides, the diphenyl ethers acifluorfen and fomesafen, and the sulphonylurea chlorimuron-ethyl are all reported in this crop. In addition to selectivity, GSTs have also been implicated in playing a role in herbicide resistance in

Table 4.1 Examples of some herbicides metabolised by glutathione S-transferases in various plant systems.

Chemical family	Examples
Chloroacetamides	Alachlor, acetochlor, metolachlor, pretilachlor
Triazines	Atrazine
Aryloxyphenoxypropionates	Fenoxaprop
Thiocarbamates	EPTC
Diphenyl ethers	Acifluorfen, fomesafen
Sulfonylureas	Clorimuron-ethyl, triflusulfuron-methyl

a variety of weeds. In black-grass, biotypes resistant to chlorotoluron and fenoxaprop-ethyl demonstrated approximately double the GST activity of susceptible biotypes. This suggests that GSTs, as well as P450s, may play a role in enhanced metabolism resistance in this species. In velvetleaf (*Abutilon theophrasti*), resistance to atrazine has also been demonstrated to be due to higher conjugation of this herbicide to glutathione.

Another commonly encountered Phase II reaction in plants is conjugation with glucose, catalysed by the glucosyltransferases (EC 2.4.1.71) utilising uridine diphosphate glucose (UDPG) as the glucose donor. Once conjugated, the glucose may undergo a further Phase II reaction by 6-O-conjugation with malonic acid, catalysed by the malonyl-CoA-dependent malonyltransferases. These Phase-II metabolites then undergo ATP-dependent transport into the vacuole.

Interestingly, Pflugmacher and Sandermann (1998) found O-, N- and S-glucosyltransferase activity to be very widely distributed throughout the plant kingdom – not solely confined to higher plants, but even in marine macroalgae. Indeed, they hypothesised that this activity in 'lower' plants may make an important contribution to the detoxification of xenobiotics in the global environment.

Finally, acidic herbicide molecules such as the synthetic auxin phenoxyacetic acids can be conjugated to the amino acids glutamine, valine, leucine, phenylalanine or tryptophan, although the enzymology of these reactions remains obscure. As an example, crop plants are able to rapidly detoxify the photosynthetic inhibitor bentazone by rapid aryl hydroxylation followed by conjugation to glucose. Susceptible weeds appear unable to metabolise the parent herbicide and phytotoxicity is observed.

A 200-fold margin of selectivity between rice and *Cyperus serotinus* has been attributed to this metabolic route. Similarly in soybean, where an 8-hydroxy derivative has been detected, Leah *et al.* (1992) isolated and purified two glucosyltransferases from tolerant soybean that were capable of glycosylating 6-hydroxybentazone (Figure 4.6). This soluble enzyme had a relative molecular mass of 44.6 kDa with binding constants for kaempferol and 6-hydroxybentazone of 0.09 and 2.45 mM, respectively. They also found a membrane-bound enzyme, whose primary substrate was *p*-hydroxyphenylpyruvic acid, with a relative molecular mass of 53 kDa and binding constants of 0.11 and 1.96 mM for *p*-hydroxyphenylpyruvic acid and 6-hydroxybentazone, respectively. These findings, and those subsequently shown by others, imply an overlapping specificity of aryl hydroxylated herbicides and the synthesis and storage of secondary metabolites.

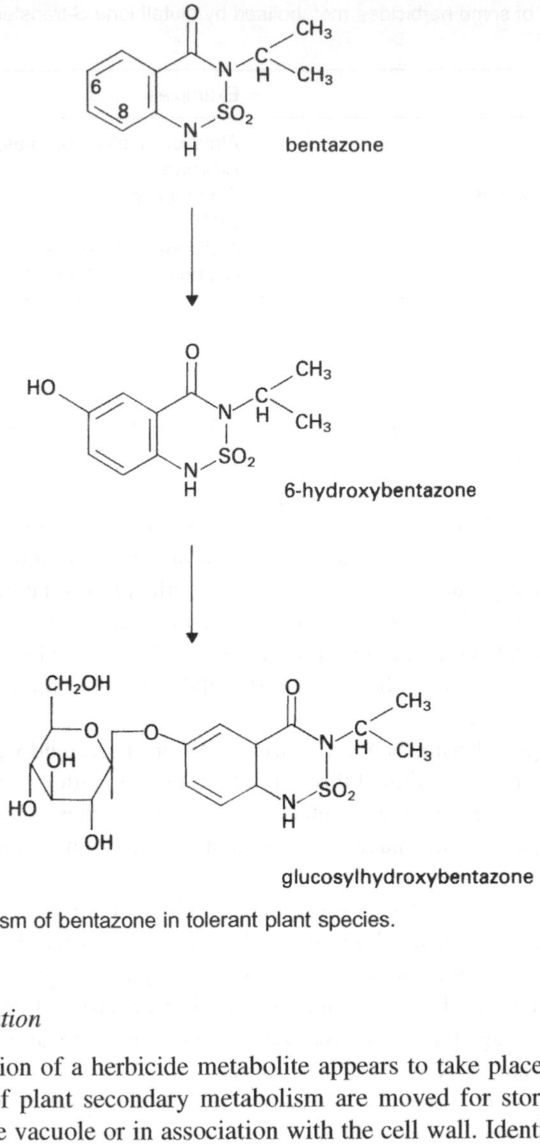

Figure 4.6 Metabolism of bentazone in tolerant plant species.

4.2.1.4 Sequestration

Compartmentalisation of a herbicide metabolite appears to take place in much the same way as products of plant secondary metabolism are moved for storage. The place for storage is either the vacuole or in association with the cell wall. Identification of a membrane-bound glutathione-dependent ABC pump in the vacuolar membrane suggests that Phase II conjugation to glutathione or malonate might serve to facilitate the movement of metabolites and could be considered as a way of 'tagging' molecules for movement into the vacuole.

The processing and vacuolar import of herbicide conjugates in plants is a two-step process, first involving glucosylation and then derivatisation of the sugar with malonic acid. The significance of this reaction is not entirely understood, though malonylation appears to act as a tag, directing the conjugates for vacuolar import.

The malonylation reaction is carried out by malonyltransferases which can conjugate compounds containing amino (*N*-malonylation) or hydroxyl (*O*-malonylation) residues.

Malonylation also appears to prevent digestion by glucosidases and may also facilitate conjugate transport across the plasma membrane. For example, glucosidic conjugates of pentachlorophenol formed in soybean and wheat undergo malonylation, and 3,4-dichloroaniline undergoes *N*-malonylation in soybean and wheat. In each case, the transport of the newly formed conjugates was shown to be routed towards the vacuole. Although malonylation of glucosides is important in directing their importation into the vacuole, it is clear that the glucosides themselves can undergo vacuolar deposition.

Once conjugates are situated in the vacuole, sequential removal of peptides from glutathione is carried out by peptidases. This results in the metabolite being conjugated to glutamylcysteine and possibly just to cysteine. It is postulated that this allows for recycling of amino acids back to the cytoplasm and in addition may prevent the conjugated metabolite from being exported back there, as it no longer is a full glutathione conjugate. This pumping mechanism may have the additional benefit of stopping the build-up of glutathione conjugates from inhibiting cytoplasmic GST activity, as some conjugates have been demonstrated to be powerful competitive inhibitors of GSTs. Once the metabolite conjugate has entered the vacuole it may be further metabolised, stored there or excreted across the plasma membrane to the extracellular matrix. Transport of glucosylated herbicides into the vacuole has also been reported. This is ATP-requiring, so is also active transport. It appears that the membrane pump carrying this out is distinct from the glutathione system (Joshua *et al.*, 2007).

Most herbicide metabolites eventually become associated with the insoluble compartments of the cell, bound to lignin or polysaccharides. In these forms they can be released by enzymic hydrolysis, and both hydroxylated and unaltered forms have been detected. It is thought that this process can occur by covalent linkage.

While conjugates usually represent the end-product of herbicide metabolism, they should not be regarded as totally inert. Some evidence exists for the hydrolysis of the glucoside to regenerate an active molecule, and so the conjugate may be regarded as a reservoir of potential activity. Such regeneration depends on the nature of the glycoside linkages involved and their proximity and susceptibility to the action of β-glucosidase.

An overview of the different phases of herbicide metabolism in a plant cell is presented in Figure 4.7.

4.3   Herbicide safeners and synergists

Herbicide safeners, also termed antidotes in some texts, are chemicals that, when applied before or with herbicides, increase the tolerance of a cereal crop to a herbicide. This activity has been known since the 1970s and the safening effect is not seen on the weeds. A list of herbicide safeners available as commercial products is given in Table 4.2.

As some safeners may show structural homology with herbicides, it was previously thought that they competed with the herbicide molecule for the target site. We now know, however, that their protective effect results from a general enhancement of detoxification processes, including the induction of:

−   P450 oxygenases
−   Glutathione *S*-transferases

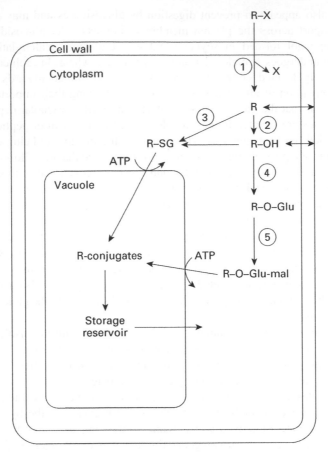

Figure 4.7 A schematic representation of the different phases of herbicide metabolism in a plant cell (R, denotes the active herbicide molecule). Key: 1. hydrolase; 2. oxido-reductase; 3. glutathione S-transferase; 4. glucosyl transferase; 5. malonyl transferase.

– Glucosyltransferases
– Vacuolar transport
– Glutathione synthesis
– Glutathione peroxidase
– Sulphate assimilation

Since a crop has to show some degree of tolerance to the herbicide for it to be safened, it would appear that the process is a 'top-up' mechanism for detoxification routes that are already operative. Despite considerable interest in safeners, their activities have mainly been demonstrated in monocotyledonous crops, notably maize, wheat, sorghum and rice.

It is likely that the identification of genes encoding safener-inducible enzymes will allow their transfer to give genetically modified crops with an enhance ability to detoxify herbicides. This approach has already shown potential (Davies and Caseley, 1999), although further development will require a greater public acceptance of GM crops in future.

Table 4.2 Herbicide safeners available as commercial products (from Davies and Caseley, 1999).

Safener	Crop	Herbicide	Application method
Benoxacor (CGA 154281)	Maize	Metolachlor	Spray as mixture with herbicide
Cloquintocet-mexyl (CGA 184927)	Wheat	Clodinafop-propargyl	Spray as mixture with herbicide
Cyometrinil (CGA 43089)	Sorghum	Metolachlor	Seed-treatment
Dichlormid (DDCA, R25788)	Maize	EPTC, butylate, vernolate	Pre-plant incorporated with herbicide
Fenchlorazole-ethyl (HOE 70542)	Wheat	Fenoxaprop-ethyl	Spray as mixture with herbicide
Fenclorim (CGA 123407)	Rice	Pretilachlor	Spray as mixture with herbicide

(Continued)

Table 4.2 (*Continued*)

Safener	Structure	Crop	Herbicide	Application method
Flurazole (MON 4606)		Sorghum	Alachlor	Seed-treatment
Fluxofenim (CGA 133205)		Sorghum	Metolachlor	Seed-treatment
Furilazole (MON 13900)		Cereals	Halosulfuron-methyl	Spray as mixture with herbicide
Mefenpyr-diethyl		Wheat, rye, triticale, barley	Fenoxaprop-ethyl	Spray as mixture with herbicide
MG 191		Maize	Thiocarbamates	Spray as mixture with herbicide
Naphthalic anhydride (NA)		Maize	EPTC, butylate, vernolate	Seed-treatment
Oxabetrinil (CGA 92194)		Sorghum	Metolachlor	Seed-treatment

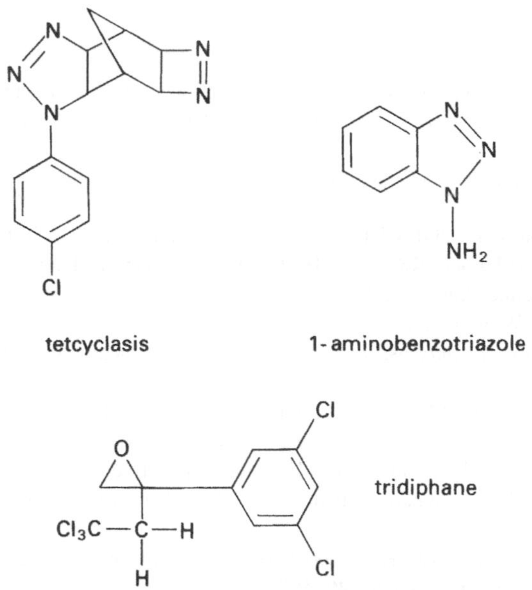

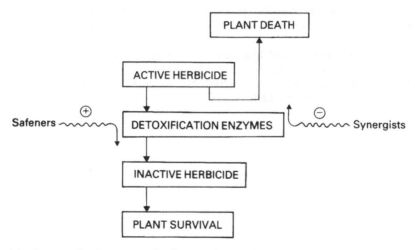

Figure 4.8   Structures of tetcyclasis, 1-aminobenzotriazole and tridiphane.

Figure 4.9   A generalised summary of safener and synergist action.

In contrast to safeners functioning by stimulating herbicide degradation, herbicide synergists inhibit these enzyme systems, so that plant defence mechanisms are overcome and phytotoxicity rapidly ensues. Aminobenzotriazole and tetcyclasis, for example, are able to bind to the haem portion of the cytochrome P450 to prevent herbicide oxidations. Furthermore, tridiphane can inhibit GSTs to synergise atrazine, may form its own phytotoxic glutathione conjugate, and may also inhibit cytochrome P450-linked monooxygenases (Moreland *et al.*, 1989) (Figure 4.8).

A generalised summary of safener and synergist action is presented in Figure 4.9.

References

Cummins, I., Burnet, M. and Edwards R. (2001) Biochemical characterisation of esterases active in hydrolysing xenobiotics in wheat and competing weeds. *Physiologia Plantarum* **113**, 477–485.

Davies, J. and Caseley, J.C. (1999) Herbicide safeners: a review. *Pesticide Science* **55**, 1043–1058.

Edwards, R. and Dixon, D.P. (2000) The role of glutathione transferases in herbicide metabolism. In: Cobb, A.H. and Kirkwood, R.C. (eds) Herbicides and their Mechanisms of Action. Sheffield, UK: Sheffield Academic Press, Ch. 3.

Leah, J.M., Worrall, T.W. and Cobb, A.H. (1992) Isolation and characterisation of two glucosyl-transferases from *Glycine max* associated with bentazone metabolism. *Pesticide Science* **34**, 81–87.

Lu, Y.P., Li, Z.S. and Rea, P.A. (1997) AtMRP1 gene of Arabidopsis encodes a glutathione S-conjugate pump. *Proceedings of the National Academy of Sciences, USA* **94**, 8243–8248.

Moreland, D.E., Novitzky, W.P. and Levi, P.E. (1989) Selective inhibition of cytochrome p450 isozymes by the herbicide synergist tridiphane. *Pesticide Biochemistry and Physiology* **35**, 42–49.

Pflugmacher, S. and Sandermann Jr, H. (1998) Taxonomic distribution of plant glucosyltransferases acting on xenobiotics. *Phytochemistry* **49**, 507–511.

Reade, J.P.H., Milner, L.J. and Cobb, A.H. (2004) A role for glutathione S-transferases in resistance to herbicides in grasses. *Weed Science* **52**, 468–474.

Wen-Sheng, X., Xiang-Jin, W., Tian-Rui, R. and Xiu-Lian, J. (2005) Expression of a wheat cyto-chrome P450 monooxygenase in yeast and its inhibition by glyphosate. *Pest Management Science* **62**, 402–406.

Werck-Reichhart, D., Hehn, A. and Didierjean, L. (2000) Cytochrome P450 for engineering herbicide tolerance. *Trends in Plant Science* **5**, 116–123.

Yuan, J.S. Tranel, P.J. and Stewart Jr, C.N. (2007) Non-target-site herbicide resistance: a family business. *Trends in Plant Science* **12**(1), 6–13.

Chapter 5
Herbicides That Inhibit Photosynthesis

5.1   Introduction

Photosynthesis is the process whereby solar energy is converted into chemical energy by plants, algae and some bacteria, and takes place in the chloroplast, an organelle unique to the plant kingdom (Figure 5.1). The generalised and highly simplified equation

$$CO_2 + H_2O \xrightarrow{\text{light}} (CH_2O) + O_2$$

summarises photosynthesis, in which, firstly, water is oxidised within the thylakoid membranes to protons, electrons and oxygen. The movement of protons and electrons through the thylakoid generates ATP and reduced $NADP^+$, and this energy is utilised in the enzymatic reduction of CO_2 to carbohydrates within the chloroplast stroma. No herbicides are known to directly interfere with carbon reduction, but many of the herbicides currently in use act by either blocking or diverting thylakoid electron flow. It is therefore not surprising that considerable research effort has been directed towards the chemical inhibition of photosynthetic electron flow.

5.2   Photosystems

In energy terms, carbon reduction and water oxidation are separated by about 1.2 V. This means that electrons have to move uphill against an energy gradient of 1.2 V in order to reduce CO_2 to carbohydrates. This unique process is driven by two inputs of light energy at huge pigment–protein complexes termed photosystems, which span the thylakoid membrane. The thylakoid itself is now known to contain two functional regions, namely appressed and non-appressed membranes (Figure 5.1). The appressed region shows close membrane interaction and is enriched in Photosystem II (PS II) and the non-appressed region contains stroma-exposed areas and is enriched in Photosystem I (PS I) and the ATP synthase complex. The relative proportions of appressed versus non-appressed regions depends on the chemical environment of the chloroplast, and is very sensitive to available photon flux density. Thus, leaves growing in shade conditions will contain chloroplasts with more appressed areas (i.e. granal stacks), whereas leaves in full sun will exhibit

Herbicides and Plant Physiology, Second Edition By Andrew H. Cobb and John P.H. Reade
© 2010 A.H. Cobb and J.P.H. Reade

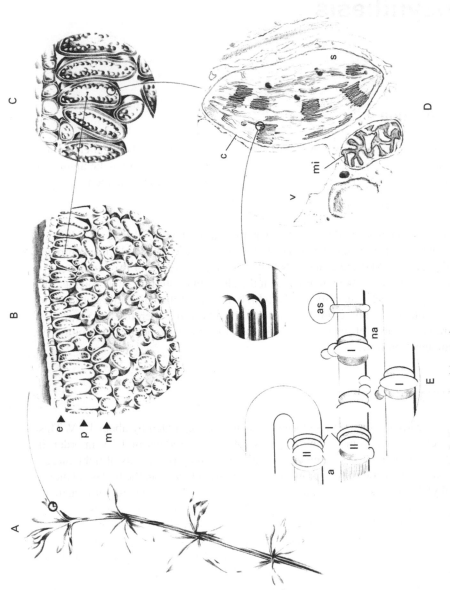

Figure 5.1 Leaf structure in *Galium aparine* (cleavers).
A. Leaf arrangement in whorls.
B. Leaf section to show epidermal (e), palisade (p) and mesophyll (m) cells.
C. Palisade cell detail.
D. Cell ultrastructure featuring chloroplast (c), stroma (s), vacuole (v) and mitochondrion (mi).
E. Detail of thylakoids. Appressed thylakoids (a) are enriched in Photosystem II (II) and light-harvesting complexes (L); non-appressed thylakoids (na) are enriched in Photosystem I (I) and ATP synthase (as).

proportionately more non-appressed regions and a decreased thylakoid to stroma ratio. Such changes in thylakoid architecture can occur within hours and suggest that this membrane is particularly fluid in nature. Indeed, thylakoids are about 50% lipid with high concentrations of electroneutral galactolipids, and this property permits the movement of electron carriers, such as plastoquinone, and light-harvesting complexes, from appressed to non-appressed areas.

Each photosystem contains specific polypeptides, pigments and electron donors/acceptors, and a unique chlorophyll termed a reaction centre (p680 or p700) at which an electron is moved from low to higher energy. Functionally, the two photosystems operate in series, such that the primary reductant generated from the photolysis of water at the oxidising side of PS II passes electrons through a series of carriers of lower reducing power to PS I. Here a second light reaction transfers electrons to their eventual natural acceptor, $NADP^+$.

When a photon of light is absorbed by the light-harvesting complex and the excitation energy transferred to the PS II reaction centre (p680), a charge separation occurs to $p680^+$ and $p680^-$. The species $p680^+$ is quenched by an electron from water via a tyrosine residue on the D1 protein, and phaeophytin, a chlorophyll molecule lacking magnesium, is the primary electron acceptor from $p680^-$. The first stable electron acceptor is a quinone, Q_A, tightly bound within a particular protein environment. Q_A acts as a single electron carrier and is closely associated with the two-electron carrier, Q_B, located on the D1 protein, which delivers pairs of electrons to the mobile plastoquinone pool (Figure 5.2). Photosystem II may therefore be regarded as a water-plastoquinone oxidoreductase. Electrons then pass via a cytochrome $b_6 - f$ complex to the copper-containing protein plastocyanin and to PS I.

The PS I complex contains the reaction centre chlorophyll p700, comprising iron–sulphur centres which act as electron acceptors, several polypeptides and the electron carriers A_0 and A_1. Light excitation causes charge separation at p700 and A_0, a specific monomeric chlorophyll a, is the first acceptor. A_1 is thought to be a phylloquinone (vitamin k_1) which donates an electron to the iron–sulphur centres and hence reduces $NADP^+$ via ferredoxin (Figure 5.3). Thus PS I operates as a plastocyanin–ferredoxin oxidoredutase.

Detailed structures are now known of the light-harvesting and photosystem complexes and the electron transport chain components in the thylakoid membrane. Indeed, some have actually been crystallised and their three-dimensional structures established. Figure 5.4 gives a schematic version of the thylakoid, demonstrating H^+ and e^- flow. The reader is referred to Blankenship (2002) and Lawlor (2001) for more details.

5.3   Inhibition at Photosystem II

The action of photosynthetic inhibitors has traditionally been monitored *in vitro* by measuring the so-called '**Hill Reaction**'. This is the ability of a thylakoid preparation to evolve oxygen in the presence of a suitable electron acceptor. The natural acceptor, $NADP^+$, is washed free from isolated thylakoids, so artificial acceptors (A) are used experimentally.

$$2H_2O + 2HA \xrightarrow[\text{thylakoids}]{\text{light}} 2AH_2 + O_2$$

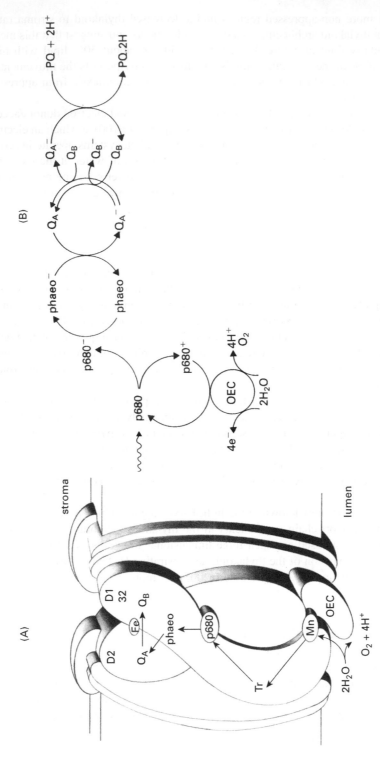

Figure 5.2 Structural (A) and functional (B) models of Photosystem II. 32 refers to the molecular mass of the D1 polypeptide in kDa. Tr, a tyrosine residue acting as an electron donor to p680[+], the bold arrow shows the direction of electron transfer; OEC, oxygen-evolving complex; p680, PS II reaction centre; phaeo, phaeophytin a; Q_A and Q_B, quinones; PQ, mobile plastoquinone pool (after Rutherford, 1989).

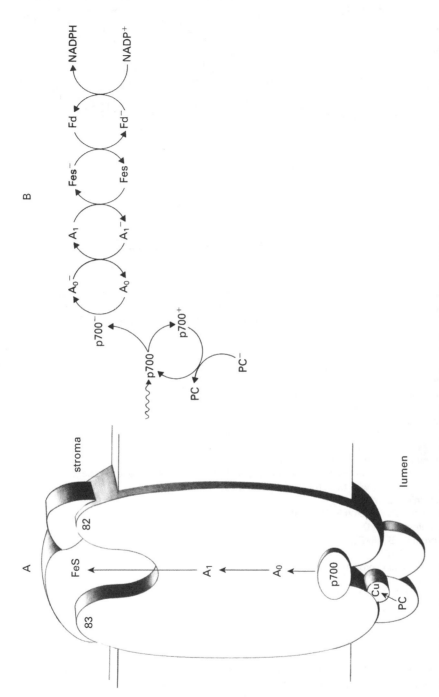

Figure 5.3 Structural (A) and functional (B) models of Photosystem I. Numbers 82 and 83 refer to molecular masses of polypeptides in kDa. PC, plastocyanin (the electron donor to p700[+]); A_0 and A_1, electron acceptors/donors; bold arrow shows direction of electron transfer. p700, PS I reaction centre; FeS, iron-sulphur centres; Fd, ferredoxin. The final electron acceptor is NADP[+] (after Knaff, 1988).

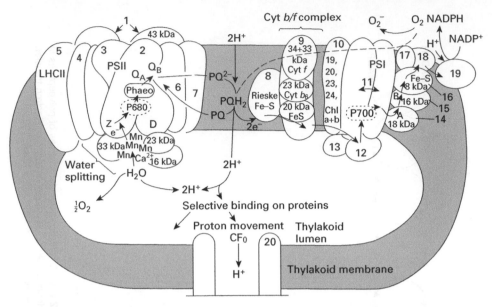

Figure 5.4 Schematic representation of the light-harvesting and photosystem complexes and electron transport chain in the thylakoid membrane (from Lawlor, 2001, by permission of Oxford University Press). The number of protein complexes and their relation is shown. The molecular mass of components is indicated. The components are listed as follows. Key:

1   Antenna protein–pigment complex
2   32 kDa, D1 herbicide binding protein of the reaction centre
3   32 kDa, D2 reaction centre protein
4   Cytochrome b_{559}, 9 kDa b_{559} type 1 and 4 kDa b_{559} type 2 proteins
5   Light-harvesting antenna
6   10 kDa docking protein
7   22 kDa stabilising protein (intrinsic membrane protein)
8   20 kDa Rieske Fe-S centre
9   Cytochrome b_6–f complex with polypeptides
10 Light-harvesting protein-pigment (chlorophyll a and b) complex of PSI and polypeptides
11 PSI reaction centre, with two 70 (?) kDa polypeptides
12 Plastocyanin, 10.5 kDa
13 Plastocyanin binding protein (10 kDa)
14 Fe–S protein A, 18 kDa
15 Fe–S proteins B, 16 kDa
16 Fe S protein 8 kDa
17 Ferredoxin binding protein
18 Ferredoxin
19 Ferredoxin, NADP oxidoreductase
20 Coupling factor, CF_0, membrane subunits

A commonly used artificial electron acceptor is potassium ferricyanide which is reduced to ferrocyanide with the evolution of oxygen. Oxygen evolution can be conveniently monitored using an oxygen electrode. Similarly, the blue dye 2,6-dichlorophenolindophenol (DCPIP) will accept electrons from the functioning thylakoid and the reduced product is colourless. Thus, photosynthetic electron flow can be measured by following DCPIP decolourisation with a spectrophotometer (Figure 5.5).

Such reactions, however, measure the activity of both photosystems rather than specific sites, and are therefore of limited value in identifying precise sites of photosynthetic inhibition. Nowadays the partial reactions of electron flow can be characterised in great detail using highly specific artificial electron donors and acceptors (e.g. Table 5.1; Trebst, 1980). Indeed, herbicides have proved invaluable in unravelling the flow of electrons in the thylakoid.

It was using DCPIP as an artificial electron acceptor that Wessels and van der Veen (1956) first demonstrated that the urea herbicide diuron could reversibly inhibit photosynthetic electron flow at micromolar concentrations. Subsequent studies in the following decades established that diuron was acting at the reducing side of PS II in the vicinity of Q_A and Q_B. This was deduced from the observations that (1) PS I activity was insensitive to diuron, (2) electron flow at the cytochrome $b_6 - f$ complex was similarly unaffected, and (3) diuron had no influence on the photolysis of water, nor on charge separation at p680, but electron flow to plastoquinone was potently inhibited.

Several classes of herbicide including the ureas, triazines and phenols are now known to inhibit PS II activity by displacing plastoquinone from the Q_B site and so preventing electron flow from Q_A^-. Although Q_A is tightly bound to the D2 protein, Q_B is not firmly bound to D1, and so herbicides successfully compete with Q_B for this site. Inhibition by the ureas and triazines is characteristically reversible and competitive. Furthermore, it was found that there was only one binding site for each PS II reaction centre, and that binding/dissociation constants were very similar to inhibition constants, hence occupancy of this site is required for photosynthetic inhibition.

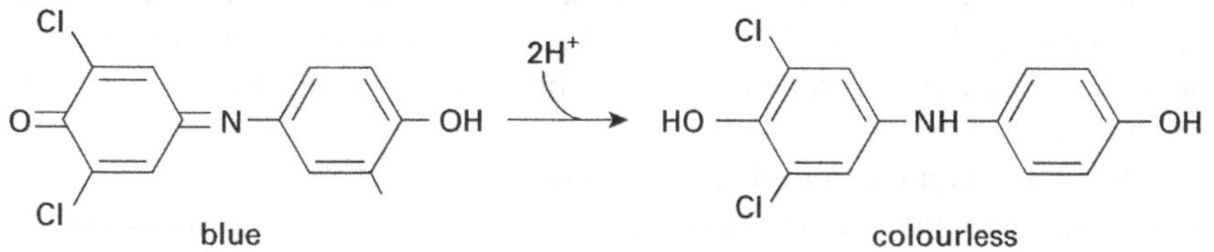

Figure 5.5 DCPIP decolourisation in the presence of functioning thylakoids.

Table 5.1 Examples of the measurement of photosystem activity *in vitro* by oxygen exchange.

Donor	Acceptor	Inhibitor	Reactions measured	Oxygen
H_2O	Ferricyanide	Absent	PS II and PS I ($H_2O \rightarrow$ FeS centres)	Evolution
H_2O	Dimethyl-benzoquinone	Absent	PS II only ($H_2O \rightarrow$ plastoquinone)	Evolution
H_2O	Silicomolybdate	Absent	PS II only ($H_2O \rightarrow$ phaeophytin)	Evolution
Ascorbate + DCPIP	Methyl viologen	Diuron	PS I only (plastocyanin $\rightarrow$ ferredoxin)	Uptake

Studies in the 1970s established that thylakoids treated with trypsin became insensitive to diuron suggesting a stromal-facing proteinaceous binding site (Renger, 1976). This protein was further characterised using radiolabelled herbicides as a rapid turnover polypeptide (molecular mass 32 kDa) that was encoded in the chloroplast *psb*A genome (Mattoo *et al.*, 1981). Conclusive evidence that this protein had a dominant role in herbicide binding came from its analysis in plants showing resistance to PS II herbicides. Hirschberg and colleagues (1984) cloned the *psb*A gene from atrazine-resistant and susceptible biotypes of *Solanum nigrum* and *Amaranthus retroflexus*, and detected a single base substitution from serine to glycine at amino acid 264 to be the basis of resistance to the herbicide. Resistance is therefore achieved by the substitution of one amino acid in a protein containing 353 amino acid residues!

Figure 5.6 presents the interaction between the quinones and atrazine at protein D1.

A separate binding site specific to the phenols (such as dinoseb) and the hydroxybenzonitriles (e.g. ioxynil) has been proposed with a molecular mass of 41–47 kDa that is less susceptible to trypsin digestion, and therefore more deeply located in the thylakoid. It is argued that these compounds bind to a different site on the D1 protein.

Our understanding of this protein was vastly enhanced when the reaction centre of the photosynthetic bacterium *Rhodopseudomonas viridis* was first crystallised and then its structure resolved by Michel and Deisenhofer in 1984 (see Michel and Deisenhofer, 1988). This seminar work, for which the Nobel Prize for Chemistry was awarded in 1988, pointed to clear similarities and sequence homologies between the bacterial reaction centre and PS II, and provided three-dimensional detail of the amino acids involved in this quinone-binding herbicide niche on the D1 protein. Trebst (1987) utilised this X-ray data and, with information gained from photoaffinity labelling of herbicides and site-directed mutagenesis, proposed a detailed model of the herbicide binding niche. This model shows that the D1 polypeptide contains five transmembrane helical spans (1–5) and two parallel helices (A and B), and that helices 4, 5 and B specifically participate in herbicide binding (Figure 5.7A). The model predicts the orientation of residues at the quinone niche (Figure 5.7B) and envisages Q_B binding via two hydrogen bridges at His 215 and close to Ser 264.

Following closer examination of the Q_B binding niche, Trebst (1987) suggested that inhibitors with a carbonyl or equivalent group (e.g. ureas, triazines, triazinones) were orientated towards the peptide bond close to Ser 264 and could form a hydrogen bridge to this peptide bond. However, the phenol group of inhibitors cannot form this bridge and are therefore thought to bind towards His 215, where they are bound more strongly to the membrane. In this way, Trebst envisages two families of PS II herbicides, namely the serine and the histidine families.

More recently, additional mutants have been identified with altered amino acids in the stroma-exposed β-helix, particularly in algae, and other substitutions noted that produce herbicide resistance (Table 5.2). Substitutions at position 264 from serine to glycine yields atrazine resistance, as demonstrated in Table 12.2, while serine to alanine or threonine produces additional resistance to diuron. This observation has led to the proposal that a serine hydroxyl group may be essential for atrazine binding. Changes at nearby residues 255 and 256 also cause triazine resistance, although changes at positions 219 and 275 give resistance to diuron and also to bromoxynil in the latter example. Resistance may then be due to conformational changes in the binding niche, an absence

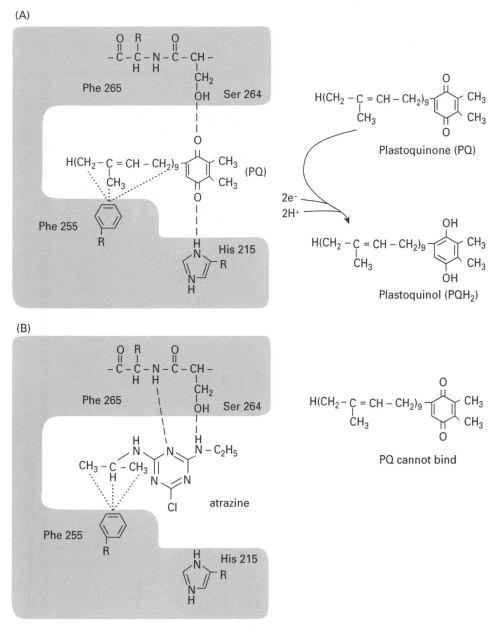

Figure 5.6 Proposed interaction between plastoquinone (A) and atrazine (B) with the Q_B site of protein D1. Dashes represent hydrogen bonds and dots represent hydrophobic interactions (after Fuerst and Norman, 1991, from the journal *Weed Science*, courtesy of the Weed Science Society of America). (A) PQ binds to the D1 protein, accepts two electrons and two protons, and is released as PQH_2. (B) Atrazine binding to the D1 protein prevents the binding of PQ.

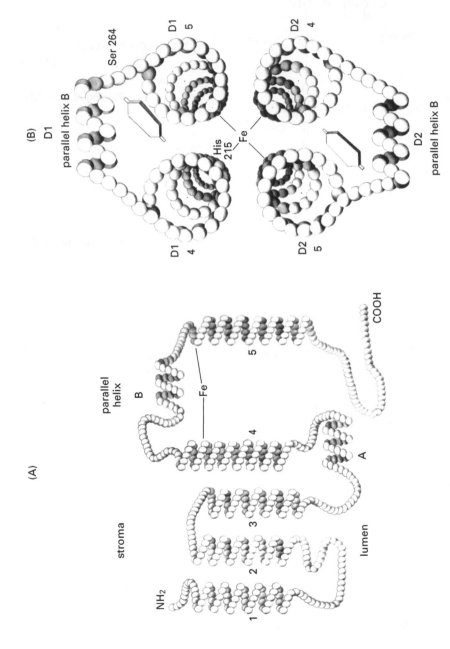

Figure 5.7 The D1 polypeptide. (a) Schematic arrangement of the protein in the thylakoid. (b) The quinone/herbicide binding niche viewed from above (after Trebst, 1987).

Table 5.2 Characteristics of mutants with amino acid substitutions in D1 (modified from Gressel, 2002).

Mutation	Organism	Primary resistances	Negative cross-resistance
Phe$_{211}$→Ser	Cyanobacteria	Atrazine ×9	
Val$_{219}$→Ile	Alga + cyano	Metribuzin ×200	Ketonitrile × 0.6
		Metabenzthiazuron ×62	
	Poa annua	Metribuzin and/or diuron	
Tyr$_{237}$→Phe	Cyanobacterium	Diuron/ioxynil ×5	BNT ×0.2
Lys$_{238}$→Val	Cyanobacterium	Ioxynil ×2.3	BNT ×0.4
Ile$_{248}$→Thr	Cyanobacterium	Metribuzin ×28	
Ala$_{250}$→Arg	Alga	Phenmedipham ×6.3	Bromoxynil ×0.3
Ala$_{250}$→Asn	Alga	Metamitron ×5	Bromoxynil ×0.2/ atrazine ×0.3
Ala$_{250}$→Asp	Alga	Phenmedipham ×5	-oxynils ×0.2/atrazine ×0.25
Ala$_{250}$→His	Alga	Phenmedipham ×10	
Ala$_{250}$→Ile	Alga	Phenmedipham ×2.5	Metribuzin ×0.25
Ala$_{250}$→Tyr	Alga	Phenmedipham ×20	Bromoxynil ×0.4
Ala$_{251}$→Cys	Alga	Metamitron/bromoxynil ×6.3	
Ala$_{251}$→Gly	Alga	Metamitron ×10	
Ala$_{251}$→Ile	Alga		Diuron ×0.6
Ala$_{251}$→Leu	Alga	Metribuzin ×108	
		Bromacil ×26	
Ala$_{251}$→Val	Alga / cyano	Metribuzin ×1000	Ketonitrile ×0.5
Phe$_{255}$→Tyr	Alga / cyano	Cyanoacrylate ×39	Metamitron ×0.3
Gly$_{256}$→Asp	Alga	Bromacil ×10	
Arg$_{257}$→Val	Cyanobacterium	Atrazine / diuron ×34	BNT ×0.3
Ala$_{263}$→Pro	Cyanobacterium	Atrazine ×2000/metribuzin ×1600	
Ser$_{264}$→Ala	Alga / cyano / *Euglena*	Metribuzin ×>3000 Chloroxuron x480 / atrazine variable	
Ser$_{264}$→Gly	Weeds / cyano	Most *s*- and *as*-triazines ×>500	-oxynils/pyridate ×<0.5
Ser$_{264}$→Asn	Tobacco cells	Terbutryn	
Ser$_{264}$→Pro	Cyanobacterium	Atrazine ×10,000	
Ser$_{264}$→Thr	Plant cells / *Euglena* *Portulaca oleracea*	Atrazine ×>50 Linuron	Dinoseb/-oxynils/BNT ×<0.3
Asn$_{266}$→Asp	Cyanobacterium	Ioxynil ×2.5	
Asn$_{266}$→Thr	Cyanobacterium	Bromoxynil ×15	
Ser$_{268}$→Pro	Soybean cells	Atrazine ×50	
Arg$_{269}$→Gly	Alga	Terbutryn ×8	Ioxynil ×0.2
Leu$_{275}$→Phe	Alga	Metamitron ×63	

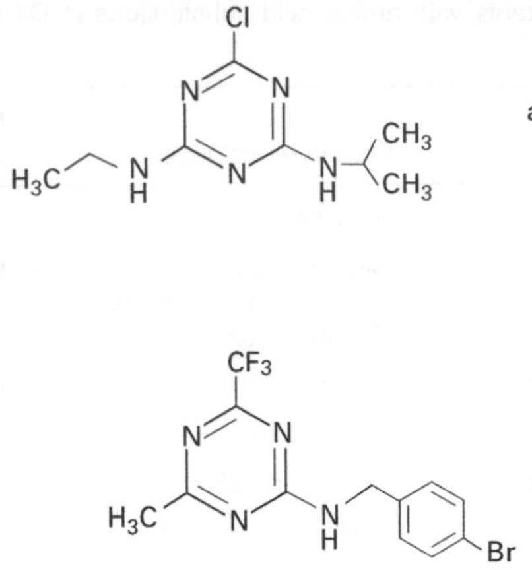

Figure 5.8   The structures of atrazine and BW314.

of a specific residue so that a herbicide is unable to bind or the introduction of steric hindrance to prevent herbicide access to the binding niche.

A new group of triazine herbicides, the 2-(4-halogenobenzyl amino)-4-methyl-6-trif-luoromethyl-1,3,5-triazines, were shown in 1998 to inhibit photosynthetic electron flow at the D1 protein. Interestingly, they have shown activity against atrazine-resistant species, such as *Chenopodium album* L. and *Solanum nigrum*, presumably by binding to different amino acids in the D1 niche (Kuboyama *et al.*, 1999; Kohno *et al.*, 2000). Figure 5.8 shows the structure of one example, BW 314, with the structure of atrazine also presented for comparison.

5.4   Photodamage and repair of Photosystem II

Plants are remarkable organisms in their ability to absorb light energy and convert it into chemical energy. Unlike animals, however, plants cannot move away from unfavourable environmental conditions and may not be able to process all the excitation energy to which they are exposed. This excess energy will lead to the generation of active oxygen species and membrane protein damage if it is not dissipated effectively. Such photodamage appears to be mainly confined to the D1 protein in the core Photosystem II reaction centre complex and inactivation can be rapidly detected following minutes of exposure to high light energies.

Fortunately, a damage–repair cycle (Figure 5.9) is present in which the PS II reaction centre is disassembled, a new D1 protein is synthesised and the complex reassembled and activity is restored (Melis, 1999). The precise mechanism of photodamage remains unclear. Perhaps an over-reduction of the quinone acceptor is the cause, although the presence of two β-carotene molecules in the reaction centre appears to be crucial to the process. Without these molecules, the breakdown of protein D1 is favoured. While the precise role of these pigments is uncertain, it is clear that the inhibition of carotenoid biosynthesis by

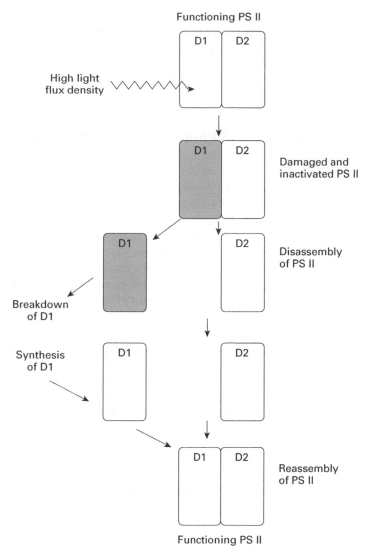

Figure 5.9 Damage–repair cycle in Photosystem II.

lycopene cyclase or phytoene desaturase inhibitors leads initially to an inhibition of photosynthetic electron flow and later to pigment bleaching (Fedtke *et al.*, 2001).

Herbicidal action as a consequence of binding to the D1 protein is now well known. Activity is effective and well characterised both in crops and weeds. One disadvantage, however, is the relatively high application rates required for phytotoxicity, typically 2–4 kg ha^{-1} with atrazine. Duff and colleagues (Fabbri *et al.*, 2005; Duff *et al.*, 2007) at Monsanto have suggested that the inhibition of D1 protein biosynthesis could be a better target for herbicide action at gram rather than kilogram doses. They have discovered that a carboxy-terminal processing protease, CtpA, is a low-abundance thylakoid enzyme that catalyses the conversion of pre-D1 into the active form by cleaving the 9-C terminal residues. They expressed a recombinant form of the enzyme in *Escherichia coli* (rCtpA), purified it and used it in a high throughput screen for CtpA inhibitors. Lead compounds were studied *in vitro* and most were shown to be competitive inhibitors with K_i values in the range of 2–50 μM. These authors conclude that CtpA inhibitors could become a new generation of effective herbicides.

5.5 Structures and uses of Photosystem II inhibitors

Observations made in recent decades have shown that the quinone–herbicide binding domain in PS II is highly conserved throughout the plant kingdom. Since this domain involves the interaction of many amino acids in the D1 protein, it is not too surprising that a large number of chemically dissimilar molecules have been found to bind at this site, thus preventing Q_B reduction. Many have become highly successful herbicides and so detailed quantitative structure–activity relationships have been attempted with the PS II inhibitors. Trebst *et al.* (1984) have suggested that the ureas and triazines, now classified in the 'serine' family, all have a lipophilic group in close association with an sp^2 hybrid and an essential positive charge (Figure 5.10). Examples of herbicide classes in the 'serine' family are shown in Figure 5.11.

Trebst and colleagues (1984) considered that the essential features of the phenol-type 'histidine' herbicides are those shown in Figure 5.12. Examples of herbicides of this type are shown in Figure 5.13.

A few major examples of PS II herbicides in commercial use today are presented in Figure 5.14. It is noteworthy that about 50% of all the pesticides cited in *The Pesticide Manual* (Tomlin, 2000) are inhibitors of photosynthetic electron transport, although recent legislation referred to in Chapter 2 may reduce this value in years to come.

5.6 Interference with electron flow at Photosystem I

The bipyridinium compounds paraquat and diquat are well known as potent, total herbicides with a contact action. Paraquat was known in the 1930s as methyl viologen and used

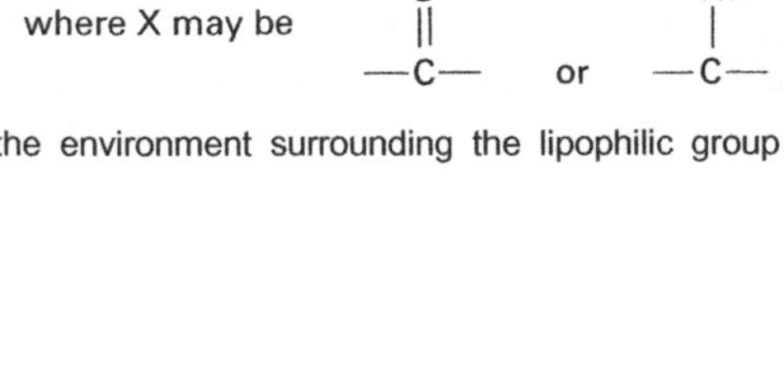

Figure 5.10 Detail of the environment surrounding the lipophilic group in the 'serine' family of herbicides.

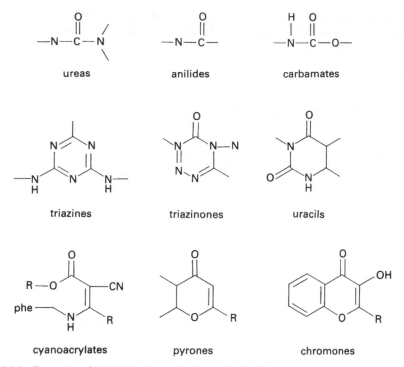

Figure 5.11   Examples of herbicide classes in the serine family.

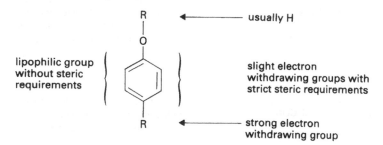

Figure 5.12   Essential features of the phenol-type 'histidine' herbicides (after Trebst *et al.*, 1984).

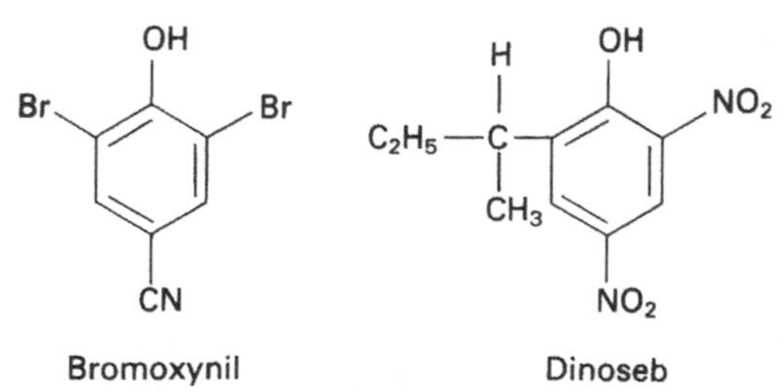

Bromoxynil   Dinoseb

Figure 5.13   Structures of the 'histidine' family of herbicides: the hydroxybenzonitriles bromoxynil and the nitrophenol dinoseb.

as an oxidation–reduction indicator, but it was not until the mid-1950s that its herbicidal properties were discovered. Paraquat and diquat, with redox potentials of $-0.446\,V$ and $-0.349\,V$ respectively, exist as dications which can accept an electron from one of the iron-sulphur proteins near ferredoxin ($-0.420\,V$) on the stromal side of PS I to form a stable free radical.

This radical is re-oxidised by molecular oxygen to produce active oxygen species, and the paraquat is then free to accept another electron from near ferredoxin (Figure 5.15).

In this way a futile cycle operates, whereby the dication is constantly regenerated in the light and active oxygen species kill the plant (see section 5.8).

Although many molecules have been tested, only paraquat and diquat have found commercial use. They appear to possess unique properties for bipyridinium-type action, namely:

(a)   they are readily soluble in water and are stable at physiological pH ranges;
(b)   the herbicidal dication can be tightly bound by plants, organic matter or soil clay minerals, and so is rapidly inactivated and unavailable to plants;
(c)   they have suitable negative oxidation–reduction potentials to accept a single electron from PS I;

Figure 5.14   A selection of herbicides that inhibit photosystem II electron flow.

UREAS	USES
Diuron	General weed control in non-crop areas. Selective, pre-emergent control of annual weeds in alfalfa, maize, cotton, pineapple, sugar cane and sorghum. Soil active and residual. Readily absorbed by roots and leaves and rapidly translocated.
Isoproturon	Control of annual grasses and broadleaf weeds in barley, rye and wheat. Soil active and residual. Can leach to groundwater.
Chlorotoluron	As isoproturon.
Linuron	Selective, pre-emergent control of annual weeds in potatoes, carrots, peas, beans, cotton, maize, soybean and winter wheat. Soil active and residual.

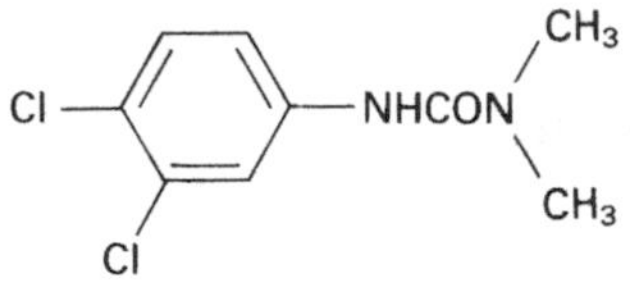

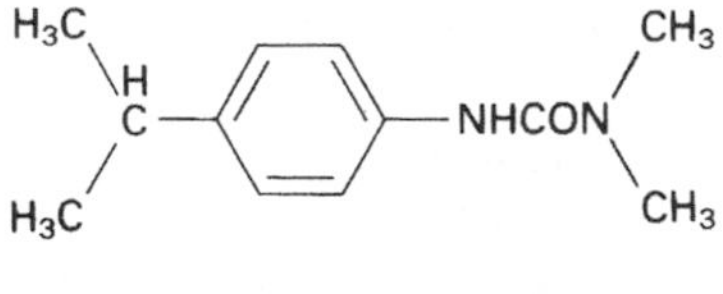

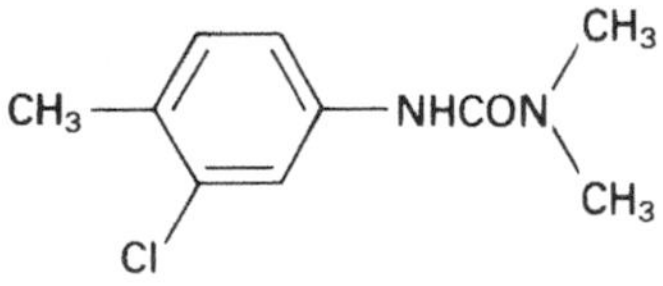

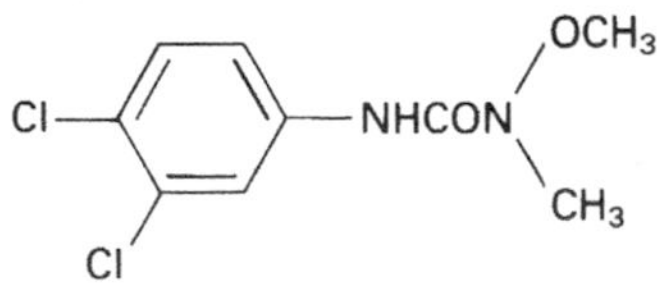

Figure 5.14 (*Continued*)

TRIAZINES	USES
Atrazine 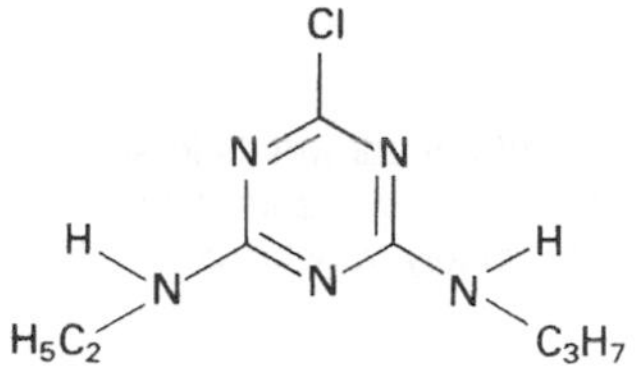	Selective pre- and post-emergent control of annual weeds in maize, sorghum, sugar cane, raspberries and roses. Residual, can persist in soils at high soil pH. Can leach to groundwater.
Simazine 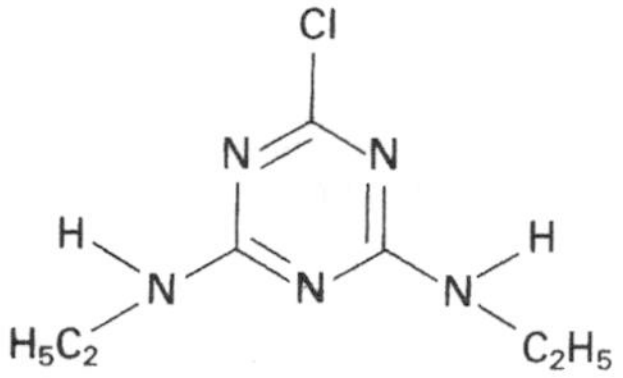	Total weed control in non-crop areas. Pre-emergent or early post-emergent control of annual weeds in beans and maize, fruit bushes, canes and trees.
Prometryn 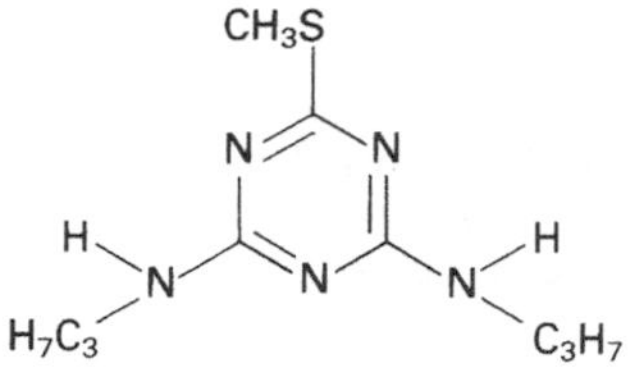	Pre-emergent control of annual weeds in cotton, peas and carrots. Post-emergent weed control in many vegetable crops and sunflowers.
Terbutryn 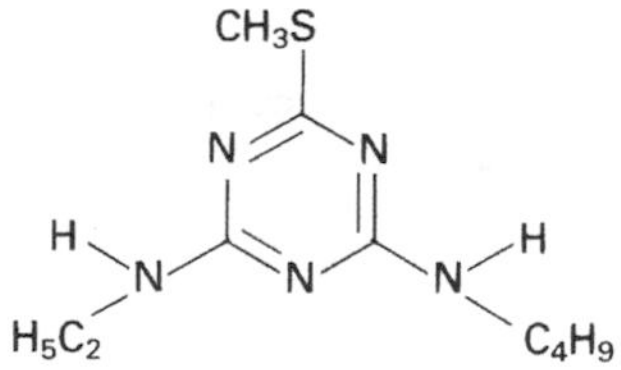	Pre-emergent control of blackgrass and broadleaf weed seedlings in winter wheat, barley, sugar cane and sunflowers.
TRIAZINONES **Metribuzin**	Pre- or post-emergent control of annual weeds in lucerne, potatoes, soybean, tomatoes, asparagus and sugar beet. Readily absorbed by roots and leaves and rapidly translocated.
Metamitron 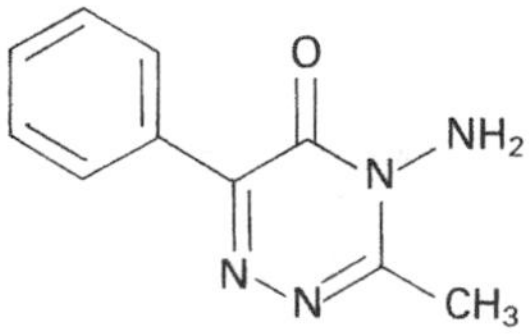	Pre- or post-emergent control of annual weeds in sugar beet and fodder beet.

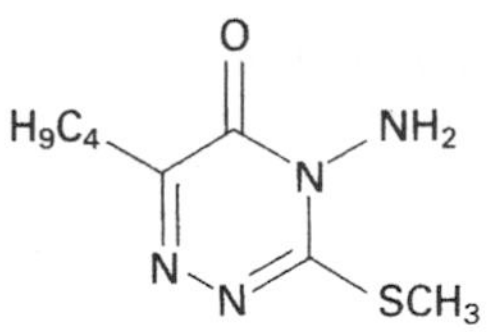

(*Continued*)

Figure 5.14 *(Continued)*

	USES
URACILS	
Lenacil	
	Pre-emergent control of annual weeds in sugar beet and fodder beet. Readily absorbed by roots and leaves and rapidly translocated.
Terbacil	
	Soil applied for weed control in sugar cane, strawberry, peach, citrus and apple.
ANILIDES	
Propanil	
	Post-emergent control of barnyard grass, sedges and some broadleaf weeds in rice, maize and wheat. Contact action.
Pentanochlor	
	Controls annual weeds in carrots, celery, strawberries, tomato and soft fruit.
PHENYLCARBAMATES	
Phenmedipham	
	Post-emergent selective control of most broadleaf weeds in sugar beet. Readily absorbed by foliage but poorly translocated.
MISCELLANEOUS	
Bentazone	
	Post-emergent selective control of broadleaf weeds in wheat, barley, maize, rice, peas, beans and soybean. Absorbed by leaves and rapidly transported in the transpiration stream.

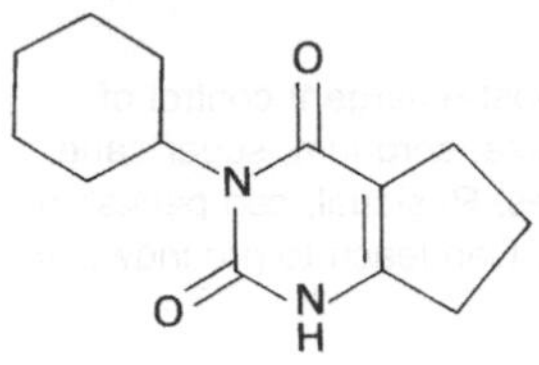

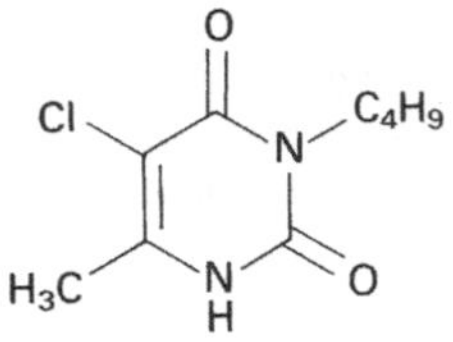

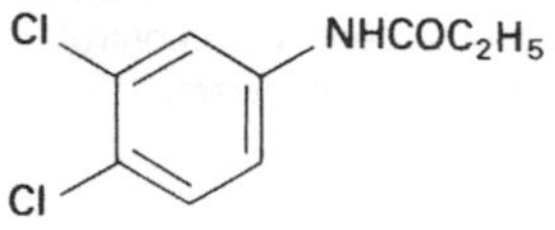

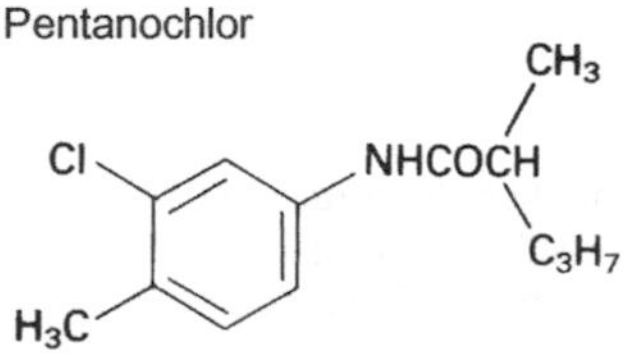

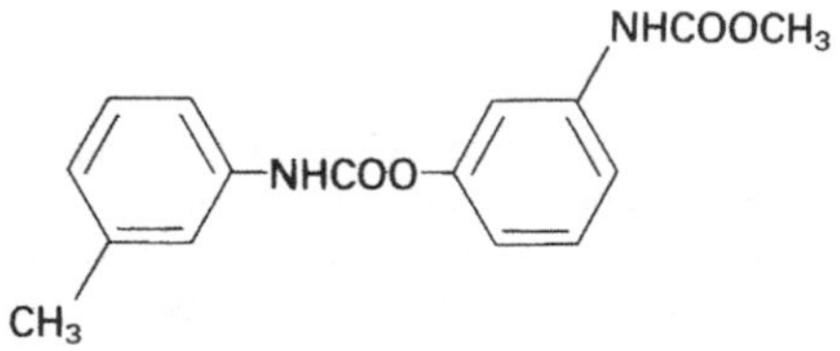

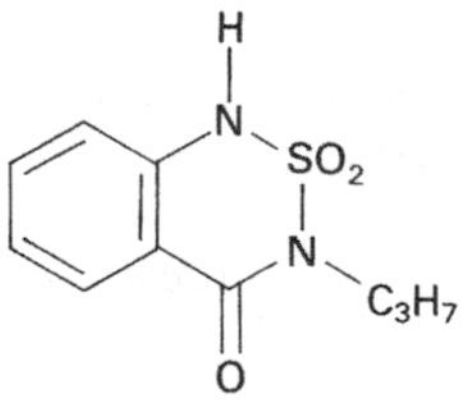

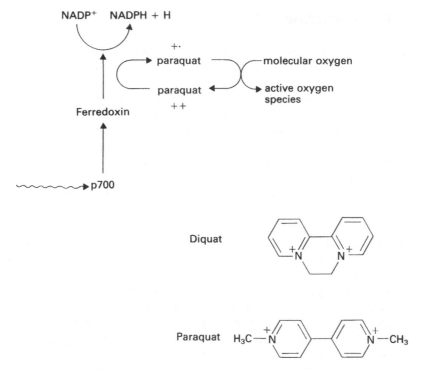

Figure 5.15 How paraquat and diquat act at Photosystem I to give rise to active oxygen species.

(d) the reduced dication is stable and readily oxidised by molecular oxygen; and

(e) the molecule is always available to accept further electrons from PS I.

On a more cautionary note, the bipyridinium herbicides are toxic to animals including mammals, as well as plants. Consequently, their use is now banned in some countries and becoming increasingly limited in others. Although the manufacturers have added bright colourants and powerful emetics to the formulation, intelligent and careful use of these molecules is called for.

There have been many attempts to develop new herbicides based on this site of action, of which a few are noted here. Itoh and Iwaki (1989) have proposed a herbicide-binding site in PS I, designated the Q_ψ or phylloquinone-binding site. They found that phylloquinone (A_1) could be reversibly extracted from PS I with diethyl ether, leaving the photochemical charge separation almost intact. Reconstitution experiments identified potential inhibitors at this site, which appears to be more hydrophobic than equivalent quinone sites in PS II, and diuron and atrazine, for example, bind weakly to it.

A recent report by Smith and colleagues (2005) demonstrated impressive *in vitro* activity of indolizine-5,8-diones, a novel class of quinone-like compounds, though volatility and photoinstability conspired to reduce foliar persistence, and so these compounds have not been developed commercially.

5.7   RuBisCo activase

While there are no known herbicides that interfere with photosynthetic carbon reduction or oxidation, a promising candidate enzyme target has emerged in recent years: RuBisCo activase.

Ribulose 1,5-bisphosphate carboxylase-oxygenase (E.C. 4.1.1.39), or RuBisCo, is one of the most important and probably the most abundant soluble proteins in Nature. It is located in the chloroplast stroma and is responsible for the net incorporation of carbon dioxide into products. It is a large and complex protein, consisting of 8 large and 8 small subunits with a total molecular mass of over 500 kDa. The large subunits, containing the active site, are encoded in the chloroplast genome, while the small subunits, which may confer stability or in some way enhance catalytic efficiency, are nuclear encoded (see Gutteridge and Gatenby, 1995 for further details).

For RuBisCo to be catalytically active, a lysine residue at the active site becomes modified by carbamylation. This is achieved by the reaction of a molecule of CO_2 with the free ε-amino lysine group. Subsequent binding of an atom of magnesium then renders RuBisCo active (Figure 5.16).

It appears, however, that the substrate RuBP is able to bind very tightly to the non-carbamylated form of RuBisCo, preventing activation of the enzyme and inhibiting carbon fixation. There are other natural RuBP analogues, such as carboxyarabinitol-1-phosphate and xylulose 1,5-bisphosphate which are equally able to inhibit the enzyme, leaving it in the inactive state. This inhibition is overcome by RuBisCo *activase* which alters the conformation of RuBisCo protein, increasing the rate at which the inhibitors dissociate from the decarbamylated enzyme, thus allowing carbamylation, magnesium attachment and a competent enzyme.

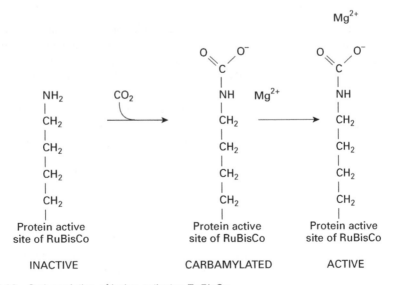

Figure 5.16   Carbamylation of lysine activates RuBisCo.

RuBisCo activase is a nuclear-encoded tetrameric protein of 47-kDa subunits which may account for 1–2% (tobacco) or 5% (*Arabidopsis*) of the total soluble protein in leaves. Enzyme activity requires ATP hydrolysis to alter the conformation of RuBisCo. The activase is an abundant stromal protein because it is a relatively slow enzyme. Decreasing the amount of activase protein by antisense methodology indicates an important role in regulating CO_2 assimilation.

The ratio of activase to RuBisCo concentration and activity appears important in the regulation of carbon fixation, with the ratio greater in fluctuating light than in constant bright light (Mott and Woodrow, 2000). Clearly, the herbicidal inhibition of RuBisCo activase would favour the inactive form of the enzyme, carbon fixation would be prevented and the cessation of plant growth would ensue.

5.8   How treated plants die

Symptoms: Following root absorption of PS II herbicides, initial symptoms of chlorosis between leaf veins on lower leaves are observed. This is followed by the death of leaf tips and margins. After foliar absorption, leaves contacted by the spray become chlorotic and then necrotic. In both cases, there is limited herbicide movement to other leaves and weed death follows within days after treatment. This is influenced by temperature and photon flux density after treatment.

All of the herbicides mentioned so far in this chapter kill plants, owing to the photo-peroxidation of thylakoid and other cell membrane lipids. Thylakoids are particularly susceptible to lipid peroxidation because they contain a high concentration of the unsaturated fatty acids, linoleic (18:2) and linolenic (18:3) acids, and molecular oxygen is always being produced by the photolysis of water, which can be used to generate active oxygen species.

5.8.1   Active oxygen species

Electrons usually exist as pairs in an atomic orbital and a free radical is defined as any species with one or more unpaired electrons (Halliwell and Gutteridge, 1984). By this definition oxygen itself is a free radical since it possesses two unpaired electrons, each located in a different π^* anti-bonding orbital, with the same parallel spin quantum number (Figure 5.17) which renders the molecule relatively unreactive. This is because when another molecule is oxidised, by accepting a pair of electrons from it, both new electrons must be of parallel spin to fit into the vacant spaces in the π^* orbitals.

However, such a pair would be expected to have anti-parallel spins ($\downarrow\uparrow$) which slow and prevent many potential oxidations. Spin restriction can be overcome in biological systems by the presence of transition metals, especially iron, found at the active site of many oxygenases. The metals can do this by their ability to accept or donate single electrons. The reactivity of oxygen is also increased by moving one of the unpaired electrons to overcome the spin restriction and form singlet oxygen. Singlet oxygen has no unpaired electrons and is therefore not a radical, but can exist for long enough (2–4 µs) to react with many biological molecules. It is commonly generated by excitation energy transfer from photosensitisers such as chlorophylls, porphyrins, flavins and retinal (Figure 5.18).

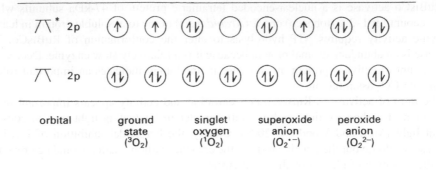

| orbital | ground state ($^3O_2$) | singlet oxygen ($^1O_2$) | superoxide anion ($O_2^{\cdot-}$) | peroxide anion (O_2^{2-}) |

Figure 5.17 Bonding orbitals in diatomic oxygen (reproduced with permission from Halliwell and Gutteridge, 1984, © The Biochemical Society).

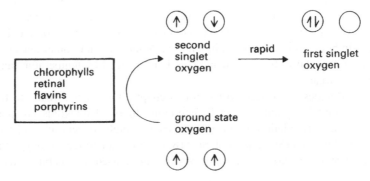

Figure 5.18 Generation of singlet oxygen by photosensitisers.

The superoxide radical ($O_2^{\cdot-}$) is formed when a single electron is accepted by ground state oxygen and the addition of a second electron, either enzymically by superoxide dismutase or non-enzymically, yields the peroxide ion (O_2^{2-}).

$$2O_2^{\cdot-} + 2H^+ \longrightarrow H_2O_2 + O_2$$

In addition, hydroxyl free-radicals ($\cdot$OH) are formed in the presence of peroxide (H_2O_2) and ferrous salts in what is termed the Fenton Reaction (Figure 5.19). The reactivity of $\cdot$OH is so great that they react immediately with whatever biological molecules are in the vicinity to produce secondary radicals.

$$H_2O_2 \xrightarrow{\quad Fe^{2+} \quad Fe^{3+} \quad} .OH + OH^-$$

All four active oxygen species ($^1O_2$, $O_2^{\cdot-}$, .OH and H_2O_2) are naturally generated at the thylakoid and are normally quenched by a series of scavengers or antioxidants within the stroma or dissipated as heat.

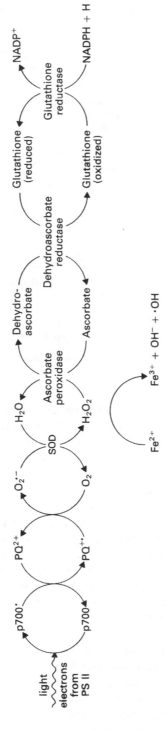

Figure 5.19 Superoxide generation and detoxification at Photosystem I. PQ, paraquat; SOD, superoxide dismutase (after Shaaltiel and Gressel, 1986, http://www.biochemj.org).

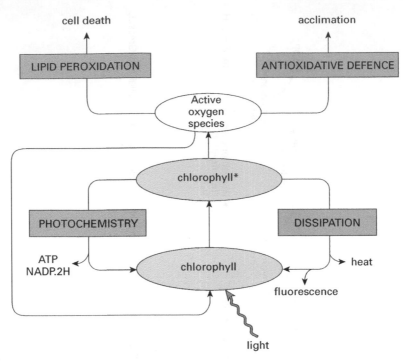

Figure 5.20 The fates of solar energy absorbed by the thylakoid light-harvesting complexes. *, denotes an excited state.

Chloroplast superoxide dismutase is a metalloprotein (apparently existing as Cu-Zn, Mn or Fe forms) partially bound to the thylakoid. Its product, hydrogen peroxide, can inhibit several stromal enzymes involved in photosynthetic carbon reduction, and is removed by the enzymes of the ascorbate–glutathione cycle (Figure 5. 19). Stromal ascorbate and glutathione concentrations fluctuate seasonally and millimolar amounts are commonly measured in the summer months.

α-Tocopherol (vitamin E) is a most effective antioxidant and can react rapidly with singlet oxygen and lipid peroxide radicals. The carotenoid pigments, particularly β-carotene, are also important quenchers of triplet chlorophyll in addition to singlet oxygen, and lead to the dissipation of excess excitation as heat. More recent studies have suggested that polyamines and some flavonols may also act as natural photoprotectants.

Plant survival therefore depends on the balance between photo-oxidative stress and the effectiveness of natural antioxidant protection systems, which have the ability to 'mop up' oxygen free-radicals (Figure 5.20).

In the presence of PS II inhibitors, excitation energy generated by p680 cannot be dissipated by normal electron flow beyond Q_A^-, and so fluorescence yield is dramatically enhanced and activated oxygen species generated. Similarly, paraquat will divert electrons at higher energy from PS I and rapidly generate oxygen free-radicals, as previously described. Under these conditions the natural protective mechanisms are rapidly overloaded, especially at increased temperatures and photon flux densities, and lipid peroxidation is initiated in thylakoids by hydrogen abstraction (Figure 5.21).

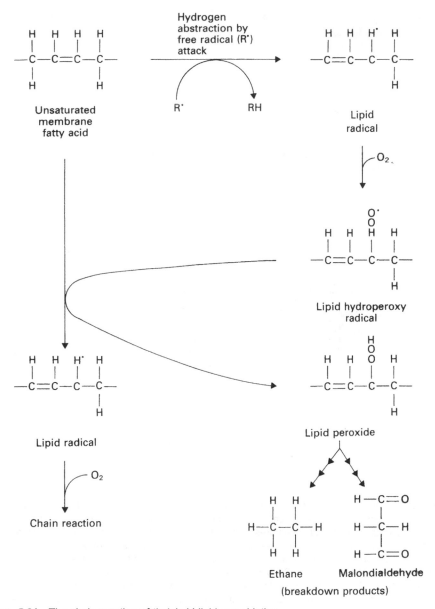

Figure 5.21   The chain reaction of thylakoid lipid peroxidation.

The free-radical attacks unsaturated membrane fatty acids and is quenched by hydrogen atom abstraction. Since a hydrogen atom has only one electron, it leaves behind an unpaired electron on a carbon atom. This carbon (lipid) radical rapidly reacts with oxygen to yield a hydroperoxy radical, which is itself able to abstract hydrogen atoms from other unsaturated lipid molecules, thus initiating a chain reaction of lipid peroxidation. Eventually, the unsaturated fatty acids of the thylakoid are totally degraded to

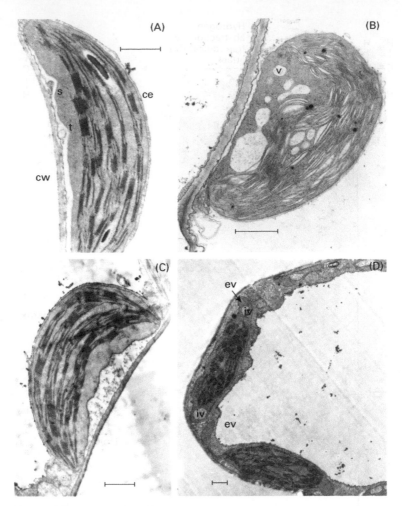

Figure 5.22 Ultrastructural symptoms following treatment with photosynthetic inhibitors. (A) Mesophyll cell chloroplast from an untreated leaf of *Tripleurospermum maritimum* subsp.*inodora* (scentless mayweed). Note appressed and non-appressed thylakoids (t), cell wall (cw), stroma (s) and chloroplast envelope (ce). (B) As (A), but seven days after treatment with ioxynil-sodium at a rate equivalent to 560 g active ingredient ha^{-1}. Note that the chloroplast is swollen and that vesicles (v) are present in both the stroma and the thylakoids. Such intergranal vacuolation is a typical symptom of photoperoxidative damage. (C, D) Mesophyll cell chloroplasts from *Galium aparine* leaves 3 hours after treatment with (C) water and (D) 100 μM of the diphenyl ether herbicide, acifluorfen. Note the invaginations (iv) and evaginations (ev) of the chloroplast envelope. Bar, 1 μM (A and B from Sanders and Pallett, 1986, C and D from Derrick *et al.*, 1988).

malondialdehyde and ethane, and the appressed thylakoid structure progressively opens up and disintegrates (Figure 5.22). Finally, cell membranes and tissues disintegrate from this chain reaction of free-radical attack.

The analysis of chlorophyll fluorescence has become a valuable experimental technique in recent years to investigate the effects of environmental stress, including herbicides, on

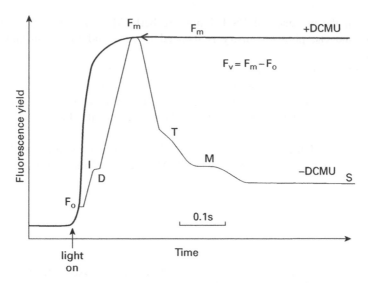

Figure 5.23 A characteristic fluorescent transient or Kautsky curve induced when a dark-adapted leaf is exposed to light in the presence and absence of DCMU (diuron).

PS II activity. Devices that are relatively simple to use and portable have become available, which enable rapid measurements in the laboratory, glasshouse and in the field. Indeed, chlorophyll fluorescence has become a useful screening technique for PS II inhibitors.

As shown in Figure 5.20, chlorophyll in an excited state can be used in one of three ways: photochemistry to drive photosynthesis, dissipation as heat, or re-emission as fluorescence. The three processes are essentially competitive and environmentally sensitive. Under low light exposure about 97% of the absorbed photons are used in photochemistry, 2.5% are transformed to heat and 0.5% emitted as fluorescence. On the other hand, if all the PS II reaction centres are closed by high light or the presence of PS II inhibitors, 95–97% of the absorbed energy may be dissipated as heat and as much as 3–5% via fluorescence.

Kautsky and colleagues (1960) were the first to report the pattern of fluorescence induction when dark-adapted leaves are exposed to light. An example of this fluorescent transient or Kautsky curve is presented in Figure 5.23.

An initial rise from a dark-adapted low value (Fo) to a maximum value (Fm) is observed in less than a second (fast phase), followed by a slower phase, lasting minutes, before a constant value is reached. The fast phase is related to PS II photochemistry, while the slower phase is determined by thylakoid function such as electron transfer away from PS II and carbon metabolism in the stroma. Specifically, at Fo all the reaction centres at PS II are fully oxidised and ready to accept electrons. At Fm, all quinone carriers, especially Q_A, are reduced and unable to accept another electron until they have passed the first one on to the next carrier, Q_B. During this time the reaction centres are thought to be closed. Thus, when reaction centres are closed, photochemical efficiency is reduced and fluorescence yield is increased.

Note that in the presence of PS II-inhibiting herbicides that prevent electron flow between Q_A and Q_B, photochemistry is reduced to zero and Fm is sustained over a prolonged period.

Variable fluorescence, Fv, (Fm − Fo) is a valuable parameter that reflects the efficiency with which incident light energy is used in, and downstream of, the photosystems. It is also a valuable measure of how a plant may tolerate a herbicide, since metabolism in a crop plant will detoxify the herbicide and the Fm value will decline with time.

For a more detailed account of the underlying theory and interpretation of fluorescence data and photochemistry, the reader is referred to Maxwell and Johnson (2000) and Lawlor (2001), respectively.

References

Blankenship, R.E. (2002) *Molecular Mechanisms of Photosynthesis*. Oxford: Blackwell Science.

Derrick, P.M., Cobb, A.H. and Pallett, K.E. (1988) Ultrastructural effects of the diphenyl ether herbicide Acifluorfen and the experimental herbicide M & B 39279. *Pesticide Biochemistry and Physiology* **32**, 153–163.

Duff, M.G.S., Chen, Y-C.S., Fabbri, B.J. *et al.* (2007) The carboxyterminal processing protease of D1 protein: Herbicidal activity of novel inhibitors of the recombinant and native spinach enzymes. *Pesticide Biochemistry and Physiology* **88**, 1–13.

Fabbri, B.J., Duff, M.G.S., Remsen, E.E. *et al.* (2005) The carboxyterminal processing protease of D1 protein: Expression, purification and enzymology of the recombinant and native spinach proteins. *Pest Management Science* **61**, 682–690.

Fedtke, C.B., Depka, O., Schallner, K., *et al.* (2001) Mode of action of new diethylamines in lycopene cyclase inhibition and in photosystem II turnover. *Pesticide Management Science* **57**, 278–282.

Fuerst, E.P. and Norman, M.A. (1991) Interactions of herbicides with photosynthetic electron transport. *Weed Science* **39**, 458–464.

Gressel, J. (2002) *Molecular Biology of Weed Control*. London and New York: Taylor & Francis Group.

Gutteridge, S., and Gatenby, A.A. (1995) Rubisco synthesis, assembly, mechanism and regulation. *Plant Cell* **7**, 809–819.

Halliwell, B. and Gutteridge, J.M.C. (1984) Oxygen toxicity, oxygen radicals, transition metals and disease. *Biochemical Journal* **219**, 1–14.

Hirschberg, J., Bleeker, D.J., Kyle, L., McIntosh, L. and Arntzen, C.J. (1984) The molecular basis of triazine-herbicide resistance in higher-plant chloroplasts. *Zeitschrift für Naturforschung* **39c**, 412–420.

Itoh, S. and Iwaki, M. (1989) Vitamin k_1 (phylloquinone) restores the turnover of FeS centers in the ether-extracted spinach Photosystem I particles. *FEBS Letters* **243**, 47–52.

Kautsky, H., Appel, W. and Amman, H. (1960) Chlorophyllfluorescenz und kohlensaureassimilation. *Biochemische Zeitschrift* **322**, 277–292.

Knaff, D.B. (1988) The Photosystem I reaction centre. *Trends in Biochemical Sciences* **13**, 460–461.

Kohno, H., Ohki, A., Ohki, S. *et al.* (2000) Low resistance against novel 2-benzylamino-1,3,5-triazine herbicides in atrazine resistant *Chenopodium album* plants. *Photosynthesis Research* **65**, 115–120.

Kuboyama, N., Koizumi, K., Ohki, A., *et al.* (1999) Photosynthetic electron transport inhibitory activity of 2-aralkylamino-4-methyl-6-trifluoromethyl-1,3,5-triazine derivatives. *Journal of Pesticide Science* **24**, 138–142.

Lawlor, D.W. (2001) *Photosynthesis*, 3rd edn. Oxford: BIOS.

Matoo, A.K., Pick, U., Hoffman-Falk, H. and Edelman, M. (1981) The rapidly metabolised 32,000 dalton polypeptide of the chloroplast is the 'proteinaceous shield' regulating Photosystem II electron transport and mediating diuron herbicide sensitivity. *Proceedings of the National Academy of Sciences, USA* **78**, 1572–1576.

Maxwell, K. and Johnson, G.N. (2000) Chlorophyll fluorescence – a practical guide. *Journal of Experimental Botany* **51**, 659–668.

Melis, A. (1999) Photosystem II damage and repair cycle in chloroplasts: What modulates the rate of photodamage *in vivo*? *Trends in Plant Science* **4**, 130–135.

Michel, H. and Deisenhofer, I. (1988) Relevance of the photosynthetic reaction center from purple bacteria to the structure of Photosystem II. *Biochemistry* **27**, 1–7.

Mott, K.A. and Woodrow, I.E. (2000) Modelling the role of rubisco activase in limiting nonsteady-state photosynthesis. *Journal of Experimental Botany* **51**, 399–406.

Renger, G. (1976) Studies on the structural and functional organisation of system II photosynthesis. The use of trypsin as a structurally selective inhibitor at the outer surface of the thylakoid membrane. *Biochimica et Biophysica Acta* **440**, 287–300.

Rutherford, A.W. (1989) Photosystem II, the water-splitting enzyme. *Trends in Biochemical Sciences* **14**, 227–232.

Sanders, G.E. and Pallett, K.E. (1986) Studies into the differential activity of the hydroxybenzonitrile herbicides. *Pesticide Biochemistry and Physiology* **26**, 116–127.

Shaaltiel, Y. and Gressel, J. (1986) Multienzyme oxygen radical detoxifying system correlated with paraquat resistance in *Conyza bonariensis*. *Pesticide Biochemistry and Physiology* **26**, 22–28.

Smith, S.C., Clarke, E.D., Ridley, S.M. *et al.* (2005) Herbicidal indolizine-5,8-diones: Photosystem I redox mediators. *Pest Management Science* **61**, 16–24.

Tomlin, C.D.S. (ed.) (2000) *The Pesticide Manual*, 12th edn. Farnham, Surrey: British Crop Protection Council.

Trebst, A. (1980) Inhibitors in electron flow: tools for the functional and structural localisation of carriers and energy conservation sites. *Methods in Enzymology* **69**, 675–715.

Trebst, A. (1987) The three-dimensional structure of the herbicide binding niche on the reaction centre polypeptides of Photosystem II. *Zeitschrift für Naturforschung* **42c**, 742–750.

Trebst, A., Donner, W. and Draber, W. (1984) Structure–activity correlation of herbicides affecting plastoquinone reduction by Photosystem II: electron density distribution in inhibitors and plastoquinone species. *Zeitschrift für Naturforschung* **39c**, 405–411.

Wessels, J.S.C. and van der Veen, R. (1956) The action of some derivatives of phenylurethan and of 3-phenyl-1,1-dimethyl urea in the Hill reaction. *Biochimica et Biophysica Acta* **19**, 548–549.

Chapter 6
Inhibitors of Pigment Biosynthesis

6.1 Introduction: structures and functions of photosynthetic pigments

Photosynthesis relies on the unique light-harvesting abilities of the chlorophyll and carotenoid pigments to trap solar energy in the chloroplast thylakoids and transfer excitation energy to the reaction centres. The principal light-harvesting pigment is chlorophyll *a* which contains a light-absorbing 'head', centred on a magnesium ion and a hydrophobic 'tail' attached to proteins and lipids, and embedded in the thylakoid membrane. Light is absorbed at the rate of about one photon per chlorophyll molecule per second and an electron is boosted to an excited state for each photon absorbed.

The light-harvesting complexes contain two other types of pigment in addition to chlorophyll *a*, namely chlorophyll *b* and the carotenoids. These so-called accessory pigments absorb light of shorter (i.e. higher-energy) wavelengths than does chlorophyll *a*, so they increase the width of the spectrum available for photosynthesis. Energy transfer occurs in the sequence: carotenoids, chlorophyll *b*, chlorophyll *a*, reaction centre.

Chlorophylls *a* and *b* are found in the leaves of all higher plants and in green algae, typically at a concentration of $0.5\,g\,m^{-2}$ of leaf area. Chlorophyll *b* differs from chlorophyll *a* in having a formyl group (–CHO) instead of a methyl group (–CH₃) on one of the N-containing rings that make up the head group, as indicated in Figure 6.1. The carotenoids are abundant in green tissues and over 600 have been found in Nature.

Herbicides that interfere with the biosynthesis of chlorophylls or carotenoids are generally termed 'bleaching herbicides' because bleaching (or whitening, i.e. lack of typically green pigmentation) is a principal symptom in treated plants. Bleaching herbicides were the most frequently patented class of herbicides in the 1980s and the 1990s.

6.2 Inhibition of chlorophyll biosynthesis

Chlorophyll is the principal pigment in photosynthesis. In addition to a light-harvesting function, chlorophyll is located within reaction centres, and so it plays a pivotal role in the movement of electrons from low energy (H_2O) to high energy ($NADP^+$). The

Herbicides and Plant Physiology, Second Edition By Andrew H. Cobb and John P.H. Reade
© 2010 A.H. Cobb and J.P.H. Reade

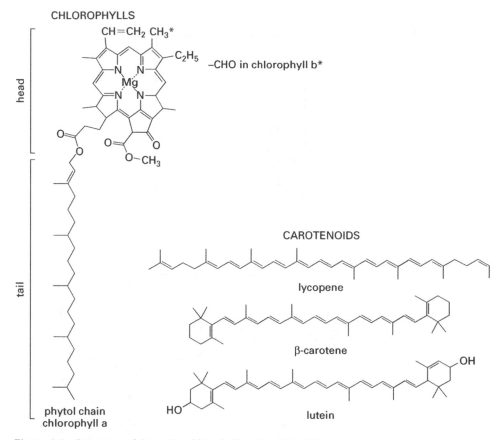

CHLOROPHYLLS

–CHO in chlorophyll b*

CAROTENOIDS

lycopene

β-carotene

lutein

phytol chain
chlorophyll a

Figure 6.1   Structures of the major chlorophylls and carotenoids.

biosynthesis of porphyrins and tetrapyrroles begins with the formation of δ-aminolaevulinic acid (δ-ALA) and its conversion to porphobilinogen (PBG) (Figure 6.2). In animals, δ-ALA is formed from glycine and succinyl-CoA and in plants δ-ALA is synthesised *via* glutamate and a pyridoxal phosphate-linked transaminase. Gabaculine inhibits the transaminase by covalent binding to the pyridoxal phosphate cofactor, whereas 4-amino-5-fluropentanoic acid (AFPA; Gardner *et al.*, 1988) can attack the transaminase active site to form a stable complex that inactivates the enzyme. Laevulinic acid, dioxovaleric acid and dioxoheptanoic acid all compete with δ-ALA to inhibit the synthesis of PBG. A succinyl moiety (COOH–$(CH_2)_2$–C=O) appears to be an essential prerequisite for this competitive inhibition of δ-ALA dehydratase (Figure 6.2).

Since these two reactions are common to the biosynthesis of all tetrapyrroles their inhibition also results in the cessation of phytochrome, cytochrome, peroxidase and catalase synthesis. The inhibitors just described clearly show some potential as **non-selective herbicides**, but since terapyrroles have a central role in the metabolism of all organisms, a lack of mammalian toxicity would need to be conclusively demonstrated.

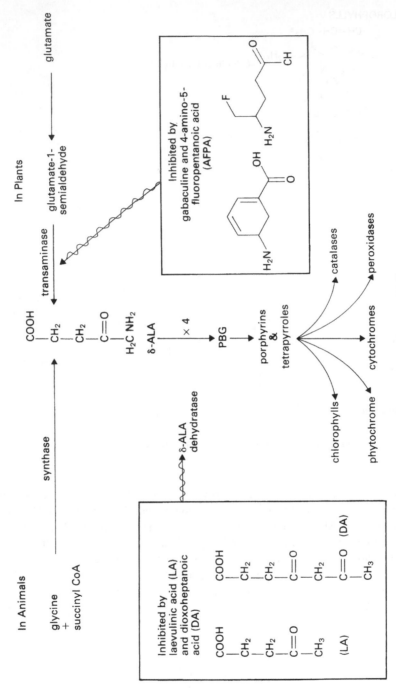

Figure 6.2 Biosynthesis of δ-aminolaevulinic acid (ALA) and porphobilinogen (PBG).

The chlorophyll biosynthesis inhibitors are also referred to as 'peroxidising herbicides' because their action results from the peroxidising effect of active oxygen species. The pioneering examples are the *p*-nitrophenyl ethers (Table 6.1) and the cyclic imides (Table 6.2), developed between 1965 and 1980, and the most recent examples (such as fluthiacet-methyl, Table 6.2) have been developed in the last decade to be active at very low rates of 1–10 g ha^{-1}.

The diphenyl ethers and cyclic imides are an important group of broad-spectrum herbicides that have been widely used for the selective control of annual grasses and broadleaf weeds in major world crops, including soybean, peanut, cotton and rice. Successful active ingredients include acifluoren, oxyflurorfen and bifenox (Table 6.1), and the generation of singlet oxygen produces phytotoxicity. It was initially thought that the phytotoxic species was generated in the thylakoid – as is the case with the inhibitors of photosynthetic electron flow – but research has conclusively demonstrated that photosynthesis is not directly involved in diphenyl ether action, and that the chloroplast envelope is an initial site of action (see Figure 5.14; Derrick *et al.*, 1988).

Observations that diphenyl ether action was particularly sensitive to low energy blue light led to the suggestion that carotenoids could mediate singlet oxygen generation. However, Matringe and Scalla (1988) have shown typical phytotoxic symptoms in carotenoid-free cell lines and instead reported the accumulation of protoporphyrin IX in treated plants. These authors have since conclusively demonstrated that protopor-phyrinogen oxidase is the target enzyme (Matringe *et al.*, 1989).

It is currently envisaged that protoporphyrinogen IX synthesis and its oxidation to protoporphyrin IX are chloroplast envelope membrane-bound reactions, and that pro-toporphyrinogen IX spontaneously and non-enzymatically oxidises to form the potent photosensitiser, protoporphyrin IX, often referred to as PROTOGEN. This molecule cannot be further metabolised to porphyrins and so accumulates, possibly in the stroma, where it generates toxic singlet oxygen and lipid photoperoxidation ensues (Figure 6.3).

Diphenyl ethers are now known to be potent inhibitors of protoporphyrinogen IX oxidase, with concentrations as low as 4 nM aciflurofen-methyl inhibiting this enzyme

Table 6.1 Structure of some diphenyl ether peroxidising herbicides.

R$_1$	R$_2$	Comman name
CF$_3$	COOH	acifluorfen
CF$_3$	OC$_2$H$_5$	oxyfluorfen
Cl	H	nitrofen
Cl	COOH	bifenox

Table 6.2 Structures of some cyclic imide peroxidising herbicides.

	Dose and use
fluthiacet-methyl	5–15 g ha^{-1} post-emergence in maize and soybean
cinidon-ethyl	30–50 g ha^{-1} post-emergence in winter wheat and winter barley
pentoxazone	150–450 g ha^{-1} pre-and early post-emergence in rice

by 50% in corn etioplasts (Matringe et al., 1989). However, this enzyme from either mammalian and yeast sources appears equally sensitive, which suggests that the toxicological properties of these herbicides may need to be closely re-examined.

Peroxidising herbicides inhibit protoporhyrinogen oxidase, commonly abbreviated to PROTOX (E.C. 1.3.3.4). Recent studies suggest that this activity is located in several parts of the cell apart from the plastids, including at the mitochondrial inner membrane for haem production, the endoplasmic reticulum and possibly at the plasmalemma, in addition to soluble forms in the chloroplast stroma.

The membrane-bound protox activity shows very high sensitivity to peroxidising herbicides with I$_{50}$ values as low as 10^{-10} M. Since all eukaryotic protox enzymes are sensitive to these inhibitors, their safety in animals and humans has been questioned. Some authors cite evidence that these compounds can alter porphyrin metabolism in animals, with an accumulation of porphyrin intermediates. The literature suggests, however, that no health problems associated with the consumption of crops treated with protox inhibitors have been reported and that these herbicides are rapidly metabolised in animals.

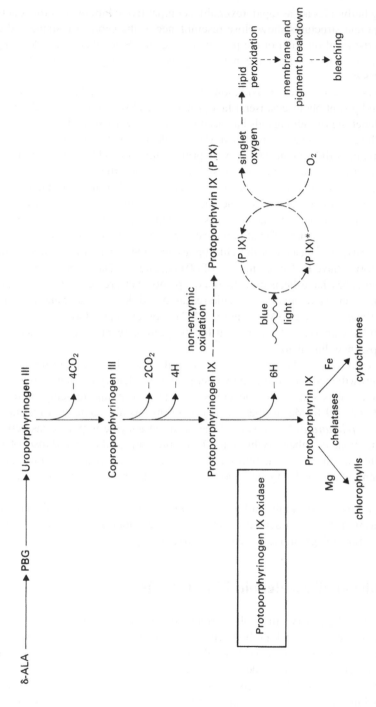

Figure 6.3 An overview of porphyrin biosynthesis in the presence (dashed line) and absence (solid line) of diphenyl ether and cyclic imide herbicides.

Peroxidising herbicides act as rapid, reversible, competitive inhibitors of protox activity and the most potent structures show close resemblance to the geometric shape and electronic characteristics of one-half of the protogen molecule. Since this molecule is a photodynamic pigment in its own right, it generates highly active oxygen species in the stroma and the cytoplasm.

Protox catalyses the six-electron oxidation of protoporphyrinogen IX to protoporphyrin IX, a known and potent photosensitiser. Flavoproteins are closely involved in the activity. Photoaffinity labelling of radioligands has been used to study herbicide binding, with one herbicide binding site suggested for each FAD molecule, with the active site at the C-terminal domain of the protein. The enzyme protein appears highly resistant to protease degradation and it is thought that acylation stabilises its conformation.

Genes involved in protox activity have now been isolated from bacteria, yeasts and higher plants. The *Arabidopsis* protox gene encodes a protein of 537 amino acid residues with little structural homology to the enzymes encoded in human or mouse genes. There have been two cDNA sequences found in tobacco and other plants, which share only 27% sequence similarity, one located in the chloroplast and the other in the mitochondrion. These two isoforms have molecular masses of 60 kDa and 55 kDa, respectively.

Several approaches have been used to obtain plants that are resistant to the protox inhibitors. This is an active area of herbicide research and development, and new protox inhibitors continue to emerge (Matsumoto, 2002). For example, Grossmann and colleagues (2010) have announced saflufenacil for the pre-emergent control of dicot weeds in several crops, including maize.

Mutant cell cultures have been obtained by several groups by incorporating the herbicides into the growth medium, followed by later studies to identify the gene sequence that confers resistance. This strategy has generated resistance to pyraflufen-ethyl in tobacco. A mutation of Val 389 to Met in the plastidic isoform has also generated resistance in *Chlamydomonas reinhardtii*. Expression of the bacterial enzyme in plants has generated tolerance to the diphenyl ether oxyfluorfen. Overexpression of the *Arabidopsis* plastidic form has resulted in resistance to acifluorfen. Clearly, several companies are now pursuing the development of peroxidising herbicide resistance in crops and further details are anticipated.

Readers are referred to the multi-authored text *Peroxidising Herbicides*, edited by Böger and Wakabayashi (1999) for a detailed treatise on these important herbicides, including their chemistry, physiology, mode of action and toxicology.

6.3   Inhibition of carotenoid biosynthesis

In addition to their light-harvesting role, carotenoids protect chlorophyll from attack by active oxygen species by quenching both triplet chlorophyll and singlet oxygen dissipating this energy as heat (Figure 6.4). How thermal energy dissipation is achieved has challenged researchers for several decades. In 2000, Li *et al.* reported a breakthrough using a mutant of the plant *Arabidopsis thaliana* that contained normal concentrations of zeaxanthin but was unable to dissipate thermal energy. They used molecular and genetic markers to show that a photosystem II (PS II) protein known as CP22 (see Figure 5.4, a chlorophyll-binding protein with a relative molecular mass of 22,000) was

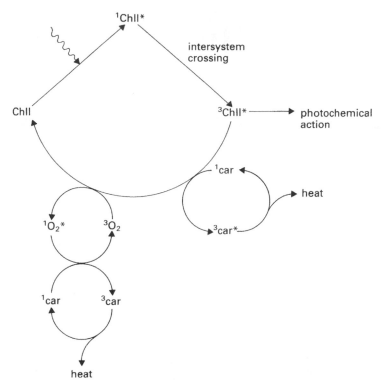

Figure 6.4   Protection of chlorophyll (Chll) by carotenoids (car) (after Britton *et al.* 1989). © Cambridge University Press.

absent in the mutant. On reintroduction into the mutant plant of a normal copy of the gene encoding CP22, the ability to dissipate thermal energy was regained. CP22 is one of over 30 proteins thought to be involved in light harvesting and its precise function remains unclear. Li and colleagues imply that a conformational change in the thylakoid takes place via CP22 to achieve the dissipation, which requires zeaxanthin and a trans-thylakoid pH gradient. Interestingly, these workers have also suggested that CP22 synthesis may increase in leaves exposed to excess light on a daily basis, implying an important physiological role for this PS II protein.

Surplus reductive capacity in the chloroplasts (i.e. production of NADP.2H in excess of that required for carbon fixation and other biosynthetic reactions) is dissipated by the xanthophyll cycle (Figure 6.5) which involves three different carotenoids: violaxanthin (a di-epoxide), antheraxanthin (a mono-epoxide) and zeaxanthin (epoxide-free). These three carotenoids are reversibly interconvertible by the addition or removal of an epoxide group, as shown in Figure 6.5. In strong light, violaxanthin is converted to zeaxanthin via antheraxanthin. This conversion, which is catalysed by a de-epoxidase enzyme, is optimal at low pH (around 5.1). The conversion of violaxanthin to zeaxanthin takes place within a few minutes in the presence of high-intensity light. During irradiation of chloroplasts, the de-epoxidase enzyme is activated by the drop in pH within the thylakoid due to photosynthetic electron transport. Under the conditions of a proton gradient across

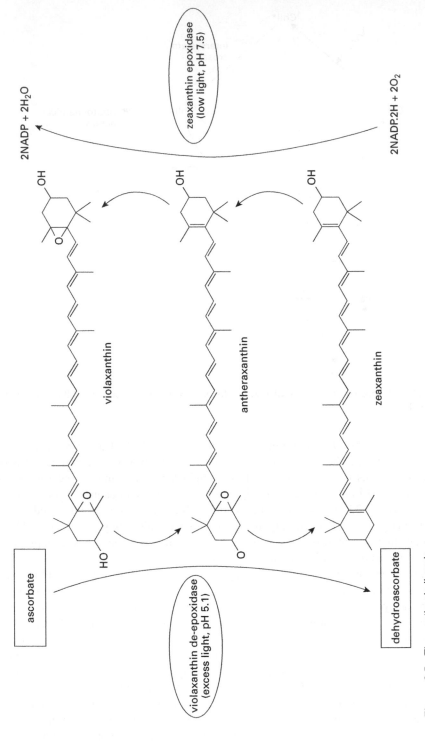

Figure 6.5 The xanthophyll cycle.

the thylakoid membrane, violaxanthin is reduced to zeaxanthin with the participation of the redox systems glutathione / oxidized glutathione and ascorbic acid / dehydroascorbic acid. Reconversion of zeaxanthin to violaxanthin is catalysed by an epoxidase enzyme, requires NADP.2H and uses oxygen. This reaction occurs rapidly in low-intensity light or darkness and is optimal at higher pH (around 7.5; see Figure 6.5). Thus, under high-intensity light conditions, the xanthophyll cycle favours production of zeaxanthin, thereby increasing the capacity for dissipating light energy as heat and protecting against photooxidative damage.

The atomic structure of the major light-harvesting antenna protein, LHC II, has been determined by X-ray crystallography. It can exist in different reversible states, operating in either light-harvesting or in energy-dissipation mode. In addition to chlorophyll molecules, the LHC II also contains the carotenoid lutein. The formation of the quenched antenna state is controlled by the carotenoids of the xanthophyll cycle. Thus de-epoxidation of violaxanthin to zeaxanthin, which stimulates energy dissipation, is thought to modulate structural changes in LHC II. In this way Pascal *et al.* (2005) consider that the LHC II molecule behaves as a 'natural nanoswitch' to control the emission or transfer of incoming light quanta.

It follows from the above that if carotenoids biosynthesis is inhibited, free-radical attack and lipid peroxidation will rapidly ensue under high light-intensity conditions, as previously described.

Carotenoids are synthesised in the chloroplast *via* the mevalonic acid pathway (Figure 6.6). In essence, this pathway involves the condensation of five carbon units (isopentenyl pyrophosphate) to yield many molecules of both physiological (e.g. natural plant growth regulators) and biotechnological (e.g. terpenes) importance. In carotenoid biosynthesis two molecules of the 20-carbon geranylgeranyl pyrophosphate combine to yield the carotenoid precursor 15-*cis*-phytoene which undergoes a series of desaturation steps to all-*trans*-lycopene. Cyclisation followed by ring hydroxylation yields the major higher plant carotenoids, namely β-carotene and lutein. The inhibition of isopentenyl pyrophosphate or geranylgeranyl pyrophosphate biosynthesis would affect the production of all plant terpenoids and could present a target for future herbicide action, although no such inhibitors are currently known. The desaturation steps between phytoene and lycopene, however, have been successfully exploited agrochemically. These enzymes are present as a multi-enzyme complex situated at the thylakoid membrane.

Any enzyme involved in the reaction sequence to β-carotene is a potential herbicide target to induce bleaching symptoms, though all commercial examples are phytoene desaturase inhibitors, with *in vitro* I_{50} values of 0.01–0.1 μM. In plants, both phytoene desaturase (PDS) and ζ-carotene desaturase (ZDS) (Figure 6.6) catalyse the desaturation sequence from phytoene to lycopene, proceeding *via* hydrogen abstraction forming double bonds with NAD and NADP as potential hydrogen acceptors, although plastoquinone has a 20-fold higher affinity than NADP. This role of plastoquinone in phytoene desaturation also explains why the presence of inhibitors of plastoquinone biosynthesis leads to phytoene accumulation.

While the herbicidal inhibition of phytoene desaturase is non-competitive with respect to the substrate phytoene, competition has been observed for the cofactors.

Phytoene desaturase inhibitors are numerous and have been intensively studied during the last two decades. Commercial examples include norflurazon, flurochloridone and

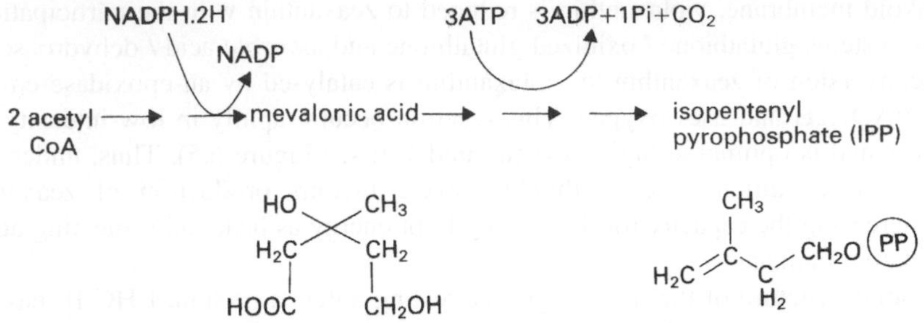

The enzyme prenyl transferase is responsible for the condensation of the 5 carbon IPP units. Thus, 4 × IPP becomes the 20-carbon compound geranylgeranyl pyrophosphate (GGPP). Two GGPPs then combine to yield the 40-carbon carotenoid precursor 15-*cis* phytoene. The stepwise desaturation of phytoene leads to all-*trans* lycopene, i.e.

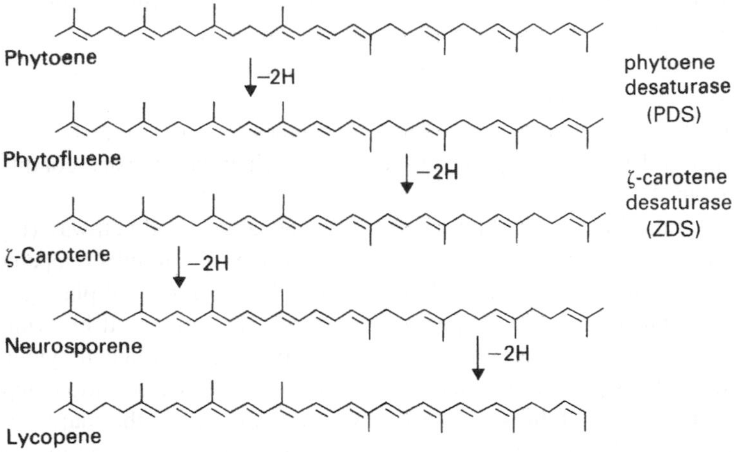

The cyclic carotenes are formed from lycopene to yield the β-, γ- or ϵ- rings, and xanthophylls are

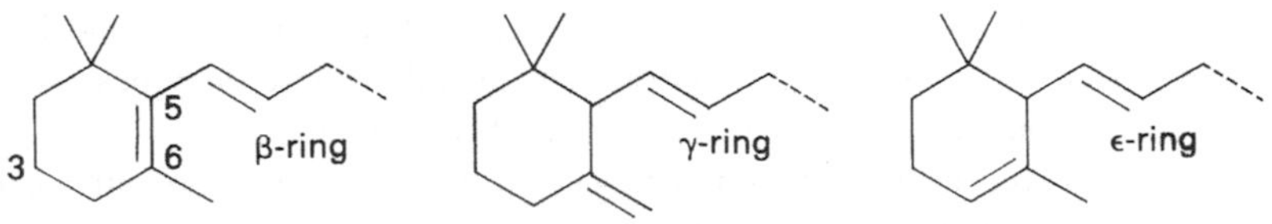

produced by the insertion of an hydroxyl group at C_3 or an epoxy group across the C_5—C_6 bond

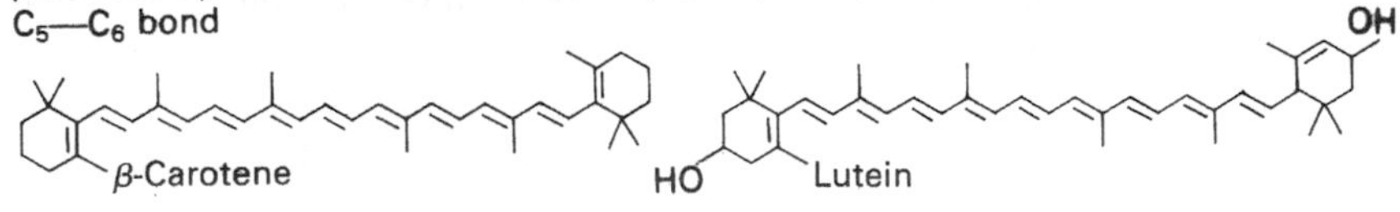

Figure 6.6 Carotenoid biosynthesis.

diflufenican (Figure 6.7), which are used as pre-planting or pre-emergent herbicides. The persistent activity of diflufenican has been effectively exploited in long-term weed control in paths and drives. Since germinating weeds lack photoprotective carotenoids, lipid peroxidation rapidly follows seedling emergence; weeds appear bleached and quickly die.

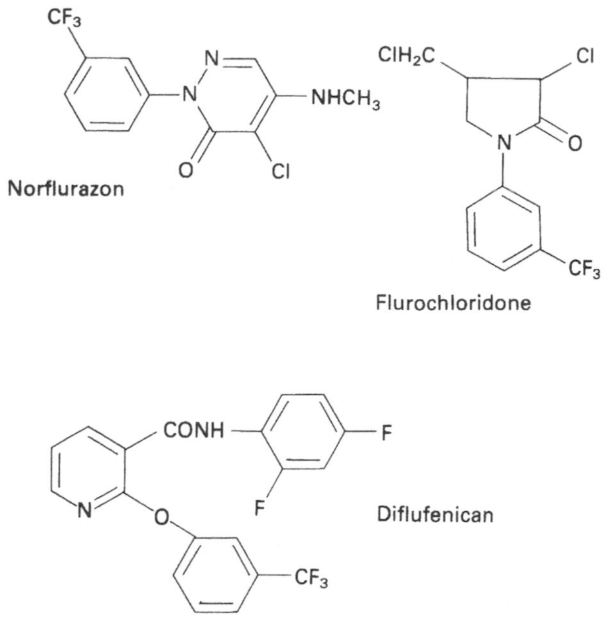

Figure 6.7   Commercial examples of herbicides that inhibit phytoene desaturase.

Fewer compounds are known to interfere with ξ-carotene desaturase or lycopene cyclase, and none appears to have been commercially developed as yet.

Efficient bleaching herbicides possess a central five-or-six-membered heterocycle, carrying one or two substituted phenyl rings, implying a common binding site at the phytoene desaturase protein (Sandmann, 2002).

Fedtke *et al.* (2001) have reported the inhibition of lycopene cyclase by some novel diethylamines. Lycopene accumulates in treated plants with marked reductions in the concentrations of neoxanthin, violaxanthin, lutein and β-carotene. Those workers demonstrated that an observed inhibition of photosynthesis was due to interference with the turnover of the D1 protein in the PS II reaction centre. The reassembly process requires the continuous biosynthesis of two reaction-centre β-carotene molecules, without which protein D1 disappears, especially at high light-flux densities. Interestingly, this depletion of PS II precedes the bleaching process, which may imply a new mechansim of herbicidal activity shared by both the lycopene cyclase and phytoene desaturase inhibitors.

6.4   Inhibition of plastoquinone biosynthesis

Plastoquinone is an electron acceptor in carotenoid biosynthesis in addition to its key role in photosynthetic electron transport (Figure 6.8). Inhibitors of plastoquinone biosynthesis are also herbicides and cause typical bleaching symptoms. Plants synthesise plastoquinone from the aromatic amino acid tyrosine via the intermediate homogentisic acid (Figure 6.9).

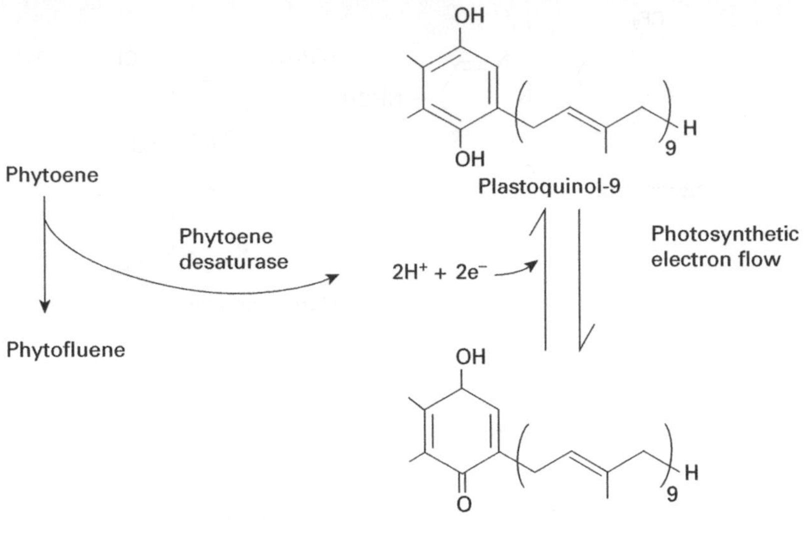

Figure 6.8 The role of plastoquinone in carotenoid biosynthesis and its regeneration by photosynthetic electron flow.

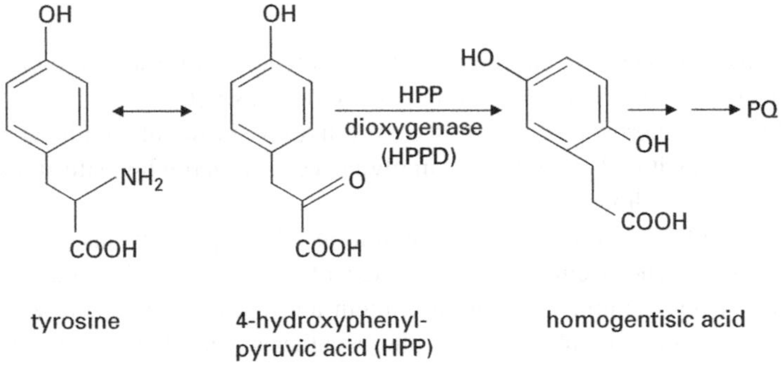

Figure 6.9 Synthesis of homogentisic acid from tyrosine.

The herbicides that inhibit the activity of hydroxyl phenylpyruvate dioxygenase (HPPD) are shown in Figure 6.10.

Isoxaflutole is a pro-herbicide and its metabolic byproduct, a diketonitrile, acts as the inhibitor. Similarly, a metabolite of the herbicide pyrazolate, termed detosyl-pyrozolate, is a potent HPPD inhibitor. Intriguingly, inhibition of HPPD activity by the graminicide sethoxydim has also been claimed, in addition to its inhibitory activity against acetyl-CoA carboxylase (Lin and Young, 1999).

The benzoyl-cyclohexanediones, sometimes termed the herbicidal triketones, were first patented in 1985 as a new group of bleaching herbicides. They are particularly effective when applied pre-emergence in maize at doses as low as $60-100\,\mathrm{g\,ha^{-1}}$ and provide an effective low-dose alternative to atrazine.

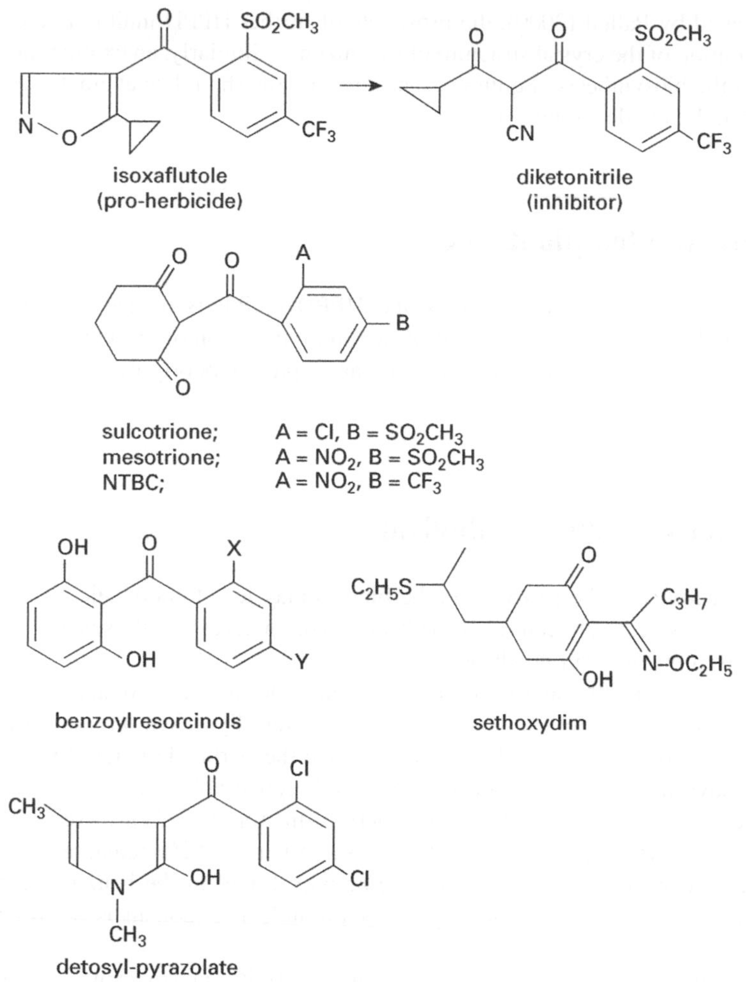

Figure 6.10 Structures of HPPD inhibitors.

Barta and Böger (1996) isolated HPPD from maize and demonstrated potent competitive inhibition by several experimental benzoyl-cyclohexanediones with I_{50} values in the range of 3–23 nm. Viviani *et al.* (1998) performed a detailed kinetic study of HPPD inhibition by diketonitrile and other triketones and concluded that the herbicides acted as tightly binding inhibitors that dissociate extremely slowly from enzyme–inhibitor complex.

HPPD (E.C. 1.13.11.27; E.C. 1.14.2.2) is a monomeric polypeptide of molecular mass 43 kDa (Barton and Böger, 1996) in maize, whereas it behaves as a homodimer of 48-kDa subunits in cultured carrot cells. Amino acid sequences are known from plant, animal and microbial HPPDs. Highly conserved regions at the C-terminus suggest an involvement of this region in the catalytic process, perhaps including sites for the binding of the substrate and an iron atom at specific histidine and glutamate residues.

As indicated by Pallett (2000), the prospects of further HPPD inhibitors are enhanced by the resolution of the crystal structure of the enzyme. Similarly, an examination of other enzymes in the biosynthesis of quinones and tocopherols should be evaluated as potential targets for herbicide development.

6.5   How treated plants die

The *p*-nitrophenyl ethers, cyclic imides and HPPD inhibitors induce rapid wilting and browning of shoots, leading to pigment bleaching, retardation of growth and plant death. They are applied pre- or post-emergence and are typically poorly translocated in treated leaves.

6.6   Selectivity and metabolism

The contact activity of the peroxidising herbicides may lead to poor selectivity. Cinidon-ethyl, however, selectively controls broadleaf weeds in cereals, with selectivity resulting from increased metabolism in wheat.

The basis of isoxaflutole selectivity in weeds and maize also appears to be differential rates of metabolism. The herbicide is rapidly taken up and translocated following both soil and leaf applications. In maize the herbicide is rapidly metabolised, so that 6 days after a root application 59% of recoverable activity was found in a benzoic acid derivative and 29% in the active diketonitrile. Conversely, in the susceptible weed *Abutilon theophrasti*, after the same period, 82% remained as the active diketonitrile, with only 12% of recoverable activity found in the benzoic acid metabolite (Pallet *et al.*, 1998). The pathway of isoxaflutole metabolism is shown in Figure 6.11.

Sulfentrazone shows selective pre-emergence activity in soybeans and peanuts at doses of $125–500\,g\,ai\,ha^{-1}$. Its selective action is explained by oxidative metabolism (Figure 6.12), specifically demethylation. Interestingly, the tolerant weed sicklepod (*Cassia obtusifolia*) also metabolises the herbicide in the same way (Dayan *et al.*, 1996).

The diphenyl ether herbicides have shown commercially successful selectivity in soybean and rice. Soybean contains homoglutathione (γ-glutamyl–cysteinyl–β-alanine) instead of glutathione and uses this alternative thiol in herbicide metabolism. As examples, acifluorfen and fomesafen are rapidly detoxified by homoglutathione conjugation in soybean. Interestingly, diphenyl ether herbicides can increase the expression of glutathione S-transferases (GSTs) in soybean. The activity of the GST in question (*GM* GSTU1-1) was selectively enhanced by homoglutathione rather than by glutathione (Skipsey *et al.*, 1997).

Tolerance by peas to the diphenyl ether fluorodifen is due to rapid conjugation with glutathione and the GST responsible has been further characterised (Edwards, 1996).

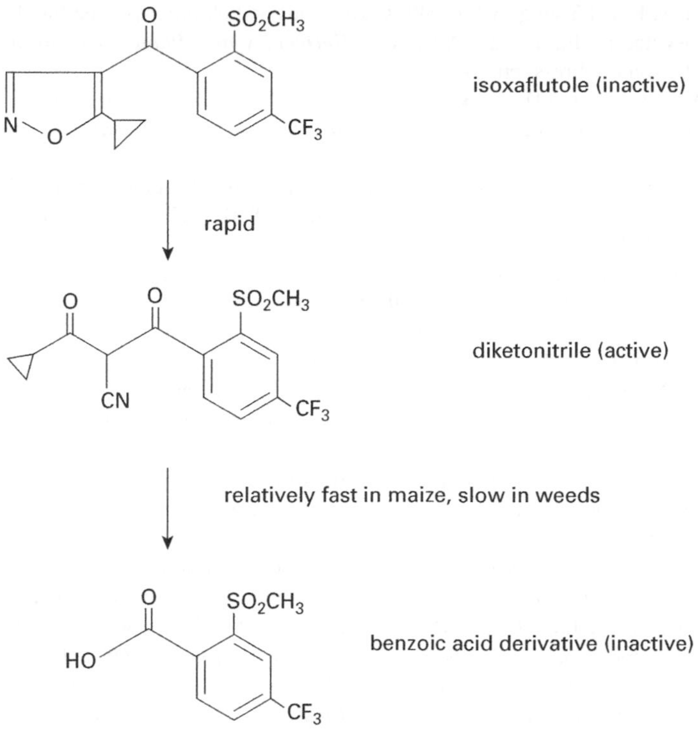

Figure 6.11   Metabolism of isoxaflutole in crops and weeds.

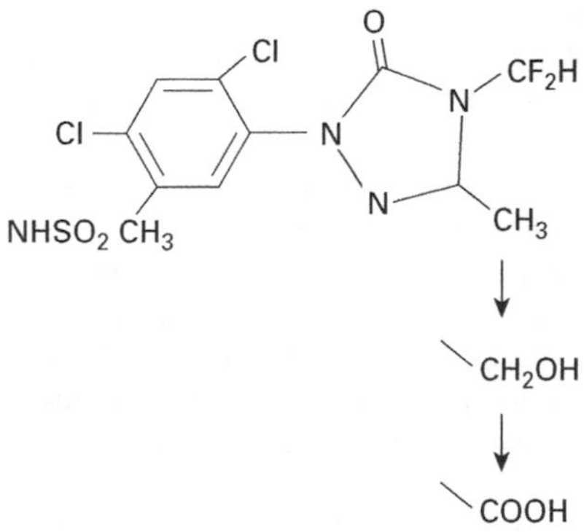

Figure 6.12   Metabolism of sulfentrazone in soybean.

References

Barta, I.C. and Böger, P. (1996) Purification and characterisation of 4-hydroxyl phenylpyruvate dioxygenese from maize. *Pesticide Science* **48**, 109–116.

Bőger, P. and Wakabayashi, K. (eds) (1999) *Peroxidising Herbicides*. Berlin and Heidelberg: Springer-Verlag.

Britton, G., Barry, P. and Young, A.J. (1989) Carotenoids and chlorophylls: herbicide inhibition of pigment biosynthesis. In: Dodge, A.D. (ed.) *Herbicides and Plant Metabolism*. Cambridge: Cambridge University Press, pp. 51-72.

Dayan, F.E., Weete, J.D. and Hancock, H.G. (1996) Physiological basis for differential sensitivity to sulfentrazone by sicklepod (*Cassia obtusifolia*) and coffee senna (*Senna occidentalis*). *Weed Science* **44**, 12–17.

Derrick, P.M., Cobb, A.H. and Pallett, K.E. (1988) Ultrastructural effects of the diphenyl ether herbicide Acifluorfen and the experimental herbicide M & B 39279. *Pesticide Biochemistry and Physiology* **32**, 153–163.

Edwards, R. (1996) Characterisation of glutathione transferases and glutathione peroxidases in pea (*Pisum sativum*). *Physiologia Plantarum* **98**, 594–604.

Fedtke, C., Depka, B., Schallner, O., *et al.* (2001) Mode of action of new diethylamines in lycopene cyclase inhibition and in photosystem II turnover. *Pest Management Science* **57**, 278–282.

Grossmann, K., Niggeweg, R., Christiansen, N., Looser R. and Ehrhardt, T. (2010) The herbicide saflufenacil (Kixor TM) is a new inhibitor of protoporphyrinogen IX activity. *Weed Science* **58**, 1–9.

Lin, S.W. and Young, D.Y. (1999) Inhibition of 4-hydroxyphenylpyruvate dioxygenase by sethoxydim, a potent inhibitor of acetyl-coenzyme A carboxylase. *Bioorganic and Medicinal Chemistry Letters* **9**, 551–554.

Matringe, M. and Scalla, R. (1988) Studies on the mode of action of acifluorfen-methyl in non-chlorophyllous soybean cells: accumulation of tetrapyrroles. *Plant Physiology* **86**, 619–622.

Matringe, M., Camadro, J.M., Labbe, P. and Scalla, R. (1989) Protoporphyrinogen oxidase as a molecular target for diphenyl-ether herbicides. *Biochemical Journal* **260**, 231–235.

Matsumoto, H. (2002) Inhibitors of protoporphyrinogen oxidase: a brief update. In: Bőger, P. Wakabayashi, K. and Hirai K. (eds) *Herbicide Classes in Development*. Berlin and Heidelberg: Springer-Verlag, Ch. 8.

Pallett, K.E. (2000) The mode of action of isoxaflutole: a case study of an emerging target site. In: Cobb, A.H. and Kirkwood, R.C. (eds) *Herbicides and Their Mechanisms of Action*, Sheffield, UK: Sheffield Academic Press.

Pallett, K.E., Little, J.P., Sheekey, M. and Veerasakaran, P. (1998) The mode of action of isoxaflutole: 1. Physiological effects, metabolism and selectivity. *Pesticide Biochemistry and Physiology* **62**, 113–124.

Pascal, A.A., Liu, Z., Broess, K., *et al.* (2005) Molecular basis of photoprotection and control of photosynthetic light harvesting. *Nature* **436**, 134–137.

Sandmann, G. (2002) Bleaching herbicides: action mechanism in carotenoid biosynthesis, structural requirements and engineering of resistance. In: Bőger, P. Wakabayashi, K. and Hirai K. (eds) *Herbicide Classes in Development*. Berlin and Heidelberg: Springer-Verlag, Ch. 2.

Skipsey, M., Andrews, C.J., Townson, J.K., Jepson, I. and Edwards, R. (1997) Substrate and thiol specificity of a stress-inducible glutathione transferase from soybean. *FEBS Letters* **409**, 370–374.

Viviani, F., Little, J.P. and Pallett, K.E. (1998) The mode of action of isoxaflutole: 2. Characterisation of the inhibition of the carrot 4-hydroxyphenyl pyruvate dioxygenase by the diketonitrile derivative of isoxaflutole. *Pesticide Biochemistry and Physiology* **62**, 125–134.

Chapter 7
Auxin-Type Herbicides

7.1 Introduction

The phenoxyacetic acids MCPA and 2,4-D were discovered apparently independently in the UK and the USA in 1941. Templeman and colleagues at ICI and Nutman and collaborators at Rothamsted Experimental Station first demonstrated the herbicidal activity of MCPA, while in the same year in the USA, Porkorny synthesised 2,4-D, and its growth regulatory properties were characterised by Zimmerman and Hitchcock in 1942 (see Kirby 1980 for further details). Both compounds were developed in secret during the war years as potential chemical warfare agents. Fortunately, they were not used in this context at that time and their agricultural potential was soon realised with the marketing of 2,4-D by the American Chemical Paint Company as 'Weedone' in 1945 and by the launch of MCPA by ICI as 'Agroxone' in 1946. Since these molecules could kill many broadleaf weeds in narrowleaf crops, they were the first, truly selective, non-toxic organic herbicides that were effective at low doses. They were also cheap to produce and became available at a time when maximum food production was essential and labour on farms was very scarce. Consequently, they completely altered traditional approaches to weed control and so provided a chemical replacement for the hoe. Furthermore, their success stimulated the chemical industry to invest in the research that led to the discovery of the wide range of herbicides now available. Further information on the discovery and development of the phenoxyalkanoic acid herbicides can be found in Kirby (1980).

In the following decades many structural analogues were developed to broaden the spectrum of weed control and selectivity and these herbicides remain some of the most widely used pesticides in the world. Indeed, estimates indicate that more than 32 million kilograms were produced annually in the USA by six manufacturers, leading to more than 1500 formulated products and 35 different esters and salts as active ingredients (Ware, 1983).

Since the registration of 2,4-D for herbicide use in 1945, there has been worldwide acceptance and widespread use of the phenoxy herbicides. Even now, they are probably the most widely used family of herbicides in the world and play a major role in weed management when used either alone or in combination with other herbicides. They have

Herbicides and Plant Physiology, Second Edition By Andrew H. Cobb and John P.H. Reade
© 2010 A.H. Cobb and J.P.H. Reade

been used for over 60 years with few, if any, reports of acute or chronic toxicity to humans, and have an outstanding record of environmental safety.

Burnside and colleagues (1996) have estimated the benefits of phenoxy herbicide use in the USA. They found that these herbicides were used in over 65 crops and in many non-crop situations, equivalent in 1992 to a value of $171 million. However, they estimated that if phenoxyherbicides were banned in the USA, this would result in an annual loss of $2,559 million: 37% of this loss would be due to increased costs in weed control from the use of more expensive herbicides or other control methods, 36% would result from decreased crop yield and the remaining 27% would arise from higher retail commodity prices.

Morphological symptoms produced by these herbicides are indicative of an exaggerated auxin response, leading to disorganised growth and death in susceptible species. However, although these molecules have been used for over 60 years and have generated a copious amount of scientific literature, the molecular basis of their activity and selectivity has until recently remained obscure. This ignorance is being slowly remedied as our understanding of the molecular biology of natural auxin improves.

7.2   Structures and uses of auxin-type herbicides

The auxin-type herbicides currently in use are categorised into five groups (Table 7.1). These are the phenoxyalkanoic acids, benzoic acids (which includes natural auxin), pyridines and the more recently announced quinoline carboxylic acids. All active auxins appear to possess a free carboxyl group, which suggests that a negatively charged group is essential for activity, although other common chemical characteristics are less obvious. The presence and position of halogens has a profound effect on both activity and selectivity since, for example, 2,3-D and 3,5-D have no auxin activity but 2,4-D does. Furthermore, the addition of an extra chlorine creates 2,4,5-T which can control woody plants that can tolerate 2,4-D.

Table 7.2 provides a brief overview of the weed spectrum controlled by auxin-type herbicides. Although only some examples are included, Table 7.2 clearly illustrates the reasons for the development of these herbicides and the introduction of mixtures to give an even broader spectrum of weed control. Thus, typical MCPA-susceptible weeds include charlock, shepherd's purse and fat hen, mecoprop was introduced to obtain control of cleavers and chickweed, and dichlorprop was developed to control *Polygonum* spp. In the following decade clopyralid became available for the additional control of scentless mayweed and creeping thistle and, more recently, soil-applied quinmerac has been developed for further control of speedwells, red deadnettle and cleavers. Mixtures first appeared in the 1960s and dicamba, for example, proved a useful addition for the control of mayweed. More recently, the addition of clopyralid, benazolin and the hydroxybenzonitriles ioxynil and bromoxynil has widened even further the spectrum of weeds controlled, and so these herbicide 'cocktails' have become widely and routinely used, particularly in cereals.

New uses for auxin-type herbicides continue to be reported. For example, in 2005 Brinkworth and colleagues reported that the mixture of two pyridine carboxylic acids, fluroxypyr and aminopyralid, was effective for the long-term selective control of annual and perennial broadleaf weeds in grassland.

Table 7.1 Structures of auxin-type herbicides.

(a) Phenoxyalkanoic Acids

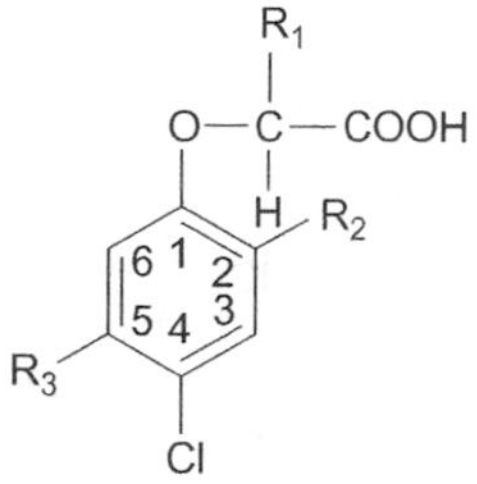

R_1	R_2	R_3	Common name
H	CH_3	H	MCPA
H	Cl	H	2,4-D
H	Cl	Cl	2,4,5-T
CH_3	CH_3	H	mecoprop
CH_3	Cl	H	dichlorprop
CH_3	Cl	Cl	fenoprop

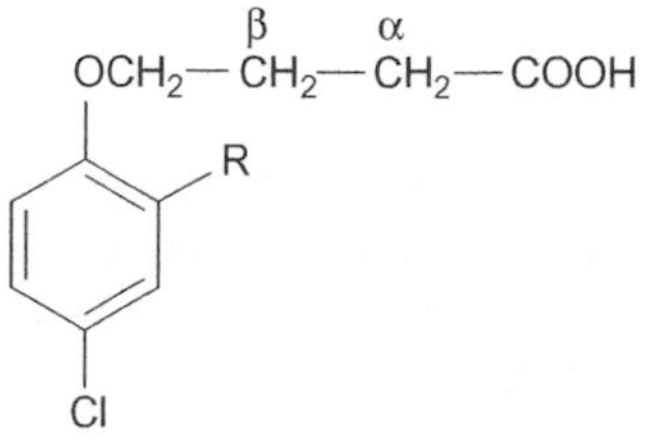

R	Common name
CH_3	MCPB
Cl	2,4-DB

(b) Benzoic Acids

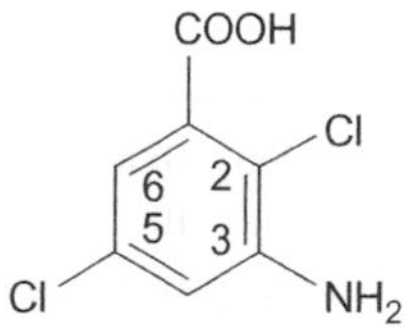

chloramben 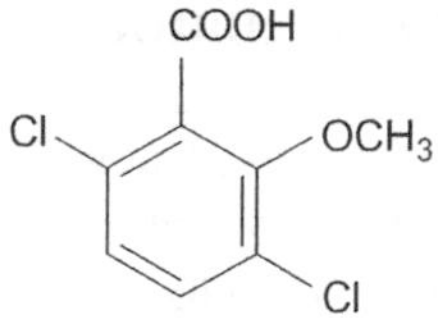   dicamba 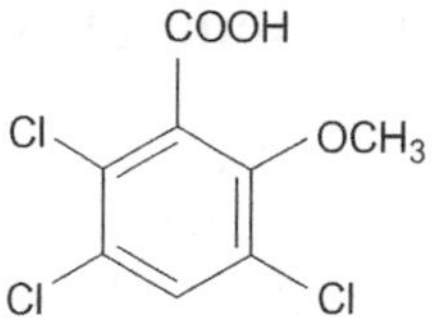   tricamba 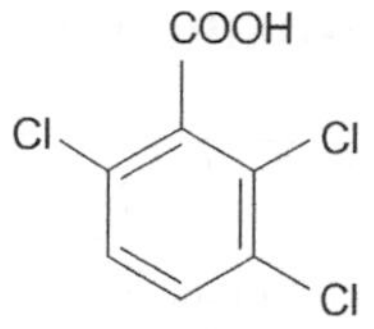   2,3,6-TBA

(c) Aromatic Carboxymethyl Derivatives

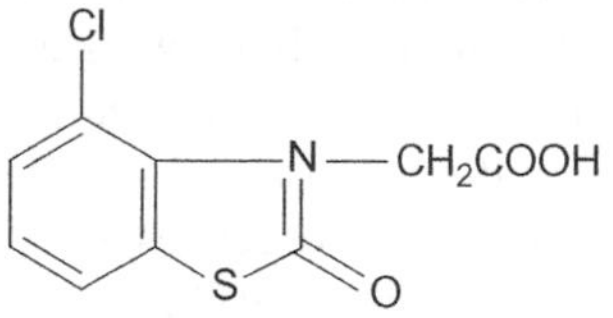

benazolin 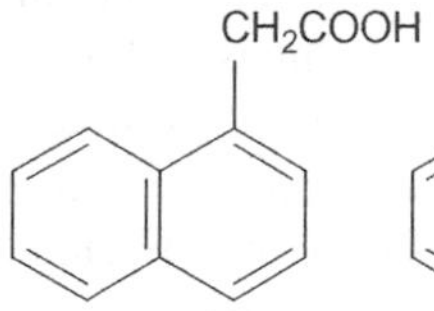   naphthylacetic acid (NAA) 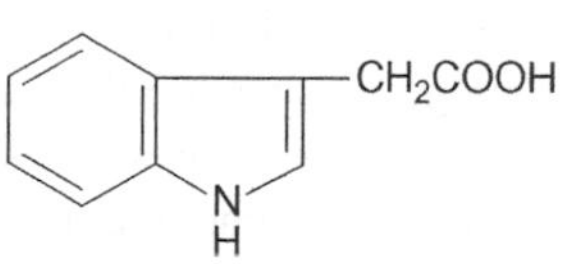   indol-3yl-acetic acid (IAA)

(*Continued*)

Table 7.1 (*Continued*)

(d) Pyridine Derivatives

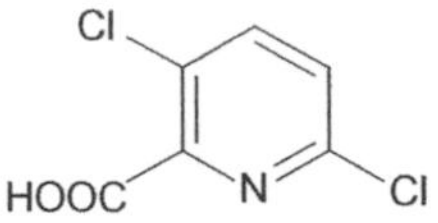

clopyralid

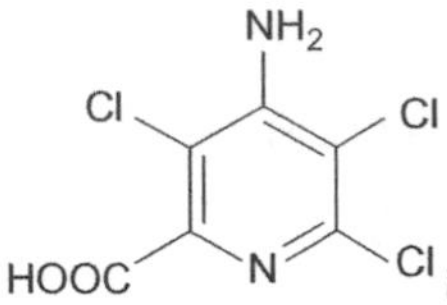

picloram

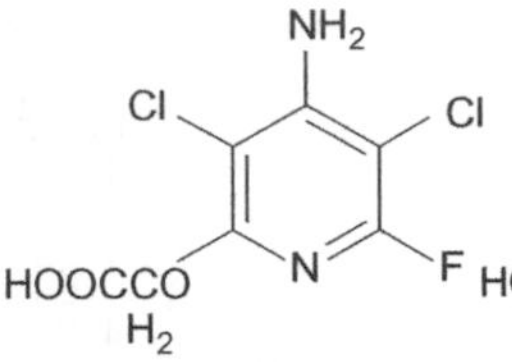

fluroxypyr

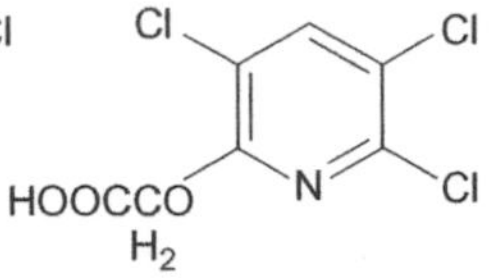

triclopyr

(e) Quinoline Carboxylic Acids

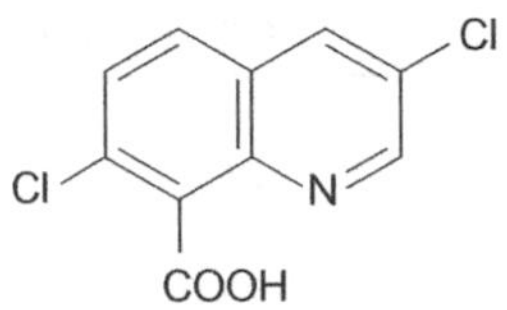

quinclorac

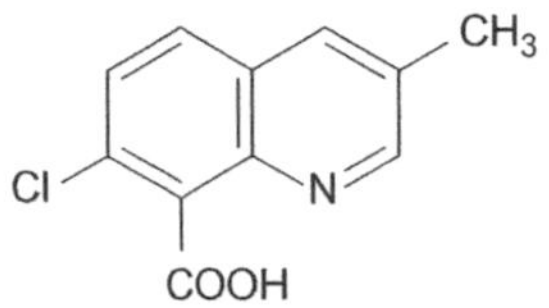

quinmerac

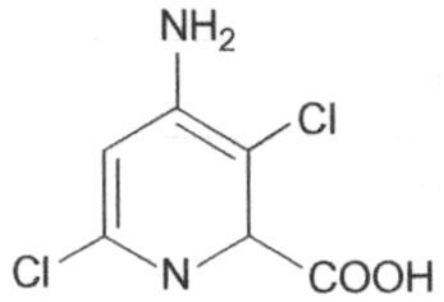

aminopyralid

The auxin-type herbicides have stood the test of time and remain a very effective means of weed control, more than 60 years after their introduction. Those in use today are generally of low persistence, are environmentally benign and are considered unlikely to result in major problems with weed resistance in the foreseeable future.

7.3 Auxin, a natural plant growth regulator

Auxin, or indol-3-yl-acetic acid (IAA), is an endogenous plant growth regulator that plays a crucial role in the division, differentiation and elongation of plant cells. At the organ and whole plant level it has a profound influence on many aspects of plant physiology, including seedling morphology, geotropism, phototropism, apical dominance, leaf senescence and abscission, flowering, and fruit setting and ripening. It is synthesised via tryptophan-dependent and tryptophan-independent pathways in meristematic, actively growing tissues and is found throughout the plant body in concentrations ranging from 1 to $100\,\mu g\,IAA\,kg^{-1}$ fresh weight. Young seedlings and tissues that are rapidly growing and elongating contain relatively higher concentrations of auxin than mature tissues and it is believed that younger tissues are the most sensitive to this growth regulator.

During the last 60 years many studies have investigated the effects of exogenous auxins on plant growth, with the general conclusions that (a) auxins can inhibit as well as stimulate plant growth in a concentration-dependent manner; and (b) different tissues show differential sensitivity to applied auxin. Growth inhibition caused by supra-optimal auxin concentrations is largely attributable to auxin-induced ethylene evolution. Thus, once a critical level of auxin is reached which is tissue-specific, ethylene is produced, and this causes relative inhibition of growth (Figures 7.1 and 7.2).

Table 7.2 Some examples of weed seedlings controlled by auxin-type herbicides (modified from Martin, 1987 and various proceedings from the British Crop Protection Conferences).

Weed	MCPA (1945)	Mecoprop (1957)	Dichlorprop (1961)	Clopyralid (1975)	Dicamba + mecoprop + MCPA	Benazolin + clopyralid	Quinmerac (1985)
Sinapis arvensis (charlock)	S	S	S		S	S	
Capsella bursa-pastoris (shepherd's purse)	S	S	S		S	S	
Chenopodium album (fat hen)	S	S	S		S	S	
Galium aparine (cleavers)	R	S	S		S	S	S
Stellaria media (chickweed)	R	S	S		S	S	
Polygonum lapathifolium (pale persicaria)	R	R	S		S	S	
Polygonum persicaria (redshank)	R	R	S		S	S	
Bilderdykia convolvulus (black bindweed)	R	R	S		S	S	
Tripleurospermum maritimum (scentless mayweed)	R	R	R	S	S	S	
Cirsium arvense (creeping thistle)	R	R	S	S	R	S	
Veronica hederifolia (ivy-leaved speedwell)	R	R	R		R	R	S
Lamium purpureum (red deadnettle)	R	R	R		R	S	S

R, Resistant; S, susceptible.

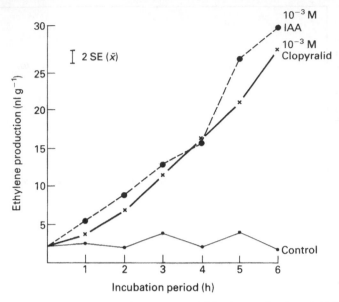

Figure 7.1 Ethylene evolution by scentless mayweed following application of 10^{-3} M clopyralid or indol-3-yl-acetic acid (IAA) (after Thompson and Cobb, 1987).

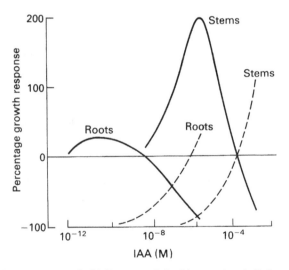

Figure 7.2 Effect of exogenous auxin (IAA) on growth (solid curves) and ethylene production (dashed curves) by roots and stems (from Goodwin and Mercer, 1983).

Klaus Grossmann and colleagues at BASF have proposed a link between hydrogen peroxide production and tissue damage in *Galium aparine* when treated by auxin-type herbicides (Grossmann *et al.*, 2001). They envisage that as a consequence of auxin-herbicide treatment, ethylene synthesis is stimulated, accompanied by an increase in the

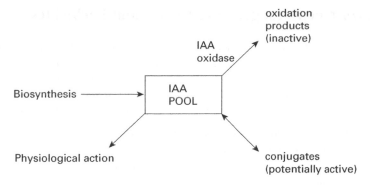

Figure 7.3   The control of auxin concentration *in vivo* (from Goodwin and Mercer, 1983).

biosynthesis of the hormone abscisic acid (ABA). Ethylene induces senescence, while the ABA induces stomatal closure and hence the cessation of carbon assimilation by photosynthesis. Since the treated plant is still exposed to light, these workers consider that H_2O_2 accumulates resulting in oxidative damage that also contributes to weed phytotoxicity.

Auxin concentration *in vivo* is tightly controlled by the relative rates of biosynthesis and degradation, with a further layer of complexity evident when conjugation is taken into account (Figure 7.3). Auxin synthesis is complex and the pool size governed by oxidation and/or conjugation. It has been known since 1947 that plant tissues are capable of the oxidative degradation of IAA by a so-called IAA oxidase and that this enzyme activity is rapid and widespread in plant tissues. However, the characterisation of this activity is awaited and it remains to be convincingly demonstrated that it can be separated from plant peroxidases. Certainly in elongating tissues, low oxidase activity is thought to ensure a relatively high auxin concentration (approximately 10^{-6} M), and in roots lower auxin concentrations (approximately 10^{-10} M) result from measurably higher oxidase activity. Many *in vitro* studies with peroxidases, especially those isolated from horseradish, have suggested that the oxidation of auxin is under the control of many naturally occurring substances, including phenols and other growth regulators, but supporting data *in vivo* is lacking.

Auxin conjugation to glucose, amino acids and myo-inositol may serve as storage forms or auxin-reservoirs, which may be hydrolysed to free auxin when necessary, especially following seed germination. This has recently assumed major physiological significance and importance with the finding that concentrations of conjugated auxins can be much higher *in vivo* than that of free IAA. The principal amino acid conjugate in vegetative tissues appears to be IAA–aspartate, which is formed by L-aspartate-*N*-acylase, an enzyme induced by all natural and synthetic auxins. Glucose esters are also common auxin con-jugates and they are formed from pre-existing glucosyltransferases. Thus, although much rigorous work remains to be done, auxin synthesis, degradation and conjugation appear to interact, with the result that natural auxin concentrations appear to be tightly controlled *in vivo*. More detailed reviews of auxin metabolism may be found in Normanly (1997) and Woodward and Bartel (2005).

7.4 Auxin receptors, gene expression and herbicides

Before an auxin can alter cell metabolism and tissue growth, it must first bind to a receptor and the signal be transmitted to the metabolic machinery of the cell. An auxin receptor may then be defined as a precise, cellular site of molecular recognition from which a series of reactions results in growth changes. Thus, the primary mechanism of auxin action is binding to an auxin receptor. According to Venis (1985) binding is predicted to be:

(a)   reversible, since reactions slow or stop when auxin is removed;
(b)   of high affinity, because endogenous concentrations of auxin are so low;
(c)   saturable, at concentrations similar to the saturation of physiological processes sensitive to auxin;
(d)   specific, to active auxins only;
(e)   confined, to a tissue sensitive to auxin; and
(f)   linked, to a biological response.

The search for plant hormone receptors has lagged far behind our understanding of animal and bacterial systems. Rapid advances have been made in recent years, although much complexity remains. The two main candidates as auxin receptors are the ABP1 (auxin binding protein) and the TIR1 (transport inhibitor response 1) protein.

7.4.1   Auxin binding protein 1

ABP1 was first detected in 1972 in crude membrane preparations of etiolated maize coleoptiles and purified in 1985. It has since been found in all green plants, including the bryophytes and pteridophytes, and in many tissues. The maize ABP1 cDNA encodes a 163-amino-acid protein and the mature protein has a molecular weight of 22 kDa containing a high-mannose oligosaccharide. Interestingly, it is localised at the endoplasmic reticulum (ER) and, like all proteins destined for delivery to the ER, carries a signal sequence 38 residues long at the C-terminus. To function as a receptor it is thought to associate with a membrane-bound 'docking' protein.

A comparison of ABP1 sequence data from a range of species indicated three highly conserved sequences, termed boxes A, B and C. Antibodies raised to box A were shown to have auxin-like activity, indicating an important role at the binding site of these 15 amino acids. The exact roles of boxes B and C remain uncertain.

Edgerton and colleagues (1994) studied auxin binding to ABP1 in isolated maize microsomes and found the characteristics very similar to those predicted solely on biological data by Katekar (1979) (Figure 7.4).

More recently (2001), the protein has been crystallised and a role for a metal ion has been suggested as an ideal carboxylic acid coordination group. The metal ion, likely to be Zn^{2+}, is complexed to three histidine and one glutamic acid residue in box A.

All available data, reviewed by Napier *et al.* (2002), have shown ABP1 to be active to auxins at the surface of the plasma membrane, despite carrying the ER sequence. These include the early responses to auxin action, such as promoting ion fluxes.

The suggestion that herbicides might bind to this site and alter plasma membrane function has been proposed by Hull and Cobb (1998). In their study, highly purified plasma

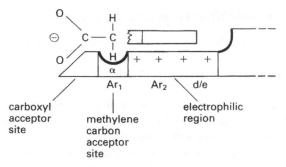

Figure 7.4 A topographic model of the auxin receptor viewed from the side. This model proposes that the auxin receptor possesses regions to accept a carboxyl group, the methylene carbon of IAA (α), the indole ring (Ar$_1$, Ar$_2$) and adjacent areas to the indole ring (d/e) (after Katekar, 1979).

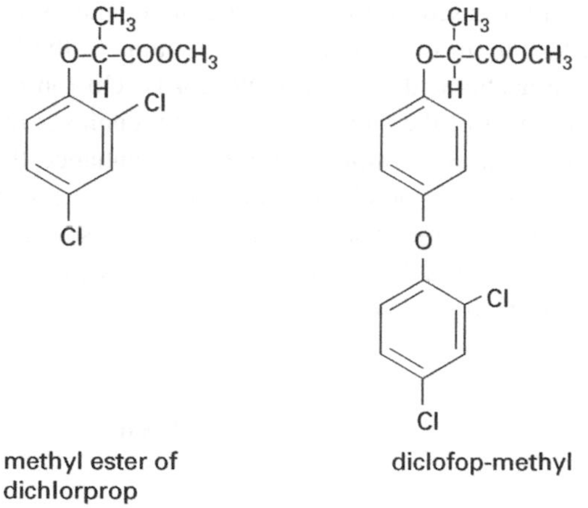

methyl ester of
dichlorprop

diclofop-methyl

Figure 7.5 Structures of the methyl ester of dichlorprop (which controls dicotyledonous weeds in monocotyledonous crops) and diclofop-methyl (which controls monocotyledonous weeds in dicotyledonous crops).

membrane vesicles were isolated from the monocotyledonous weed black-grass (*Alopecurus myosuroides* Huds.) and the dicotyledonous crop sugar beet (*Beta vulgaris*) and H$^+$-efflux measured in the presence and absence of herbicides. They found that while auxin-type herbicides in general did not affect H$^+$-efflux, the aryloxyphenoxypropionate diclofop-methyl was highly inhibitory and 2,4-D gave a slight increase in activity. Auxins are known to antagonise the action of these 'fop' herbicides in the field when present in mixtures. Since the fops are able to depolarise the plasma membrane potential by inhibiting ATP-ase, perhaps the auxin repolarises the potential by stimulating ATP-ase activity and so restores cytoplasmic homeostasis. It may also be speculated that an interaction of these herbicidal molecules could occur at ABP1. To speculate further, perhaps this interaction could also account for the differences in selectivity observed in the field between two similar herbicides (Figure 7.5).

7.4.2   Transport inhibitor response 1 protein

Since 2005 the TIR1 protein has been closely associated with auxin binding and linked to auxin-induced gene expression. The *TIR1* gene was first identified in 1997 in plants tolerant to the auxin transport inhibitor naphthylphthalamic acid (NPA) and named 'transport inhibitor response 1'. Since then, research has shown that auxin-mediated control of gene expression is achieved by the derepression of genes due to the ubiquitination and subsequent degradation of transcription factors. It is now believed that TIR1 is an auxin receptor in its own right, operating within the nucleus of the plant cell.

Sequence comparisons have found no similarities between TIR1 and ABP1, though binding affinities and pH optima indicate their different locations within the cell, that is pH 5 at the cell surface for ABP1 binding and pH 7 in the nucleus for auxin binding to TIR1. It is tempting to speculate that the rapid effects of auxins at the plasma membrane, such as H^+ efflux and ion movement, are linked to ABP1, while gene expression responses result from binding to TIR1 (Figure 7.6). Tan and colleagues in 2007 reported a structural model of the auxin receptor based on the crystallographic analysis of the TIR1 complex from *Arabidopsis*. According to these authors, auxin binds to the base of a site that can also bind 1-naphthalene acetic acid (1-NAA) and 2,4-D. On top of this, the auxin/IAA polypeptide occupies the rest of the site and completely encloses the auxin binding site. They consider that IAA acts as a "molecular glue" to enhance the TIR1-auxin / IAA protein interaction. This results in the ubiquitination of the auxin/IAA proteins leading to their degradation at the proteosome. The loss of these repressor proteins then allows the auxin response factor proteins (ARFs) to activate the transcription of the auxin-response genes, according to the prevailing auxin concentration.

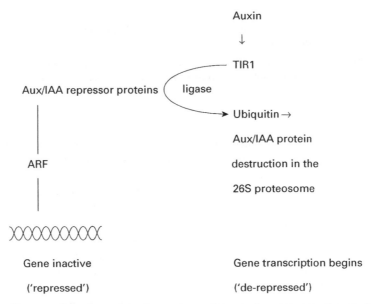

Figure 7.6   Diagram of de-repression of gene transcription in the presence of auxin (see text for details).

Through a combination of genetic, molecular and biochemical approaches we are at last beginning to unravel aspects of how auxins interact with gene expression. Auxin rapidly and transiently stimulates the transcription of three gene families, known as the primary auxin-responsive genes.

1 *Auxin/IAA proteins:* These short-lived nuclear proteins function as transcriptional regulators. They do not interact directly with DNA but exert regulation through another group of proteins known as the auxin response factors (ARFs). The *Arabidopsis* genome contains 29 AUX/IAA genes and 23 ARF genes. ARFs bind to conserved DNA sequences in the promoter regions of early auxin response genes, acting as transcriptional repressors. The AUX/IAA proteins turnover rapidly, over 10–80 minutes, and this degradation is controlled by a ubinquitin protein ligase activated by auxin binding.

2 *SAUR genes:* Small Auxin Up/RNA transcripts accumulate rapidly after auxin exposure. They appear to have short half-lives, although their function remains unknown.

3 *GH3 genes:* Auxin-induced GH3 gene transcription appears to encode IAA–amino acid conjugating enzymes. This may serve to dampen the auxin signal by inactivating the auxin itself by conjugation.

Kelley and colleagues (2006) have used auxin-induced gene expression as an indicator of auxin-type herbicide injury in soybean leaves. They found that GH3 expression was highly induced by dicamba and clopyralid, within 8 hours after application, reaching a maximum of 1–3 days after treatment. GH3 expression was not affected by environmental stress or by viral attack, indicating its potential as a diagnostic assay of herbicide injury. It will be interesting to observe what use is made of such a test, since the crop inevitably recovers via the selective metabolism of these herbicides.

The reader is referred to the recent reviews by Woodward and Bartel (2005) and by Badescu and Napier (2006) for updates in this rapidly developing field. As already demonstrated with NPA, auxin-type herbicides can have an important role as probes in unravelling these complex but important aspects controlling plant growth and development. In addition, these newly discovered genes and gene products may become the targets for future generations of herbicides and plant growth regulators.

7.5   Signal transduction

Once bound to the ABP1 receptor, a signal must be transmitted to the rest of the cell to cause the specific changes in metabolism that result in altered growth. Brummell and Hall (1987) consider that for a relatively small number of auxin molecules binding to a relatively small number of receptors to produce a major effect, some sort of rapid amplification reaction is necessary. They suggest that both signal amplification and transduction are achieved through changes in intracellular concentrations of calcium ions with important roles for inositol triphosphate (IP_3) and diacylglycerol (DG) as additional secondary messengers.

The cytoplasmic concentration of free calcium ions in a 'resting' cell is kept low, in the region of 10^{-7}–10^{-8}M, due to its continual active removal into the cell wall or into organelles, such as the endoplasmic reticulum and vacuole. However, when the cell is stimulated by auxin binding at receptors, a transient increase in cytoplasmic calcium ions (to 10^{-5}–10^{-6} M) results, which is then able to bind to a calcium-binding protein (calmodulin) and this complex may stimulate protein kinases. These in turn phosphorylate, and hence activate, key enzymes.

The link between auxin receptor occupancy and cytoplasmic calcium ion concentration is thought to involve the hydrolysis of phosphatidyl inositol bisphosphate (PIP_2) at the plasmalemma, which results in the production of IP_3 and DG. IP_3 can mobilise calcium ions from the endoplasmic reticulum, and is certainly metabolised in the presence of auxin (Ettlinger and Lehle, 1988), and DG may also activate calcium-dependent protein kinases.

A unifying, but still highly speculative, view of the consequences of binding to an auxin receptor in young tissues is presented in Figure 7.7. Auxin binding to the plasmalemmal receptor causes hydrolysis of PIP_2 to IP_3 and DG. IP_3 causes enhanced mobilisation and cytoplasmic accumulation of calcium ions, which together with DG activates protein kinases, and the activation of other key enzymes follows. The elevated cytoplasmic concentration of calcium ions is reduced to resting state levels by transport into the vacuole or endoplasmic reticulum. Transport into the vacuole is in exchange for protons (H^+) which are then exported out of the cell by an H^+-ATPase.

We now know that auxin also binds to a soluble receptor in the nucleus where specific genomes become derepressed and specific mRNA sequences are synthesised. These messengers are translated at the endoplasmic reticulum to products, for example, involved in new cell wall synthesis (Figure 7.7). This view is supported by the observation by many workers that specific mRNA synthesis can be detected within minutes after auxin application. These models are of real value for an appreciation of auxin-type herbicide action because they identify primary and measurable responses to auxin receptor occupancy, namely selective gene expression and enhanced proton (H^+)-efflux.

Useful information has been gained from studies of auxin-induced H^+-efflux. It is now well established that plant cells maintain large differences in electrical potential and pH across their intracellular compartments (Figure 7.8). Indeed, the electrogenic gradient generated by H^+ pumping provides the driving force for the transport of various solutes, including anions, cations, amino acids, sugars and auxins. These ATPases are associated with the plasmalemma and tonoplast and are the subjects of much current research. It has been known for some time that auxins are able to induce cell elongation in young tissues that is associated with H^+-efflux. Indeed, the traditional bioassays of measuring the elongation or curvature of intact tissues or tissue segments after a 24-h incubation was a cornerstone of auxin discovey. However, the relationship between auxin concentration and growth in these bioassays is poorly defined, such that a very large change in exogenous auxin is often necessary to cause a measurable difference in, for example, growth rate. Furthermore, most auxin bioassays have poor, or at least variable, sensitivity usually caused by poor or slow penetration of the exogenous auxin. Indeed, mecoprop was initially overlooked as a potential auxin herbicide because of its relative inactivity in the oat straight growth test (as cited in Kirby 1980). On the other hand, the measurement of rates of H^+-efflux from sensitive tissues is rapid, linear with time, and highly dependent on auxin concentration. In oat coleoptile tissues at least, it can be considered a primary auxin response in ion transport, which occurs as a direct consequence of auxin–receptor complex formation. By measuring H^+-efflux from oat segments, Fitzsimons (1989) and Fitzsimons *et al.* (1988) derived sensitivity parameters for a wide range of auxin-type herbicides

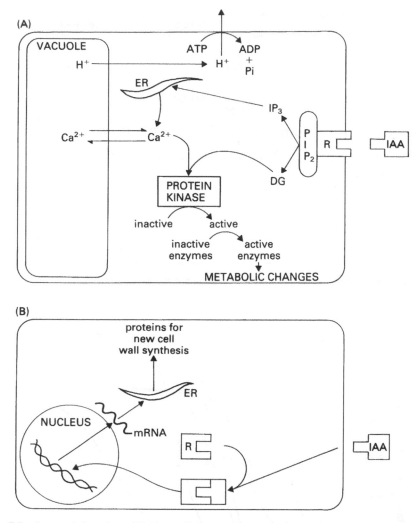

Figure 7.7 A speculative view of the immediate consequences of auxin binding to a plasmalemmal (A) or cytoplasmic (B) receptor. PIP$_2$, phosphatidyl inositol 4,5-bisphosphate; ER, endoplasmic reticulum; R, receptor; IP$_3$, inositol 1,4,5-triphosphate; DG, diacylglycerol (from Brummell and Hall, 1987).

(Figure 7.9). In this example, and with many other auxins tested, the maximum value of the rate of efflux was always similar, with the unexplained exception of benazolin. Large-scale differences in the activity observed were attributed to the relative affinity of each herbicide for the auxin receptor in this species, which were computed by reference to a model of hormone–receptor interaction. These H_{50} values (the auxin concentration giving half maximal response) enable a more accurate and relevant comparison of auxin activity than data from growth bioassays, and may offer a means for a more detailed understanding of the auxin receptor in monocotyledonous species in the presence of herbicides.

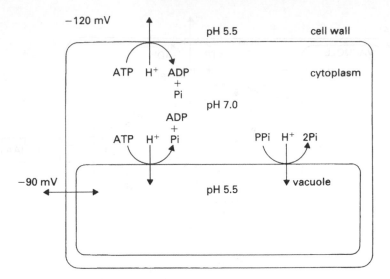

Figure 7.8 Approximate differences in electrical potential and pH in a plant cell.

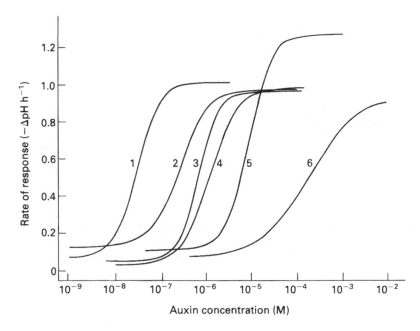

Auxin	*H_{50} (M)	Activity relative to IAA
1. IAA	2.8×10^{-8}	1
2. Mecoprop	2.4×10^{-7}	9
3. Fluroxypyr	6.5×10^{-7}	23
4. 2, 4-D	1.1×10^{-6}	39
5. Benazolin	8.3×10^{-6}	307
6. Clopyralid	1.7×10^{-4}	6071

*H_{50}, the auxin concentration giving half maximal response.

Figure 7.9 Dose–response curves for auxin-induced proton-efflux (modified from Fitzsimons *et al.*, 1988).

7.6   Auxin transport

Auxin appears to regulate a wide range of plant developmental processes owing to its asymmetrical distribution across adjacent cells and tissues during important stages of growth and development. Auxin is synthesised in meristematic tissues and is distributed throughout the plant either through the phloem or by a more controlled cell-to-cell, polar transport system. A now well-accepted model for polar auxin transport is known as the chemiosmotic hypothesis.

As already shown in Figure 7.8, there is a pH gradient across the cell, from about pH 5.5 in the matrix of the cell wall to pH 7.0 in the cytoplasm. In the cell wall the IAA molecule exists primarily in the protonated form (IAAH) and enters the cell passively through the plasma membrane. It dissociates once inside the cell and the IAA$^-$ ion becomes trapped inside the cytoplasm. Specific auxin efflux carriers are required to transport the dissociated form out of the cell. These are predicted to be asymmetrically located in the cell to account for unidirectional auxin transport. We now know that an auxin carrier molecule also exists to further enable IAA$^-$ uptake (Figure 7.10).

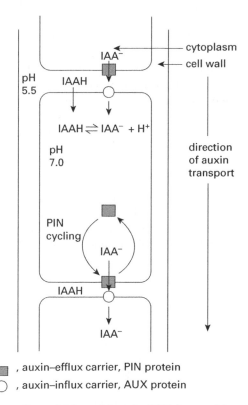

Figure 7.10   The chemiosmotic model for polar auxin (IAA) transport in xylem parenchyma cells (from Vieten *et al.*, 2007). See text for further details.

The nature of the auxin-efflux carrier was identified from studies with the pin-1 mutant of *Arabidopsis thaliana*. These mutants have needle-like stems that lack flowers and molecular analysis of the *PIN1* gene revealed in 1998 that it encoded a transmembrane carrier protein. A further seven *PIN* genes have been found and all phenotypes related to aberrant auxin accumulation. All PIN proteins are localised in a polar fashion, according to the predictions of the chemiosmotic hypothesis.

Arabidopsis mutants were also used to discover the auxin influx carrier in 1998. The AUX-1 mutant was identified as having resistance to movement of 2,4-D. The *AUX1* gene was cloned and found to encode a protein with similarity to amino acid permeases. The AUX1 protein is also localised asymmetrically in some cells.

An intriguing question is 'how do these auxin carriers end up at distinct sides of the cell?'. Current thinking, as reviewed by Vieten *et al.* (2007), suggests that their movement and targeting results from the vesicle trafficking pathways in the cytoplasm. An accessory protein AXR4 is also thought to ensure that the protein is localised in the plasma membrane, although how this is achieved is not known.

In *Arabidopsis*, PIN proteins are thought to rapidly cycle between the plasma membrane and the endoplasmic reticulum via endosomes. This process is termed constitutive cycling and, according to Paciorek and colleagues (2005), is inhibited by auxin. This novel finding, shared by biologically active auxins, leads to increased auxin-efflux carriers at the cell surface where auxin concentration is least. Thus, auxins may promote their own efflux.

Inhibition of auxin transport has been often thought to be a potentially effective mechanism for herbicidal exploitation. Interestingly, one compound has emerged in recent years with these properties, diflufenzopyr (BAS 662H) (Figure 7.11).

When applied alone, diflufenzopyr stunts weed growth, but when combined with an auxin-type herbicide it appears to result in enhanced translocation of the auxin-herbicide to the weed apices to give a more effective broadleaf weed control. The combination of dicamba with diflufenzopyr appears to be particularly effective for a broad spectrum of weed control and tolerance in maize (Bowe *et al.*, 1999), with relatively low dose rates of 100–300 g ai ha^{-1}.

Diflufenzopyr acts by binding to a specific protein involved in transporting auxin away from the meristematic apices. It has a high affinity for this site, with an I_{50} of 19 nM diflufenzopyr . Could this be a PIN protein? Thus, both natural and synthetic auxins accumulate at these apices to induce an 'auxin-overdose' response. Interestingly, root geotropism is also inhibited by this treatment, with an I_{50} of 0.6 nM.

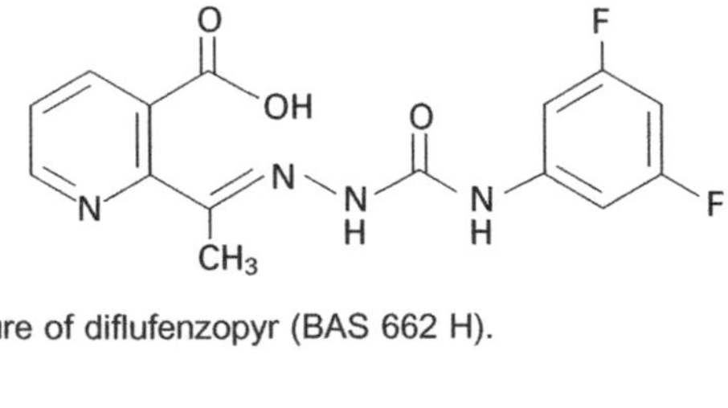

Figure 7.11 The structure of diflufenzopyr (BAS 662 H).

7.7   An 'auxin' overdose

It is evident from the preceding sections that natural auxins are present in very low, but controlled, concentrations in plant tissues, and that they can have a profound influence on plant growth and development. Studies since the early 1970s using combined gas chromatography–mass spectrometry (GC-MS) have demonstrated a range of natural auxin concentrations in plant tissues from 1 to $100\,\mu g\,kg^{-1}$ fresh weight. The lower values are commonly found in fleshy tissues and the higher amounts are reported in seeds. It follows that if a broadleaf weed seedling weighs $10\,g$, then the approximate amount of auxin present in the plant will be in the range of 10–$100\,ng$. Given that a field rate of 2,4-D is 0.2–$2.0\,kg$ active ingredient ha^{-1}, we can safely assume that a broadleaf weed will intercept at least $100\,\mu g$ of this synthetic auxin, which is at least 1000 times more auxin than is already present in the plant. Since auxins are rapidly absorbed into the leaf and translocated throughout the plant, an imbalance of growth regulator is clearly evident, the control systems are overloaded, and growth is drastically altered by supra-optimal auxin concentrations.

7.8   How treated plants die

Weed death occurs as a consequence of an auxin-overdose and is due to uncontrolled growth. The exact sequence of events depends on the age and physiological state of the tissues affected and varies considerably between species. Nonetheless, three phases of symptom development are commonly observed.

Phase 1, the first day. Profound changes in membrane permeability to cations can be discerned within minutes of auxin application. For example, rapid and sustained proton-efflux results in measurable cell elongation within an hour, and an enhanced accumulation of potassium ions in guard cells causes increased stomatal apertures and a transient stimulation of photosynthesis (Figure 7.12). Furthermore, a rapid mobilisation of cellular carbohydrate and protein reserves is commonly observed as large rises in soluble reducing sugars and amino acids. This coincides with enhanced mRNA synthesis and large increases in rates of protein synthesis. In addition, ethylene evolution is typically detected from treated plants (see Figure 7.1).

Phase 2, within a week. In the days following herbicide treatment major growth changes become apparent as visible symptoms initiated by new genome expression and powered by reserve mobilisation in Phase 1. For example, increased cell division and differentiation in the cambium leads to adventitious root formation at stem nodes in some species (Sanders and Pallett, 1987), and general tissue swelling caused by the division and elongation of the cortical parenchyma is typically observed in other tissues, such as the petiole. Classic symptoms of stem, petiole, and leaf **epinasty** are now observed in young tissues in response to ethylene evolution, and abnormal apical growth is sometimes observed (Figure 7.13). Lateral buds may also be released from apical dominance and all other meristems increase in activity.

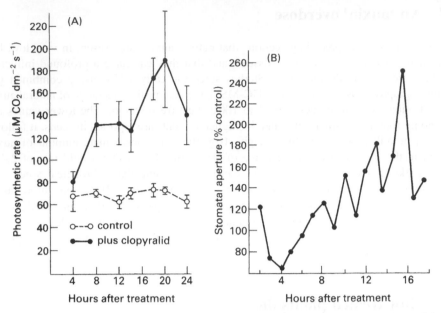

Figure 7.12   Effect of clopyralid (100 g active ingredient ha[-1]) on rate of photosynthesis (A) and stomatal aperture (B) in leaves of scentless mayweed (modified from Thompson, 1989).

Figure 7.13   Typical symptom development of auxin-type herbicides. Plants of *Tripleurospermum maritimum* subsp. *Inodora* (scentless mayweed) were treated with 0, 50, 100, 200 and 400 g a.i. ha[-1] clopyralid, and foliar symptoms were photographed after 7 days.

Phase 3, within ten days. Ultrastructural studies reveal progressive disruption of intracellular membranes culminating in the disruption of the plasmalemma, organelle breakdown and tissue collapse. Root disintegration, leaf chlorosis, and senescence are rapidly followed by plant death.

7.9   Selectivity and metabolism

The success of the auxin-type herbicides is principally due to their highly selective action. However, since these herbicides behave differently in many species, it is thought that many factors interact and contribute to selectivity in various ways. Several examples of this variation are cited by Pillmoor and Gaunt (1981) in their extensive review of phenoxyacetic acid herbicides. At an extreme level, a few micrograms may kill a susceptible species, but a tolerant crop may withstand milligram doses, although increasing the dose will eventually induce phytotoxicity in the crop. Furthermore, species sensitivity clearly varies with plant and tissue morphology and age. For example, young seedlings of cucumber (*Cucumis sativus*) and wild carrot (*Daucus carota*) are sensitive to 2, 4-D but develop tolerance as tissues mature.

Generally, herbicide selectivity is achieved either by differences in herbicide concentration reaching an active site or by differences in sensitivity at an active site. The former involves a consideration of herbicide uptake, movement and metabolism, since the amount of herbicide in a sensitive tissue is determined by its import and transport from the site of application and its metabolic fate in the largest tissue. A full balance sheet is therefore ideally needed for all these factors in both resistant and susceptible species to satisfactorily account for selectivity. However, such detailed information is invariably lacking with auxin-type herbicides. In general, differences in uptake and movement have been reported in many species, but no correlation between uptake, movement, and selectivity has been convincingly demonstrated. On the contrary, uptake and movement are sometimes faster in tolerant species! Studies with radiolabelled herbicides show that foliar uptake is typically rapid, and that active ingredients accumulate at major growth regions, especially the apical meristem, as a result of phloem transport (Figure 7.14). Slow but significant rates of root excretion of auxin-type herbicides has also been reported in some species, but how this is achieved, and to what extent it contributes to selectivity, remains unclear.

The pattern and extent of metabolism of auxin-type herbicides is also highly variable. Conjugation, hydroxylation, and side-chain cleavage are the principal routes for the metabolism of phenoxyalkanoic acids, the eventual products depending on the sequence of these processes.

Thus, the 2-methyl group of MCPA is highly susceptible to oxidation, and the hydroxyl group so formed is rapidly used in glycoside formation (Figure 7.15A). Direct conjugation of phenoxyacetic acids to form glucose and aspartate esters has also been widely reported, as has side-chain degradation to a corresponding phenol. This oxidation also yields glycolic acid, which is subsequently metabolised in photorespiration to carbon dioxide.

In addition, a reaction unique to the metabolism of 2,4-D, known as the NIH shift, is commonly observed in many species. Here, the migration of a chloride atom, usually from the 4- to the 5-carbon position, is probable evidence for an epoxide intermediate in ring hydroxylation, and the product is a substrate for further glucosylation (Figure 7.15B).

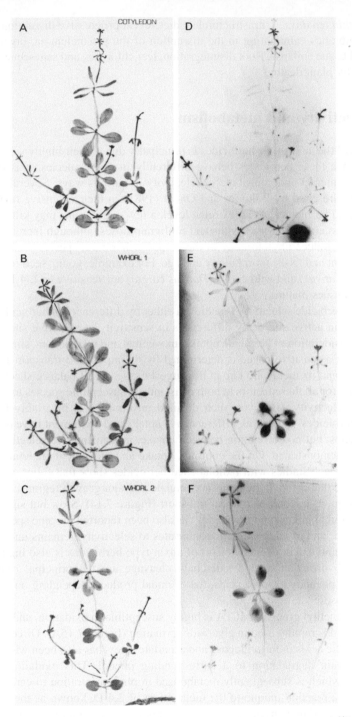

Figure 7.14 The use of autoradiography to study the translocation patterns of ^{14}C-clopyralid in *Galium aparine* (cleavers). (A–C) Photographs of treated plants showing position of ^{14}C-clopyralid application (arrowed). (D–F) Autoradiographs of plants A, B and C, respectively, showing patterns of ^{14}C-translocation from sites of application to regions of active growth (from Thompson, 1989).

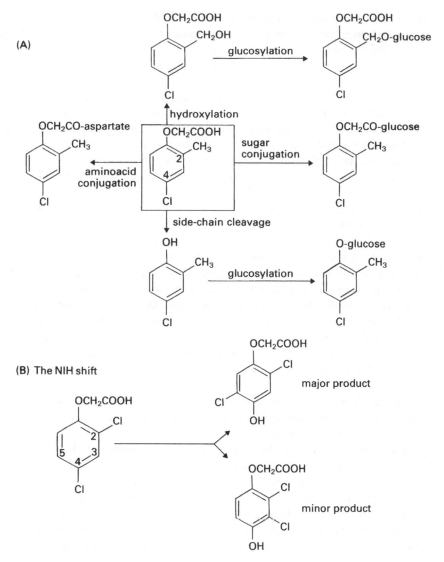

Figure 7.15   Pathways of metabolism of (A) MCPA and (B) 2,4-D.

Further sugar conjugates that are more polar than either the glucose esters or glycosides mentioned earlier can also be formed. Monocotyledonous species in particular appear able to form glycosides with two or more sugar residues, although their possible contribution to selectivity is uncertain (Pillmoor and Gaunt, 1981). In addition, studies using radiola-belled herbicides have shown that some phenoxyacetic acids and their metabolites can also become bound to insoluble fractions in monocotyledons. Structural polymers in the cell wall are often implicated and lignin, pectin and cellulose have all been suggested to bind auxin metabolites. Further studies are now needed to identify the ligands and characterise these binding phenomena. Such information is clearly needed since these bound residues are seldom found in dicotyledonous plants. Indeed, resistant monocots generally contain very low levels of free auxin-type herbicide in contrast to susceptible dicots, and this may be an important feature of selectivity to these herbicides. In most instances the products of metabolism are more hydrophilic, non-phytotoxic and polar than the parent herbicide, and can be stored, sequestered in the vacuole, or become bound to structural polymers. Each factor will contribute to the lowering of the cytoplasmic pool of free herbicide, reducing the level of auxin-receptor occupancy in sensitive tissues.

These observations on metabolism are not confined to the phenoxyalkanoic acids. In a study of triclopyr selectivity in wheat, barley and chickweed, Lewer and Owen (1990) were able to correlate rates of metabolism with species selectivity. They found that resistant wheat plants rapidly metabolised triclopyr to a glucose ester within 12 h, but susceptible chickweed slowly converted the herbicide to triclopyr-aspartate over a 48-h period. In addition, levels of free herbicide remained higher in the weed than in the crop plants.

Metabolism forms the basis of selectivity of the phenoxybutyric acid herbicides, MCPB and 2, 4-DB. MCPA and 2, 4-D cannot be used in legume crops because they kill both legumes and weeds, but their butyric acid derivatives are selective in these crops. Selectivity is achieved by the conversation of the inactive phenoxybutyric acid derivative to an active phenoxyacetic acid *only* in broadleaf weeds by the process of β-oxidation, which successfully removes two CH_2 residues from the side-chain so that an active auxin-herbicide is only produced when an odd number of CH_2 residues is originally present (Figures 7.16 and 7.17).

In this way the phenoxycaproic acids ($n = 5$) may also have theoretical use as selective herbicides in legume crops, although only the phenoxybutyric acid derivatives ($n = 3$) have been commercially developed.

The possibility that selectivity is due to differential receptor sensitivity has already been raised in Section 7.4. It is tempting to suggest that the auxin receptor in dicots is more accessible to auxin-type herbicides than the receptor in monocots. However, no direct supporting evidence is currently available. Alternatively, it may be argued that monocots are less sensitive to auxin-type herbicides because their leaves intercept and retain less herbicide, and that since their mature vascular tissues lack a layer of cambium (cells capable of cell division), they may not possess sensitive cells capable of auxin reception.

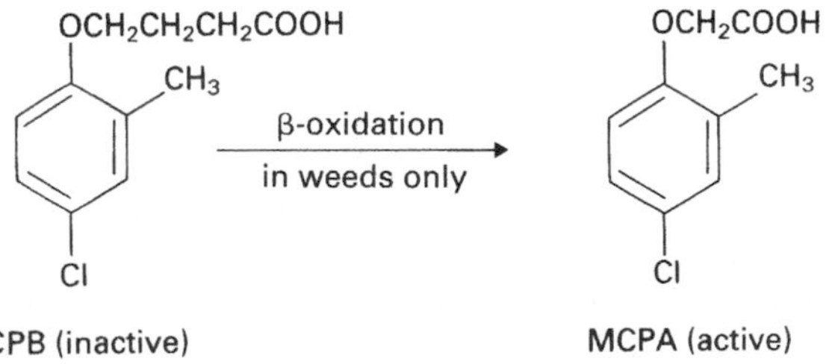

Figure 7.16 Bioactivation of MCPB in susceptible broadleaf weeds.

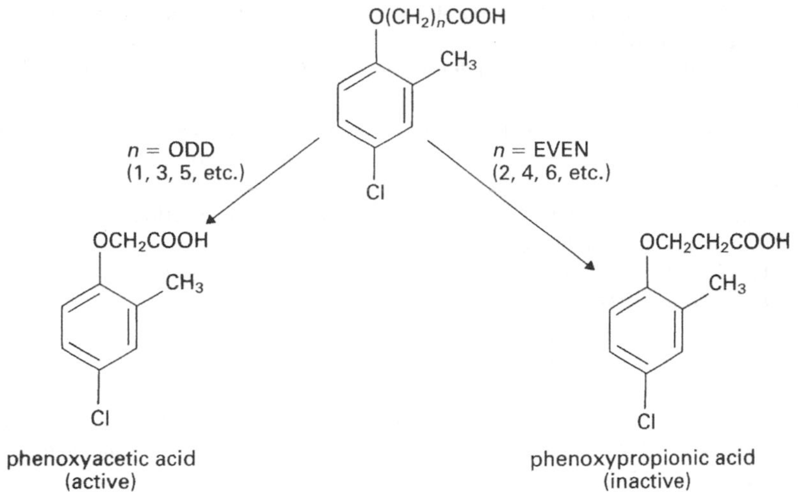

$O(CH_2)_nCOOH$

CH_3

$n = ODD$
$(1, 3, 5, etc.)$

Cl

$n = EVEN$
$(2, 4, 6, etc.)$

OCH_2COOH

CH_3

Cl

phenoxyacetic acid
(active)

OCH_2CH_2COOH

CH_3

Cl

phenoxypropionic acid
(inactive)

Figure 7.17 The effect of β-oxidation on phenoxyacids containing an odd number and an even number of CH_2 residues. Only the former gives rise to active auxin-herbicides.

In conclusion, the selectivity of the auxin-type herbicides is clearly a complex topic, dependent on many interacting aspects of herbicide behaviour and plant physiology.

References

Badescu, G.O. and Napier, R.M. (2006) Receptors of auxin: will it all end in TIRs? *Trends in Plant Science* **11**, 217–223.

Bowe, S., Landes, M., Best, J., Schmitz, G. and Graben, M. (1999) BAS 662 H: an innovative herbicide for weed control in corn. *Proceedings of the Brighton Crop Protection Conference, Weeds* **1**, 35–40.

Brinkworth, L.A, Egerton, S.A, Bailey, A.D. and Bernhard, U. (2005) Aminopyralid, a new active substance for long-term control of annual and perennial broad-leaved weeds in grassland. Proceedings of the BCPC International Congress. *Crop Science and Technology* **1**, 43–48.

Brummell, D.A. and Hall, J.L. (1987) Rapid cellular responses to auxin and the regulation of growth. *Plant, Cell and Environment* **10**, 523–543.

Burnside, O.C. *et al.* (eds) (1996) Biologic and economic assessment of benefits from use of phenoxy herbicides in the United States. USDA National Agricultural Pesticide Impact Assessment Program (NAPIAP) Report number 1-PA-96.

Edgerton, M.D., Tropsha, A. and Jones, A.M. (1994) Modelling the auxin-binding site of auxin-binding protein 1 of maize. *Phytochemistry* **35**, 111–1123.

Ettlinger, C. and Lehle, L. (1988) Auxin induces rapid changes in phosphatidyl inoitol metabolites. *Nature* **331**, 176–178.

Fitzsimons, P.J. (1989) The determination of sensitivity parameters for auxin-induced H^+-efflux from *Avena* coleoptile segments. *Plant, Cell and Environment* **12**, 737–746.

Fitzsimons, P.J., Barnwell, B. and Cobb, A.H. (1988) A study of auxin-type herbicide action based on dose–response analysis of H^+-efflux. In: Proceedings of the European weed research society symposium, *Factors Affecting Herbicidal Activity and Selectivity*. Wageningen: European Weed Research Society, pp. 63–68.

Goodwin, T.W. and Mercer, E.I. (1983) *Introduction to Plant Biochemistry*, 2nd edn. Oxford: Pergamon Press.

Grossmann, K., Kwiatkowski, A. and Tresch, S. (2001) Auxin herbicides induce H_2O_2 overproduction and tissue damage in cleavers (*Galium aparine* L.). *Journal of Experimental Botany* **52**, 1811–1816.

Hull, M.R. and Cobb, A.H. (1998) An investigation of herbicide interaction with the H^+-ATPase activity of plant plasma membranes. *Pesticide Science* **53**, 155–164.

Katekar, G.F. (1979) Auxins: on the nature of the receptor site and molecular requirements for auxin activity. *Phytochemistry* **18**, 223–233.

Kelley, K.B., Zhang, Q., Lambert, C.N. and Riechers, D.E. (2006) Evaluation of auxin-responsive genes in soybean for detection of off-target plant growth regulator herbicides. *Weed Science* **54**, 220–229.

Kirby, C. (1980) *The Hormone Weedkillers*. Farnham, Surrey: British Crop Protection Council.

Lewer, P. and Owen, W.J. (1990) Selective action of the herbicide triclopyr. *Pesticide Biochemistry and Physiology* **36**, 187–200.

Martin, T.J. (1987) Broad versus narrow-spectrum herbicides and the future of mixtures. *Pesticide Science* **20**, 289–299.

Napier, R.M., David, K.M. and Perrot-Rechenmann, C. (2002) A short history of auxin-binding proteins. *Plant Molecular Biology* **49**, 339–348.

Normanly, J. (1997) Auxin metabolism. *Physiologia Plantarum* **100**, 431–442.

Paciorek, T., Zazimalova, E., Ruthardt, N. *et al.* (2005) Auxin inhibits endocytosis and promotes its own efflux from cells. *Nature* **435**, 1251–1256.

Pillmoor, J.B. and Gaunt, J.K. (1981) The behaviour and mode of action of the phenoxyacetic acids in plants. In: Hutson, D.H. and Roberts, T.R. (eds) *Progress in Pesticide Biochemistry*, vol. 1. Chichester: Wiley, pp. 147–218.

Sanders, G.E. and Pallett, K.E. (1987) Physiological and ultrastructural changes in *Stellaria media* following treatment with fluroxypyr. *Annals of Applied Biology* **111**, 385–398.

Tan, X., Calderon-Villalobos, L.I.A, Sharon, M., *et al.* (2007) Mechanism of auxin perception by the TIR1 ubiquitin ligase. *Nature* **446**, 640–645.

Thompson, L.M.L. (1989) An investigation into the mode of action and selecticity of 3,6-dichloropicolinic acid. PhD Thesis, Nottingham Polytechnic, UK.

Thompson, L.M.L. and Cobb, A.H. (1987) The selectivity of clopyralid in sugar beet; studies on ethylene evolution. *British Crop Protection Conference, Weeds* **3**, 1097–1104.

Venis, M. (1985) *Hormone Binding Sites in Plants*. Harlow, UK: Longman.

Vieten, A., Sauer, M., Brewer, P.B. and Friml, J. (2007) Molecular and cellular aspects of auxin – transport-mediated development. *Trends in Plant Science* **12**, 160–168.

Ware, G.W. (1983) *Pesticides. Theory and Application*. San Francisco, CA: W.H. Freeman.

Woodward, A.W. and Bartel, B. (2005) Auxin: regulation, action and interaction. *Annals of Botany* **95**, 707–735.

Chapter 8
Inhibitors of Lipid Biosynthesis

8.1   Introduction

Current agricultural practices of reduced cultivation, cereal crop monoculture and the widespread use of broadleaf weedkillers has resulted in an increased spread of grass weeds. The scale of the problem became evident in the 1970s, and by 1981 it was estimated that, for example, 55% of the 209,000 ha of sugar beet in the UK was infested with wild oats (*Avena* spp.) and black-grass (*A. myosuroides*), and approximately 10% with couch grass (*Elytrigia repens*) (Siddall and Cousins, 1982). Wild oats and black-grass are highly competitive and invasive weeds, especially in cereals, and various models predict significant yield losses when as few as 10 (*Avena* spp.) or 50 (*A. myosuroides*) weed plants are present per square metre. Furthermore, an increasingly reported problem is the abundance of volunteer wheat and barley populations in succeeding broadleaf crops, such as sugar beet, potatoes, or oilseed rape. Consequently, considerable effort has been directed towards the development of selective grass weedkillers, or graminicides, in both narrowleaf and broadleaf crops. Two important families of herbicides, aryloxyphenoxypropionates (AOPPs; fops) and cyclohexanediones (CHDs; dims), demonstrate good grass-weed control and have been used extensively since their introduction in the mid-1970s. In addition to these herbicides, which inhibit lipid synthesis at the acetyl-CoA carboxylase (ACCase) step, a number of other lipid biosynthesis inhibitors are available that prevent plants from synthesising very-long-chain fatty acids (VLCFAs) and subsequently also result in disruption to plasma membranes and the development of the waxy cuticle that covers aerial surfaces.

8.2   Structures and uses of graminicides

Graminicides may be categorised into six chemical families, namely the thiocarbamates, chloroacetamides, alaninopropionates, aryloxyphenoxypropionates (AOPPs), cyclohexanediones (CHDs) and the phenylpyrazolines (Table 8.1).

The selective control of wild oats in cereals first became possible in the late 1950s with the introduction of the chlorophenylcarbamate, barban and the thiocarbamate, diallate.

Herbicides and Plant Physiology, Second Edition By Andrew H. Cobb and John P.H. Reade
© 2010 A.H. Cobb and J.P.H. Reade

Table 8.1   Structures of graminicides.

(a) thiocarbamates

$$C_3H_7\!\!\diagdown \atop C_3H_7\!\!\diagup N - \overset{\displaystyle O}{\overset{\displaystyle \|}{C}} - S - CH_2 - R$$

R	Common name
—C(Cl)=C(Cl)—H	diallate
—C(Cl)=C(Cl)—Cl	triallate
—CH₃	EPTC

3-Cl-C₆H₄—N(H)—C(=O)—O—CH₂C≡C—CH₂Cl   barban

(b) chloroacetamides

2,6-(R₁)(R₂)-C₆H₃—N(R₃)—CO—CH₂Cl

R₁	R₂	R₃	Common name
CH₃	C₂H₅	CH₂OC₂H₅	acetochlor
C₂H₅	C₂H₅	CH₂OCH₃	alachlor
CH₃	CH₃	CH₂—N—N (pyrazole)	metazachlor
CH₃	C₂H₅	CH(CH₃)—CH₂OCH₃	metolachlor

Table 8.1   (*Continued*)

R₁	R₂	R₃	Common name
H	H	$\overset{\displaystyle H}{\underset{\displaystyle CH_3}{\overset{\displaystyle |}{\underset{\displaystyle |}{C}}}}$—CH₃	propachlor

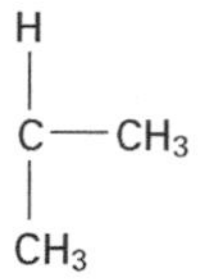

(c) alaninopropionates

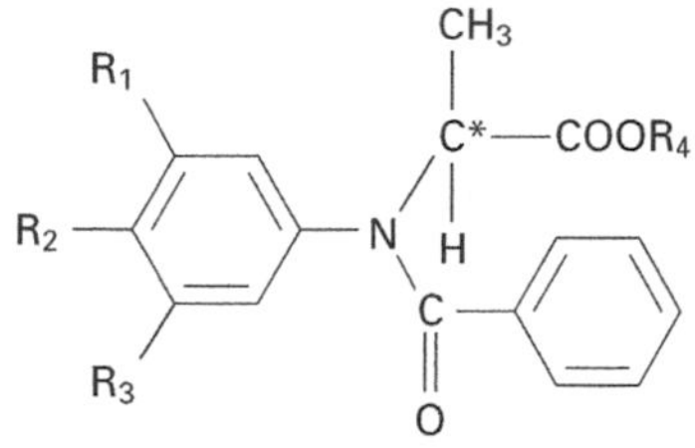

R₁	R₂	R₃	R₄	Common name
Cl	Cl	H	C₂H₃	benzoylprop-ethyl
H	F	Cl	C₃H₇	flamprop-isopropyl
H	F	Cl	CH₃	flamprop-methyl

(d) aryloxyphenoxypropionates

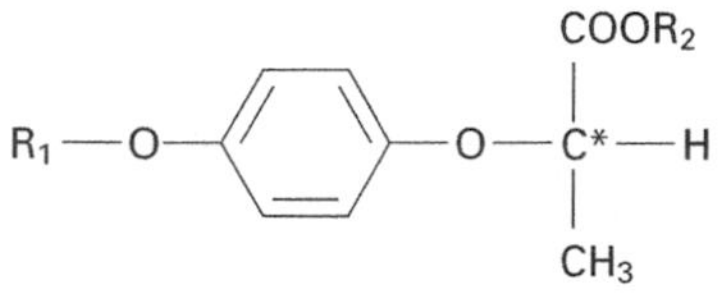

R₁	R₂	Common name
	CH₃	diclofop-methyl
	C₄H₉	fluazifop-butyl
	CH₃	haloxyfop-methyl
	C₂H₅	quizalofop-ethyl

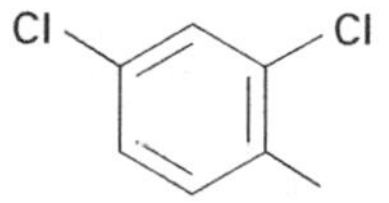

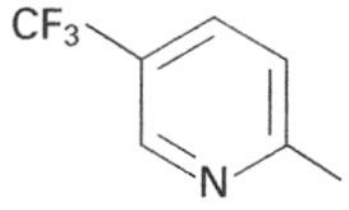

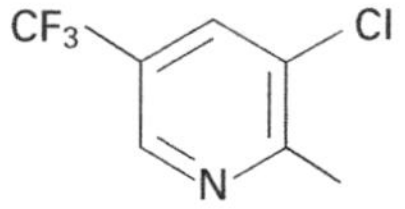

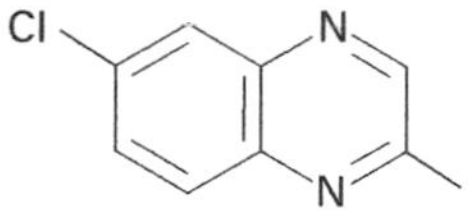

(*Continued*)

Table 8.1 (*Continued*)

R_1	R_2	Common name
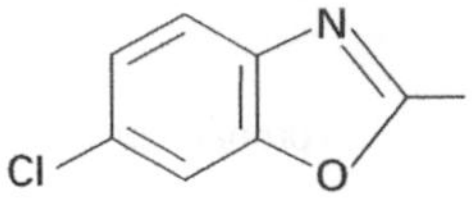	C_2H_5	fenoxaprop-ethyl
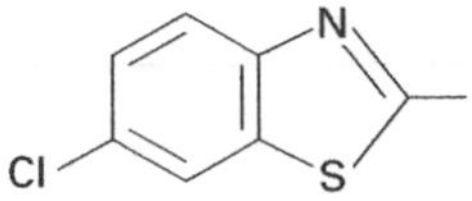	C_2H_5	fenthiaprop-ethyl
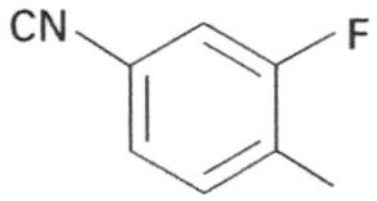	C_4H_9	cyhalofop-butyl
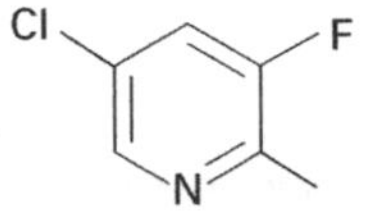	propargyl	clodinofop-propargyl
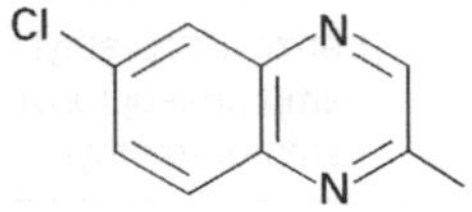	$(CH_3)_2C{=}NOCH_2CH_2$	propaquizafop

(e) cyclohexanediones

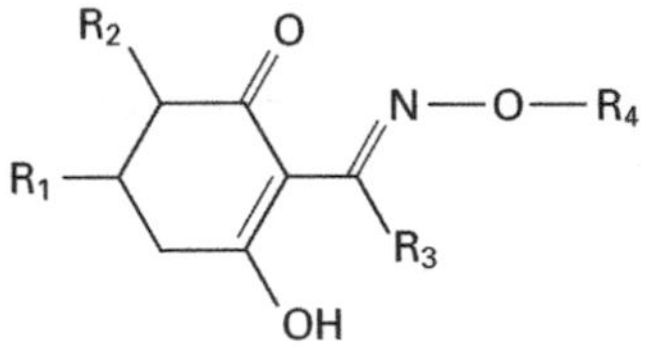

R_1	R_2	R_3	R_4	Common name
$(CH_3)_2>$	$-C(=O)CH_3$	C_3H_7	$CH_2CH{=}CH_2$	alloxydim
$C_2H_5S-C(CH_3)(H)-CH_2-$	H_2	C_2H_5	$CH_2CH{=}CH(Cl)$	clethodim
$C_2H_5S-C(CH_3)(H)-CH_2-$	H_2	C_3H_7	C_2H_5	sethoxydim

Table 8.1 (*Continued*)

R₁	R₂	R₃	R₄	Common name
S-containing tetrahydrothiopyran ring	H₂	C₃H₇	C₂H₅	cycloxydim
CH₃, CH₃, CH₃ substituted benzene	H₂	C₂H₅	C₂H₅	tralkoxydim
O-containing tetrahydropyran ring	H₂	C₂H₅	CH₂CH=CH with Cl	tepraloxydim
PrCO substituted benzene	H₂	C₂H₅	C₂H₅	butroxydim
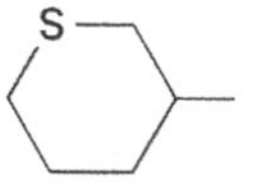 S-containing ring	H₂	C₃H₇	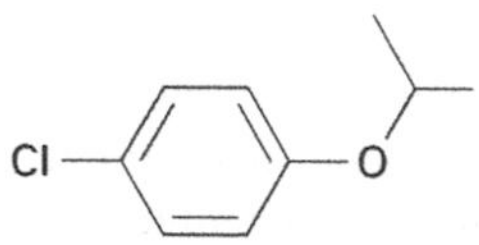 Cl—⟨ ⟩—O—CH(CH₃)₂	profoxydim

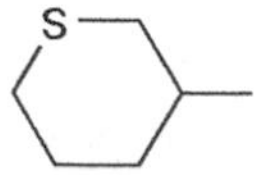

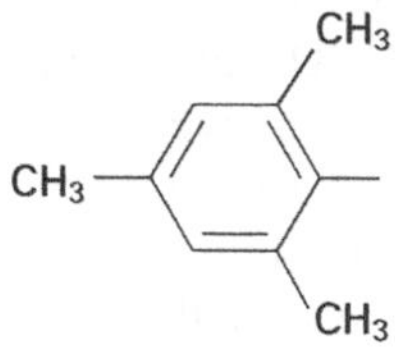

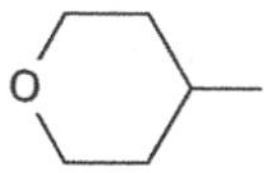

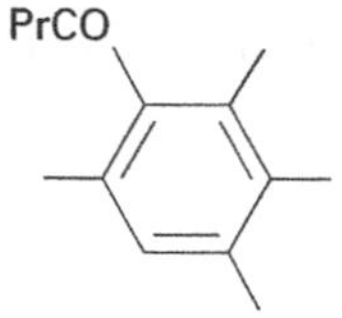

(f) phenylpyrazolines

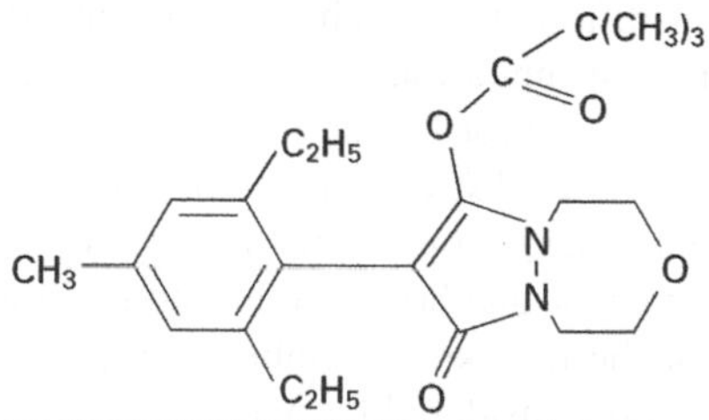

pinoxaden

However, barban has proven to be phytotoxic to certain barley cultivars, and the activity of the soil-applied diallate varies with soil type, condition and placement. Triallate has superseded diallate in cereals, since wheat and barley are more tolerant to this compound. These volatile herbicides must be rapidly incorporated into the top 2.5 cm of soil for maximum efficacy. Grasses which germinate from seed grow through this treated layer and are killed, although surface germinating seeds or those at greater depth are not controlled. No soil incorporation is required for the pre-emergent chloroacetamides, since these herbicides have a residual action in the soil, and so can kill annual broadleaf weeds and annual grasses for up to eight weeks after application.

The post-emergent alaninopropionates were introduced between 1969 and 1972 for the control of wild oats already present in cereals, and the introduction of the AOPPs and the CHDs in the mid-1970s has enabled the control of a broad spectrum of annual and perennial grasses in a wider range of crops. Furthermore, the development of the AOPPs (also termed the 'fops', such as diclofop, fluazifop and quizalofop) and the CHDs (also termed the 'dims', such as sethoxydim, tralkoxydim and clethodim) has produced active ingredients with widespread and successful grass-weed control at doses as low as 100–200 g ha^{-1}. Recent commercialisation of pinoxaden, a phenylpyrazolin that acts at the same target site as fops and dims but is chemically distinct from these groups, has resulted in a third class of ACCase-inhibiting graminicides which are termed the 'dens'.

In the cases of the alaninopropionates and the AOPPs , the C2 of the propanoic acid group is a chiral centre so that both (R) and (S) enantiomers exist. Interestingly, only the (R) enantiomer is biologically active, so the removal of the inactive (S) half of an isomeric mix represents a doubling of activity. Thus, (R) isomers are commercially produced and used at half the dose rate of the racemic mixture.

8.3 Inhibition of lipid biosynthesis

The first hint of a specific target process for graminicide inhibition came from the studies of Hoppe (1980). He found that although diclofop-methyl did not interfere with photosynthesis, respiration, protein synthesis or nucleic acid synthesis, the inhibition of acetate incorporation into fatty acids could be demonstrated in susceptible species. Lichtenthaler and Meier (1984) later reported that the CHDs also disrupted *de novo* lipid biosynthesis in developing barley seedlings, and in 1987–88 four different research teams independently reported that the enzyme ACCase was the target of inhibition by both graminicide classes (see Lichtenthaler *et al.*, 1989 and Secor *et al.*, 1989 for further details).

Plant membranes contain unique fatty acids that have crucial structural and biochemical roles. For example, at least 70% of the total leaf fatty acids consist of the unsaturated α-linolenic acid (18:3), which itself makes up between 40% and 80% of the lipid fraction in the chloroplast. Indeed, the unique functioning of the thylakoid membrane to permit the movement of protons, electrons, and their carriers is considered by many workers to result from the property of membrane fluidity conferred by this unsaturated fatty acid. Trans-Δ_3-hexadecanoic acid and linoleic acid (18:2) are additional examples of important thylakoid fatty acids. The synthesis of fatty acids in plants involves two major enzymes; acetyl-CoA carboxylase (ACCase) and fatty acid synthase. Fatty acids are synthesised both in the chloroplast stroma and the cytoplasm (Figure 8.1). Essentially, malonyl-CoA is formed from acetyl-CoA and converted to the saturated palmitate (16:0) by the action of a soluble stromal enzyme complex, termed fatty acid synthetase. This complex contains seven enzymes covalently bound to an acyl carrier protein (ACP), which transfers intermediates between the seven enzymes. Thus, seven enzyme cycles are needed for the condensation of seven additional C$_2$ units into one palmitate. Two metabolic routes are now possible from palmitate. On the one hand, a soluble condensing enzyme and

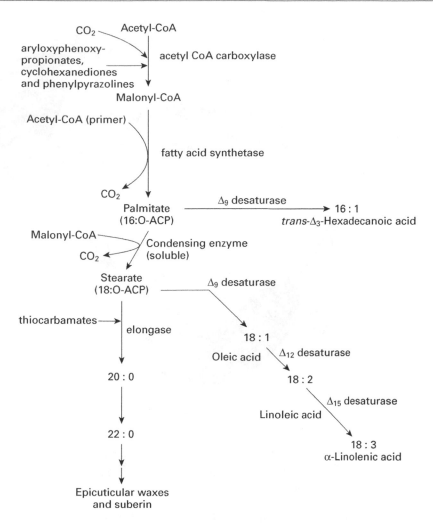

Figure 8.1   Fatty acid biosynthesis in plants. ACP, acyl carrier protein.

elongases bound to the endoplasmic reticulum are able to add further C_2 units in the cytoplasm to yield the long-chain saturated fatty acids found in suberin and the epicuticular waxes on plant surfaces, and desaturases are present in the chloroplast to form the unsaturated fatty acids mentioned earlier (Harwood, 1988).

Acetyl-CoA carboxylase (acetyl-coenzyme A: bicarbonate ligase [ATP], E.C. 6.4.1.2.), or ACCase, is the first committed step for fatty acid biosynthesis in plants, and catalyses the formation of malonyl-CoA (Harwood, 1989). ACCase is a high molecular weight, multifunctional protein with three distinct functional regions (two enzymic regions and one a carrier protein region), which involves biotin as an essential cofactor that functions as a CO_2 carrier (Figure 8.2a). Initially, a carboxyl group is donated from a bicarbonate anion, and ATP hydrolysis is used to allow the formation of a carboxybiotin intermediate

(a)

(b)

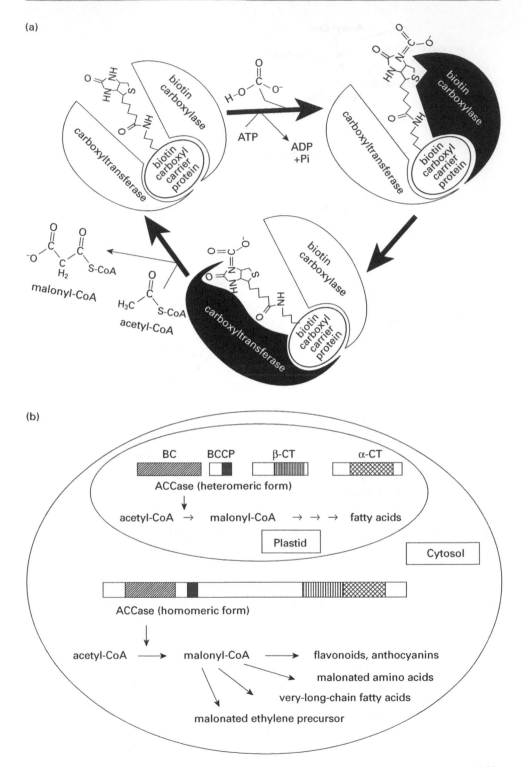

Figure 8.2 (a) Diagrammatic representation of the three functional domains of ACCase (BC, BCCP, α-CT/β-CT) (reproduced from Ohlrogge and Browse, 1995). Reproduced with permission of American Society of Plant Biologists via Copyright Clearance Center. (b) Composition and compartmentalisation of the two forms of ACCase found in higher plants. See text for abbreviations (reproduced from Sasaki and Nagano, 2004, with permission).

by biotin carboxylase (BC). Carboxybiotin is attached to an ε-amino group of a lysine residue on the biotin carboxyl carrier protein (BCCP). Carboxybiotin then functions as a CO_2 donor in malonyl-CoA formation (Figure 8.2a). Malonyl-CoA is a substrate for fatty acid synthetase and also for fatty acid elongation (and the subsequent production of a number of important secondary metabolites, including flavonoids and phytoalexins). It is therefore likely that inhibition of ACCase affects a number of malonyl-CoA-requiring metabolic pathways. ACCase activity can be regulated at the transcription and post-transcription level but is also regulated by a number of metabolic factors. ACCase is most active when a plant is in light. During photosynthesis the stromal pH rises from 7 to 8 and Mg^{2+} concentration rises from approximately 1 mM to 3 mM. Plastid ACCase activity reaches a maximum at pH 8 and at Mg^{2+} concentrations of 2–5 mM. This ensures that ACCase is most active in the light, when photosynthesis is producing ATP, reductant and photosynthate, all of which are necessary for lipid biosynthesis. Laboratory studies by Kozaki and Sasaki (1999) have demonstrated that reducing agents will also activate ACCase, specifically the carboxytransferase (CT) domain of the enzyme. This appears to be due to the formation of a disulfide bridge between 2 cysteine residues located in the α and β-CT regions of ACCase (Kozaki et al., 2001). During photosynthesis concentration of reductant rises in the plastid and this represents a further light-mediated regulation of ACCase activity. It has also been postulated that ACCase may undergo phosphorylation, and this may also regulate its activity, although the mechanisms involved have yet to be fully elucidated (Savage and Ohlrogge, 1999).

Rendina and colleagues (1990) have characterised the kinetics of ACCase inhibition by graminicides. AOPPs and CHDs appear to be non-competitive inhibitors with respect to Mg^{2+} ATP, HCO_3^- and acetyl-CoA. It therefore seems likely that this class of herbicides inhibits the carboxytransferase rather than the carboxylation step of ACCase activity. It also appears that both AOPPs and CHDs compete for the same site on the enzyme and that the inhibition is reversible. The inhibition of ACCase is both rapid and concentration dependent so that, for example, about 1 μm haloxyfop or tralkoxydim can inhibit the enzyme *in vitro* by 50% within 20 min. Furthermore, haloxyfop-acid is more than 100-fold more potent than the methyl-ester and only the R(+) enantiomer is herbicidally active (Secor et al., 1989). Recent observations that amino acids in the CT region are responsible for ACCase inhibition may lead to further information on the mechanism of herbicide binding to the enzyme. Indeed, the observation that Ile1781→Leu results in resistance to all fops and most dims but that Ile2041→Asn only results in resistance to fops, clearly indicates the importance of Ile2041 in fop but not dim binding (Délye et al., 2003). Additionally, a homozygous herbicide-resistant black-grass biotype that contains Gly at position 2078 shows decreased fitness in the absence of herbicide, highlighting the importance of position 2078 in ACCase activity (Menchari et al., 2008).

Two forms of ACCase are found in higher plants (Figure 8.2b) and this plays an important role in the selectivity demonstrated by AOPPs and CHDs. In dicotyledons a heteromeric form (termed the prokaryotic form) of ACCase, located in plastids, is insensitive to AOPPs and CHDs. The heteromeric form is composed of BCCP, BC, α-CT and β-CT polypeptides. It has been postulated that this form of ACCase is $BCCP_4 BC_2$ α-CT_2 β-CT_2, similar to bacterial ACCase (Choi-Rhee and Cronan, 2003). This form of ACCase is absent from the monocotyledon grasses and evidence suggests that this is due to the absence of the *accD* gene that encodes the β-CT polypeptide (Konishi and Sasaki, 1994). A homomeric

Table 8.2 Forms of ACCase found in higher plants (after Sasaki *et al.*, 1995). Reproduced with permission of American Institute of Biological Sciences via Copyright Clearance Center.

	Prokaryotic form	Eukaryotic form
Structure	Heteromeric (separate BCC, BC and CT subunits)	Homomer; single multifunctional polypeptide
Grasses	Absent	Plastids and cytosol
Dicotyledonous species	Plastids	Cytosol
Sensitivity to 'fops' and 'dims'	Insensitive	Sensitive (plastidic) Insensitive (cytosolic)

BCC, biotin carboxyl carrier; BC, biotin carboxylase; CT, carboxytransferase.

Table 8.3 Sensitivity of ACCase I and II from *Lolium multiflorum* biotypes to diclofop (from Evenson *et al.*, 1997).

Biotype	Isoform	Source	Diclofop conc. (μM)	Inhibition (%)
Susceptible	ACCase I	Plastid	0.2	50
	ACCase II	Cytosol	125	42
Resistance	ACCase I	Plastid	7	50
	ACCase II	Cytosol	127	31

form of ACCase, located in plastids, is found in grasses and is sensitive to AOPPs and CHDs. This homomeric form is also found in the cytosol of both dicotyledons and grasses. It is a large polypeptide (~ 250 kDa) which contains BCCP, BC, α-CT and β-CT regions. It appears to be active as a dimer. These forms of ACCase are summarised in Table 8.2 and Figure 8.2b. This explains the selectivity of ACCase inhibitors between dicotyledons and grasses, as the presence of the herbicide-insensitive heterodimeric ACCase in dicotyledons allows the synthesis of fatty acids in the presence of these herbicides.

In the monocot maize two isoforms of ACCase are reported. The plastid form (ACCase I) predominates and is inhibited by AOPPs and CHDs. A cytosolic form (ACCase II) is 2000-fold less sensitive to these herbicides. Further studies have found a similar situations in other grasses, and it appears that some naturally occurring resistant biotypes of *Lolium multiflorum* possess an altered form of the plastid ACCase that is less sensitive to herbicides (Evenson *et al.*, 1997; Table 8.3). Plant ACCases have recently been reviewed by Sasaki and Nagano (2004).

ACCase is not the only site of graminicide action. Weisshaar *et al.* (1988) have demonstrated that micromolar concentrations of the chloroacetamides also inhibit fatty acid biosynthesis by preventing the elongation of palmitate and the desaturation of oleate in the green microalga *Scenedesmus acutus*. It has subsequently been confirmed that both the elongation and desaturation steps of very long-chain fatty acid biosynthesis are targets for a number of herbicides.

Very-long-chain fatty acids (VLCFAs) are important components of the plant cell plasma membrane (plasmalemma) and are enriched in the leaf epicuticular waxes. Here they are embedded in a matrix and ensure the hydrophobicity of the leaf surface (Chapter 3.2).

Indeed, the cuticle is the main barrier against invasion by external agents and micro-organisms, while preventing the loss of water and solutes from the leaves. Disrupting the plasma membrane will lead to a loss of permeability, transport and hormone receptor functions. It therefore follows that inhibition of VLCFA biosynthesis is a valuable target for herbicide action and development.

Members of several herbicide groups are now known to act as specific inhibitors of VLCFA biosynthesis, with alkyl chains longer than C_{18}. They inhibit elongase activity in reactions taking place outside the chloroplast. Chemical groups include the chloroaceta-mides (e.g. alachlor), oxyacetamides (e.g. mefenacet), carbomylated five-membered nitro-gen heterocycles (e.g. cafenstrole), oxiranes (e.g. indanofan) and miscellaneous others (such as ethofumesate). Some examples of herbicide structures acting as inhibitors of VLCFA biosynthesis are shown in Figure 8.3.

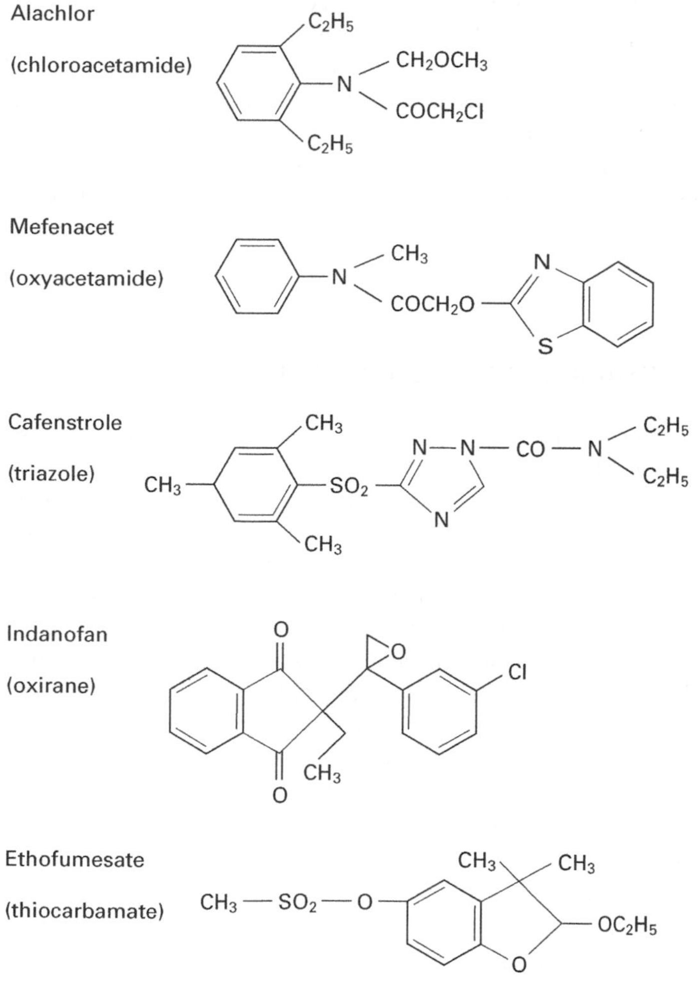

Figure 8.3 Structures of a selection of herbicides that inhibit very-long-chain fatty acid biosynthesis.

The chloracetamides have been successfully used in maize, soybean and rice for several decades, and remain today an important means of weed control, especially in maize. They are persistent herbicides, taken up from the soil, and several safeners have been developed for use in mixtures to extend their range of use. While germination is unaffected, early seedling growth is typically inhibited and the seedlings do not emerge from the soil or are severely stunted. Cell division and expansion are both inhibited.

The enzymes of acyl elongation are membrane-bound and thought to be associated with the endoplasmic reticulum and the Golgi apparatus in the cytoplasm, and perhaps with the plasma membrane itself. There are 21 genes encoding VLCFA elongases in *Arabidopsis thaliana*. Trentkamp and colleagues (2004) investigated the expression and activity of six gene products in the presence of VLCFA biosynthesis inhibiting herbicides and found wide substrate specificity. They suggest that such complex patterns of substrate specificity may explain why resistance to these herbicides is rare.

The substituted pyridazinones have been found to inhibit lipid biosynthesis in addition to their known inhibition of carotenoid biosynthesis (see section 6.3). Norflurazon can inhibit the Δ_{15}-desaturase to prevent α-linolenic acid synthesis, and metflurazon may additionally prevent the formation of trans-Δ_3-hexadecanoate by the inhibition of the Δ_9-desaturase. These herbicides are classed as carotenoid biosynthesis inhibitors and their effects on lipid biosynthesis probably represent a secondary mode of action. That they have two points of action should not be surprising as they inhibit a desaturase enzyme in both metabolic pathways.

8.4 Anti-auxin activity of graminicides

Although most authors accept the selective inhibition of ACCase as a primary target of the AOPPs and CHDs, it has also been proposed that these and other graminicides can also act as anti-auxins. Diclofop-methyl, for example, has no auxin activity alone, but can inhibit several auxin-mediated processes such as coleoptile elongation and proton-efflux (Figure 8.4), and alter cell membrane potentials, possibly by acting as a protonophore (Wright and Shimabukuro, 1987). Disruption of memberane function in the form of rapid depolarisation of the plasma membrane electrogenic potential is reported after ACCase inhibitor treatment of susceptible species. This phenomenon is not observed when the same herbicides are applied to ACCase-resistant biotypes or species. However, this alternative hypothesis of graminicide action has been rejected by many workers as a secondary consequence of the inhibition of lipid biosynthesis which would, it is argued, alter membrane function. Furthermore, for this hypothesis to gain acceptance it would be necessary that this anti-auxin activity satisfy the following four criteria:

(a) shows at least equal sensitivity to the micromolar concentrations known to inhibit ACCase,
(b) is observed within minutes,
(c) shows stereospecificity, and
(d) is selective in grasses.

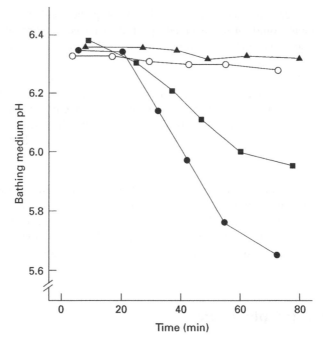

Figure 8.4   Inhibition of auxin-induced proton-efflux by diclofop-methyl (DM). ○, Control; ▲, 50 μM DM; ●, 10 μM MCPA; ■, 10 μM MCPA + 50 μM DM (from Cobb and Barnwell, 1989, with permission).

Several studies have provided supportive or circumstantial evidence to fulfil each of the above criteria, and more detailed and convincing data are slowly emerging. For example, micromolar concentrations of several AOPPs can inhibit auxin-induced proton-efflux from sensitive coleoptile segments within minutes (Figure 8.4, Cobb and Barnwell, 1989), and stereospecificity has been demonstrated (Andreev and Amrhein, 1976). Evidence is also accumulating for a selective anti-auxin activity being confined to grasses. Young, rapidly elongating monocotyledonous tissues are especially sensitive to graminicides, and a rapid retardation of grass internode elongation is commonly observed, with the result that sensitive grasses show significant stunting when compared to control plants and cereal crops. The need for a rapid rate of tissue extension has recently been identified as a prerequisite for optimal graminicide activity, since decreased extension rates associated with water stress, for example, severely reduce diclofop efficacy (Andrews *et al.*, 1989). Finally, it is surely pertinent that a foliar-applied graminicide must first come into contact with a cell membrane before its subsequent intracellular metabolism and translocation to the grass meristem. The opportunity therefore exists *in vivo* for a graminicide to demonstrate anti-auxin activity in addition to an inhibition of ACCase.

This additional activity of graminicides is thought to be of some commercial significance when herbicide mixtures are taken into account. Since an arable field will contain a mixture of both monocotyledonous and dicotyledonous weeds, it would be of obvious advantage to the farmer to mix a broadleaf weed, auxin-type herbicide with a graminicide. This single treatment could ensure a broad spectrum of weed control with additional savings in time, labour and fuel costs. However, it is well known that such tank mixtures commonly result in reduced efficacy and crop damage is often observed. Antagonism of ACCase inhibitor action by a number of herbicides including 2,4-D, dicamba, ALS inhibitors, bentazone and acifluorfen has been reported. This antagonism does not appear to be at the sites of action of the herbicide and may be due to the decreased uptake and movement of the ACCase inhibitors. Indeed, broadleaf weed herbicides such as bentazone are able to reduce the uptake of some graminicides, and so account for a reduction in graminicide efficacy.

On the other hand, if the broadleaf weedkiller is an auxin-type herbicide it may be argued that the anti-auxin activity of the graminicide is overcome by the addition of 'extra auxin'. Hence, less stunting is observed in treated grasses and plants may recover to set seed. The real significance of these observations remains to be established, although further understanding of this interaction may eventually lead to the design of more compatible graminicide/broadleaf weedkiller mixtures in the future.

8.5   How treated plants die

The foliar uptake of the AOPPs is very rapid, and studies using radiolabelled herbicides have shown that almost all of the applied dose remains at the site of application, and so contact damage is commonly observed at the treated leaf (Figure 8.5; Carr *et al.,* 1986). De-esterification occurs in the leaf tissues, and the phytotoxic acid accumulates at the apical meristem, which becomes necrotic. Active plant growth and warm temperatures encourage active transport within the phloem and passive movement in the xylem, so that the acid accumulates in all meristematic zones. However, most herbicidal injury is noted in the apical meristem due to the limited translocation of these herbicides, with far fewer symptoms noted in the root meristem. Interference with lipid biosynthesis causes an irreversible disruption in membrane synthesis, so that normal plastid development is not observed and metabolism is drastically altered. Leaf elongation ceases within two days and plastid disruption is most marked in young leaves, which appear chlorotic. The main site of action is the apical meristem where large levels of *de novo* fatty acid biosynthesis are taking place to support growth. Within two to four days new tissue at the meristem can be easily detached from the rest of the plant. Grass death follows within two to three weeks after application. Grasses treated with the alaninopropionates and the CHDs show similar symptoms to those described above, although the CHDs have slower rates of penetration into the treated leaf.

Pre-planting or pre-emergent treatments with the thiocarbamates and chloroacetamides effectively provides a chemical barrier for the growth of grass seedlings. These compounds penetrate the mesocotyl of germinating weeds to inhibit lipid biosynthesis in these young tissues. Consequently, any tissues that emerge are chlorotic and short-lived. Indeed, seedling death will occur when seed reserves of fatty acids are exhausted.

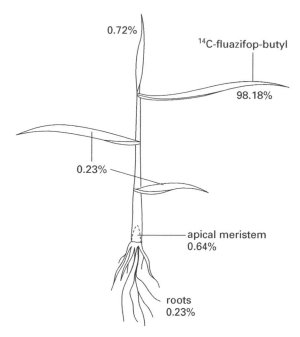

0.72%

^{14}C-fluazifop-butyl

98.18%

0.23%

apical meristem
0.64%

roots
0.23%

Figure 8.5 Percentage ^{14}C-activity recovered from plants of *Setaria viridis* 72 h after treatment with radiolabelled fluazifop-butyl (modified from Carr *et al.*, 1986).

8.6   Selectivity

Crop selectivity to the thiocarbamates is achieved through depth protection and metabolism. These herbicides are volatile and so need to be rapidly incorporated into the soil to be effective. Indeed EPTC, the most volatile example of this class, must be incorporated within 15 min of application! Such instability ensures that EPTC can be used to clear a soil of germinating couch grass (*Elytrigia repens*) or wild oats (*Avena* spp.) and that it mostly disappears before a crop is planted in the following week or so. Triallate is incorporated into the top 2.5 cm of soil and also controls grasses at germination. Cereal seeds are then drilled at a minimum 4 cm depth and their seedlings grow through this zone, the sensitive meristem being protected by the coleoptile and primodial leaves. On the other hand, germinating wild oat seedlings have an elongating mesocotyl (first internode) that extends the unprotected wild oat meristem into the phytotoxic chemical barrier. However, wild oats germinating at the soil surface or at greater depths than 4 cm may not be controlled by this treatment, and so mixtures with more persistent soil-applied herbicides such as atrazine are often used in practice. In addition, thiocarbamates appear to undergo bioactivation in susceptible species by the process of sulphoxidation. The sulphoxides so formed create an electrophilic centre in the molecule which seems to be linked to an increase in phytotoxicity. Tolerant crops such as cereals, and especially maize, appear able to detoxify the sulphoxides by conjugation with glutathione.

The chloroacetamides, such as alachlor and metolachlor, are absorbed onto soil colloids and may remain active for two or three months without major soil leaching. Selectivity

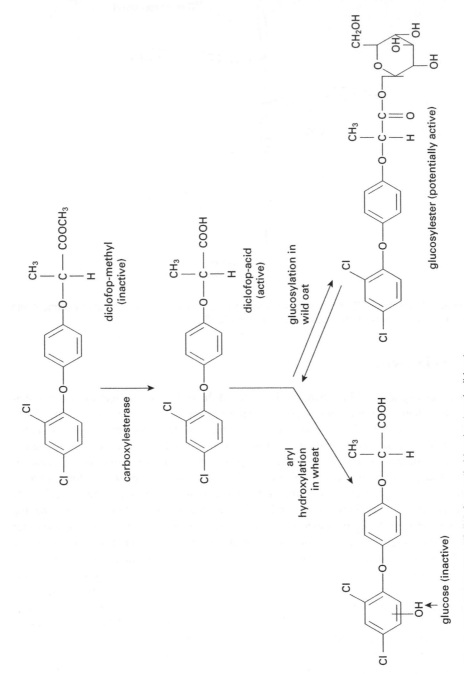

Figure 8.6   Metabolism of diclofop-methyl in wheat and wild oat.

Table 8.4 Effect of haloxyfop and tralkoxydim on plant growth (ED_{50}) and inhibition of ACCase *in vitro* (I_{50}) (from Secor *et al.*, 1989, with permission).

Species	ED_{50} (μM)		I_{50} (μM)	
	Haloxyfop	Tralkoxydim	Haloxyfop	Tralkoxydim
Maize	19 (S)	18 (S)	0.50	0.52
Wheat	83 (S)	>760 (T)	1.22	0.91
Tall fescue	133 (S)	225 (S)	0.94	0.40
Red fescue	1250 (T)	>6000 (T)	23.32	13.83
Soybean	>6000 (T)	>6000 (T)	138.50	516.72

S, susceptible; T, tolerant.

appears to be achieved by rapid metabolism to inactive glutathione conjugates in barley, sorghum, maize and sugarcane seedlings.

Ester hydrolysis appears central to the selectivity of the alaninopropionates in the control of wild oats in cereals. Thus, the inactive esters of benzoylprop and flamprop, for example, are rapidly hydrolysed in susceptible wild oats to their respective phytotoxic acids. In contrast, de-esterification is far slower in wheat, and any acid formed is rapidly inactivated by glycosylation. The carboxylesterase responsible for benzoylprop-ethyl hydrolysis in *Avena fatua* has been studied in some detail by Hill and colleagues (1978), and has also been implicated in the selective metabolism of the AOPPs. In this case Shimabukuro *et al.* (1979) found that diclofop-methyl was rapidly de-esterified in both wild oat and wheat. However, aryl hydroxylation rapidly inactivated the phytotoxic acid in the crop, and an ester glucoside was formed in the weed from which the toxic species could be easily and rapidly regenerated (Figure 8.6).

Metabolism also appears to form the basis of selectivity of the CHDs. Sulphoxidation, aryl hydroxylation, and molecular rearrangement have been observed with cycloxydim in tolerant species, and rapid conjugation of these groups leads to more polar and inactive by-products.

The striking selectivity of the ACCase inhibitors haloxyfop and tralkoxydim has been investigated in some detail by Secor *et al.* (1989). In this study, both susceptible and tolerant plants were sprayed with a range of herbicide concentrations to establish the dose at which growth was inhibited by 50% (ED_{50}) and values compared to the herbicide concentrations needed to inhibit *in vitro* ACCase activity by 50% (I_{50}). Their results, presented in Table 8.4, indicate that plant tolerance to these herbicides is clearly related to the insensitivity of ACCase. The identification of two forms of ACCase with very different sensitivities to ACCase inhibitors has confirmed that this is the major basis of selectivity between monocotyledonous and dicotyledonous crops (see section 8.3 and Table 8.4). Of the five species examined, soybean was the most tolerant at both the whole plant and the enzyme levels, although the opposite was the case in maize. Interestingly, wheat proved to be tolerant to tralkoxydim even though the isolated ACCase was sensitive to inhibition, which may imply metabolic inactivation in this crop. Further differences in grass sensitivity to the ACCase inhibitors have been characterised by Lichtenthaler and colleagues

(1989). It is therefore clear that grass tolerance to these graminicides is a consequence of both metabolism and lower sensitivity of ACCase, and that the apparent resistance of many dicotyledonous plants (e.g. soybean in Table 8.4) is due to insensitivity of the target enzyme itself.

References

Andreev, G.K and Amrhein, N. (1976) Mechanism of action of the herbicide 2-chloro-3-(4-chlorophenyl) propionate and its methyl ester: interaction with cell responses mediated by auxin. *Physiologia plantarum* **37**, 175–182.

Andrews, M., Dickson, R.L., Foreman, M.H, Dastgheib, F. and Field, R.J. (1989) The effects of different external nitrate concentrations on growth of *Avena sativa* cv. Amuri treated with diclofop-methyl. *Annals of Applied Biology* **114**, 339–348.

Carr, J.E., Davies, L.G., Cobb, A.H. and Pallett, K.E. (1986) Uptake, translocation and metabolism of fluazifop-butyl in *Setaria viridis*. *Annals of Applied Biology* **108**, 115–123.

Choi-Rhee, E. and Cronan, J.E. (2003) The biotin carboxylase–biotin carboxyl carrier protein complex of *Escherichia coli* acetyl CoA carboxylase. *Journal of Biological Chemistry* **278**, 30806–30812.

Cobb, A.H. and Barnwell, P. (1989) Anti-auxin activity of graminicides. *Brighton Crop Protection Conference, Weeds* **1**, 183–190.

Délye, C., Zhang, X.Q., Chalopin, C., Michel, S .and Powles, S.B. (2003) An isoleucine residue within the carboxyl-transferase domain of multidomain acetyl-coenzyme A carboxylase is a major determinant of sensitivity to aryloxyphenoxypropionate but not to cyclohexanedione inhibitors. *Plant Physiology* **132**, 1716–1723.

Evenson, K.J., Gronwald, J.W. and Wyse, D.L. (1997) Isoforms of acetyl-coenzyme A carboxylase in *Lolium multiflorum*. *Plant Physiology and Biochemistry* **35**, 265–272.

Harwood, J.L. (1988) Fatty acid metabolism. *Annual Review of Plant Physiology* **39**, 101–38.

Harwood, J.L. (1989) The properties and importance of acetyl-coenzyme A carboxylase in plants. *Brighton Crop Protection Conference, Weeds* **1**, 155–162.

Hill, B.D., Stobbe, E.H and Jones, B.L. (1978) Hydrolysis of the herbicide benzoylprop-ethyl by wild oat esterase. *Weed Research* **18**, 149–154.

Hoppe, H.H. (1980) Veränderungen der membranpermeabilitdt, des kohlenhydragehaltes, des lipidzusammensetzung in keimwurzel von *Zea mays* L. nach behandlung mit diclofop-methyl. *Zeitschrift für Pflanzenphysiologie* **100**, 415–426.

Konishi, T. and Sasaki, Y. (1994) Compartmentation of two forms of acetyl CoA carboxylase in plants and the origin of their tolerance towards herbicides. *Proceedings of the National Academy of Sciences, USA* **91**, 3598–3601.

Kozaki, A. and Sasaki, Y. (1999) Light-dependent changes in redox status of the plastidic acetyl-CoA carboxylase and its regulatory component. *Biochemical Journal* **339**, 541–546.

Kozaki, A., Mayumi, K. and Sasaki, Y.(2001) Thiol–disulfide exchange between nuclear-encoded and chloroplast-encoded subunits of pea acetyl-CoA carboxylase. *Journal of Biological Chemistry* **276**, 39919–39925.

Lichtenthaler, H.K., Kobek, K. and Focke, M. (1989) Differences in sensitivity and tolerance of monocotyledonous and dicotyledonous plants towards inhibitors of acetyl-coenzyme A carboxylase. *Brighton Crop Protection Conference, Weeds* **1**, 173–182.

Lichtenthaler, H.K. and D. Meier, D. (1984) Inhibition by sethoxydim of chloroplast biogenesis, development and replication in barley seedlings. *Zeitschrift für Naturforschung* **39c**, 115–122.

Menchari, Y., Chauvel, B., Darmency, H. and Delye, C. (2008) Fitness costs associated with three mutant acetyl-coenzyme A carboxylase alleles endowing herbicide resistance in black-grass *Alopecurus myosuroides*. *Journal of Applied Ecology* **45**(3), 939–947.

Ohlrogge, J. and Browse, J. (1995) Lipid biosynthesis. *Plant Cell* **7**, 957–970.

Rendina, A.R., Craig-Kennard, A.C., Beaudoin, J.D. and Breen, M.K. (1990) Inhibition of acetyl-coenzyme A carboxylase by two classes of grass-selective herbicides. *Journal of Agricultural and Food Chemistry* **38**, 1282–1287.

Sasaki, Y. and Nagano, Y. (2004) Plant acetyl-CoA carboxylase: structure, biosynthesis, regulation, and gene manipulation for plant breeding. *Bioscience, Biotechnology and Biochemstry* **68**, 1175–1184.

Sasaki, Y., Konishi, T. and Nagano Y. (1995) The compartmentation of acetyl-coenzyme A carboxylase in plants. *Plant Physiology* **108**, 445–449.

Savage, L.J. and Ohlrogge, J.B. (1999) Phosphorylation of pea chloroplast acetyl-CoA carboxylase. *Plant Journal* **18**, 521–527.

Secor, J., Csèke, C. and Owen, J. (1989) The discovery of the selective inhibition of acetyl-coenzyme A carboxylase activity by two classes of graminicides. *Brighton Crop Protection Conference, Weeds* **1**, 145–154.

Shimabukuro, R.H., Walsh, W.C and Hoerauf, R.A. (1979) Metabolism and selectivity of diclofop-methyl in wild oat and wheat. *Journal of Agricultural and Food Chemistry* **27**, 615–623.

Siddall, C. J. and Cousins, S.F.B. (1982) Annual and perennial grass weed control in sugar beet following sequential and tank-mix application of fluazifop-butyl and broad-leaf herbicides. *British Crop Protection Conference, Weeds* **2**, 827–833.

Trentkamp S., Martin, W. and Tietjen, K. (2004). Specific and differential inhibition of very long-chain fatty acid elongases from *Arabidopsis thaliana* by different herbicides. *Proceedings of the National Academy of Sciences, USA* **101**, 11903–11908.

Weisshaar, H., Retzlaff, G. and Böger, P. (1988) Chloroacetamide inhibition of fatty acid biosynthesis. *Pesticide Biochemistry and Physiology* **32**, 212–216.

Wright, J.P. and Shimabukuro, R.H. (1987) Effects of diclofop and diclofop-methyl on the membrane potentials of wheat and oat coleoptiles. *Plant Physiology* **85**, 188–193.

Chapter 9
The Inhibition of Amino Acid Biosynthesis

9.1 Introduction

Since the early 1970s a new generation of herbicides of great agronomic and commercial importance has been developed that specifically inhibit the biosynthesis of amino acids. Inhibition of plant amino acid biosynthesis has become a major target of herbicide development as only plants and microorganisms can synthesise all their amino acids. Consequently, the chances of mammalian toxicity are slight because animals cannot synthesise histidine, isoleucine, leucine, lysine, methionine, threonine, tryptophan, phenylalanine and valine, which are termed essential amino acids and which have considerable dietary significance.

Microorganisms have proved the key to our understanding of the herbicidal inhibition of amino acid biosynthesis, since our knowledge of the biochemistry and genetics of microorganisms is far greater than our understanding of plant metabolism. In particular, mutations in several bacteria and yeasts have been raised with relative ease, and the study of their biochemistry and genetics has been directly applied to green plants. Furthermore, the developed technologies to isolate genes from microorganisms responsible for herbicide resistance and to transfer them to crop plants have provided exciting opportunities in crop protection, as discussed in Chapter 13. Indeed, the major success of glyphosate, and the more recent development of the very low dosage acetolactate synthase inhibitors and glufosinate, has led to an increased interest in this area of plant metabolism, and further discoveries may be anticipated in the coming years.

9.2 Overview of amino acid biosynthesis in plants

Our understanding of the biosynthesis of amino acids in plants is fragmentary. Although the general pathways are known with some certainty, their cellular location and details of regulation often remain obscure. Similarly, information of amino acid metabolism during plant development and in different tissues is also lacking, as are details of the structure and function of many key enzymes. In fact, herbicides have proved immensely powerful and useful tools to probe this aspect of plant metabolism. One certainty, however, is the

Herbicides and Plant Physiology, Second Edition By Andrew H. Cobb and John P.H. Reade
© 2010 A.H. Cobb and J.P.H. Reade

central role played by the chloroplast in amino acid biosynthesis. Although most enzymes are nuclear-encoded, they are synthesised in the cytoplasm and imported into the chloroplast where carbon skeletons from photosynthesis provide the starting materials for amino acid biosynthesis. All amino acids derive their nitrogen from glutamate by the combined action of glutamine synthase and glutamate synthase in the chloroplast stroma. These enzymes are also central to the recycling of nitrogen within the plant from the storage amides, glutamine and asparagine. In addition, the photosynthetic carbon oxidation cycle generates phosphoglycollate, and the ammonia released from photorespiration is efficiently reassimilated by glutamine synthase. Arginine and proline are synthesised from glutamate, and aspartate, produced by a glutamate:oxaloacetate aminotransferase, is the starting point for the synthesis of methionine, threonine, lysine, and the branched-chain amino acids leucine, isoleucine and valine. Pyruvate plays a central role in the synthesis of alanine and the branched-chain amino acids. On the other hand, the aromatic amino acids tryptophan, phenylalanine and tyrosine are derived from erythrose 4-phosphate in the photosynthetic carbon reduction cycle, and histidine is similarly formed via ribose 5-phosphate. These major reactions and interconversions are summarised in Figure 9.1. Individual pathways susceptible to herbicidal inhibition will be considered in more detail in the following sections. For further details the reader is referred to Kishore and Shah (1988) and Wallsgrove (1989).

9.3   Inhibition of glutamine synthase

The ammonium salt of glufosinate (DL-homoalanin-4-yl [methyl] phosphonic acid) is a major non-selective, post-emergence herbicide introduced in 1981 for the rapid control of vegetation. It is a potent inhibitor of glutamine synthase and, in common with methionine sulfoximine and tabtoxinine-β-lactam, is a structural analogue of glutamic acid (Figure 9.2).

The discovery and development of glufosinate are of particular interest since this herbicide may be regarded as having a natural origin. Glufosinate, also termed phosphinothricin, and a tripeptide containing glufosinate bound to two molecules of L-alanine (phosphinothricyl-alanyl-alanine) were first described in 1972 as products of the soil bacterium *Streptomyces viridichromogenes*. Independent studies in Japan in 1973 also discovered this tripeptide in culture filtrates of another soil bacterium, *Streptomyces hygroscopicus*, and this compound was introduced commercially as bialaphos in the 1980s. Bialaphos itself is inactive, but it is rapidly hydrolysed in plants to form the phytotoxic glufosinate (Wild and Ziegler, 1989). An additional tripeptide, phosalacine, also produced by a *Streptomyces* sp. has been reported by Omura *et al.* (1984). This tripeptide differs from bialaphos in that one alanine moiety is replaced by leucine. Evidence for the inhibition of plant glutamine synthase by glufosinate was first provided by Leason and colleagues (1982) who demonstrated a K_i value of 0.073 mM, and this potent inhibition has since been confirmed in a wide variety of plants and algae.

Glutamine synthase (E.C. 6.3.1.2) catalyses the conversion of L-glutamate to L-glutamine in the presence of ATP and ammonia. This enzyme has a molecular weight of about 400 kDa and has eight subunits, each with an active site. It exists as separate isozymes in the leaf cytoplasm, chloroplast and roots, and in legumes as an isoform specific to root

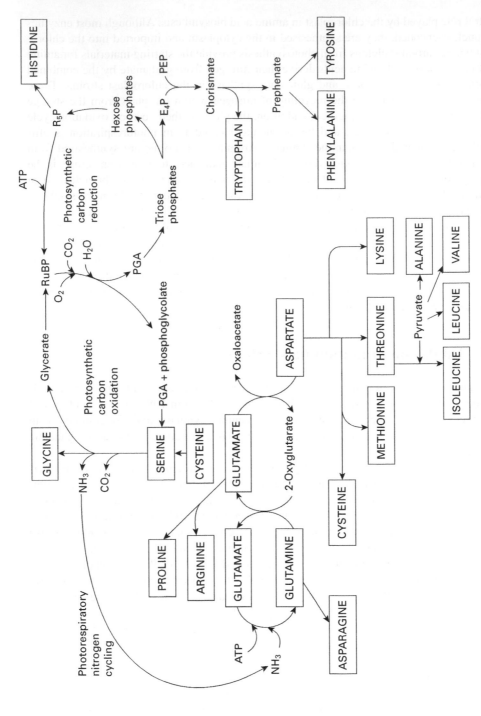

Figure 9.1 Amino acid biosynthesis. R_5P, ribulose 5-phosphate; PEP, phosphoenoylpyruvate; E_4P, erythrose 4-phosphate; RuBP, ribulose 1,5-bisphosphate; PGA, 3-phosphoglyceric acid.

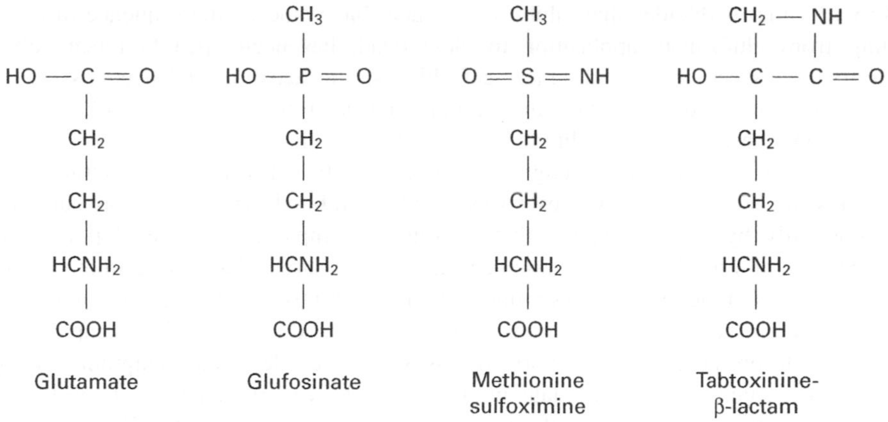

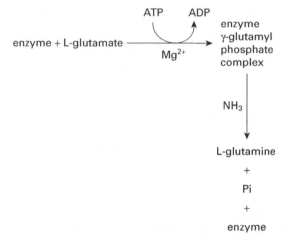

Figure 9.2   The structures of glutamate and the glutamine synthases inhibitors glufosinate, methionine sulfoximine and tabtoxinine-β-lactam.

Figure 9.3   The two step process in the catalysis of glutamine synthesis by glutamine synthase.

nodules. The isoform found in the chloroplast is expressed in greater amounts in the presence of light and at high sucrose concentrations, an indicator of high levels of photosynthetic activity. Glutamine synthase forms glutamine in a two-step process (Figure 9.3).

Normally, γ-glutamyl phosphate is first produced from ATP and L-glutamate, and then ammonia reacts with this complex to release Pi and L-glutamine. However, glufosinate, as an L-glutamate analogue, can also be phosphorylated to produce an enzyme–glufosinate-phosphate complex to which ammonia cannot bind and the enzyme is irreversibly inhibited (Manderscheid and Wild, 1986). Glufosinate can inhibit bacterial, plant and mammalian glutamine synthase *in vitro*, but is non-toxic to mammals, apparently because of its inability to cross the blood–brain barrier and its rapid clearance by the kidneys (Kishore and Shah, 1988).

As with other herbicides that inhibit amino acid biosynthesis, the sequence of events leading from glufosinate application to plant death has been open to much debate. Phytotoxicity is certainly rapid, since leaf chlorosis, desiccation and necrosis may be observed within two days after treatment, and plant death results three days later. Since glutamine synthase is potently inhibited, the rapid accumulation of ammonia was the presumed toxic species which was thought to uncouple electron flow from proton transport in the thylakoid, so that photosynthesis was rapidly inhibited. This view was supported experimentally by the observation that phytotoxic symptoms only develop in plants exposed to light, especially when photorespiration is favoured. Thus, the ammonia generated in photorespiration is not reassimilated and may directly uncouple proton gradients.

Further and more recent observations clearly imply an additional and very rapid action of glufosinate on membrane transport processes that precede visual symptom development. Thus, treated plants accumulate ammonia and show increased rates of cell leakage of potassium ions, and the uptake of nitrate and phosphate are adversely affected. Indeed, Ullrich and colleagues (1990) have convincingly demonstrated that glufosinate, like glutamate, can directly cause an electrical depolarisation of the plasmalemma, although unlike glutamate, recovery is often incomplete and a secondary depolarisation is evident. These decreased membrane potentials inhibit or alter transport processes so that, for example, the cotransport of glutamate/H^+ was irreversibly inhibited but potassium flux was increased. In this way, the accumulation of ammonia may be seen to uncouple or interfere with membrane function and transport at the plasmalemma as well as at the thylakoid.

Increases in ammonia concentrations in treated tissues have been reported as 10 times higher than in untreated tissues 4 hours after treatment and as high as 100 times within 1 day. Free ammonia is toxic to biological systems and it is for this reason that a number of processes are found in biological systems that aim to maintain free ammonia at very low concentrations.

An associated decrease in the concentration of a number of amino acids is also noted where treated plants remain in the light. A decrease in photosynthetic carbon fixation is also noted under the same conditions. These symptoms develop far slower in the dark and this is mirrored in field observations that glufosinate activity is observed far quicker under conditions of full sunlight.

Free ammonia has a severe affect on the pH gradient across biological membranes, collapsing membrane potentials. Glufosinate does not have a direct inhibitory effect on photosynthetic carbon fixation but was assumed to inhibit this process via the reduction in ATP production by photophosphorylation, due to the effect of ammonia on membrane potentials. However, Wild *et al.* (1987) reported that under conditions in which photorespiration was not taking place, photosynthesis was not inhibited by the increased ammonia concentrations resulting from glufosinate treatment. Photorespiration is an alternative metabolic pathway where ribulose 1,5-bisphosphate carboxylase-oxygenase (RuBisCo) – the enzyme responsible for carbon fixation in the Calvin cycle using the products of the light stage of photosynthesis – uses oxygen in place of carbon dioxide as a substrate to react with ribulose 1,5-bisphosphate. Photorespiration takes place in conditions of elevated oxygen and, although it does produce glyceraldehyde 3-phosphate for the carbon reduction cycle, it does this at an energetically far less economical rate compared with the carboxylase activity of RuBisCo. In addition, photorespiration produces free ammonia that must be detoxified by reassimilation into organic molecules. Glufosinate will inhibit this reassimilation and will therefore reduce or inhibit photorespiration, due to reduced concentrations of glutamate as an amino donor for glyoxylate. This will result in a reduction in photosynthetic carbon fixation, due to increased glyoxylate concentrations, and may also result in an increase in active oxygen species from triplet state chlorophyll in the light stage of photosynthesis. The result of this would be lipid peroxidation as reported for a number of other herbicide classes (Chapter 5).

It therefore appears that glufosinate can affect a number of metabolic pathways in plants by its indirect affect on membrane polarisation, reduced peptide, protein and nucleotide concentration, increased protein degradation (to release free amino acids) and the inhibition of photosynthetic carbon assimilation via the inhibition of photorespiration.

Introduction of genes encoding a specific glufosinate-metabolising enzyme has allowed the successful development of glufosinate-resistant, genetically modified crops. These are discussed in detail in Chapter 13.

9.4   Inhibition of EPSP synthase

The herbicide glyphosate (*N*-phosphonomethyl glycine) (Figure 9.4) is a major non-selective, post-emergence herbicide used in circumstances where the total control of vegetation is required. Its success lies in very low soil residual activity, broad spectrum of activity, low non-target organism toxicity and great systemicity in plants, so that even the most troublesome rhizomatous weeds can be controlled. Monsanto received a patent for use of phosphoric acid derivatives, including glyphosate, as non-selective herbicides in 1974. The Stauffer Chemical Company, who had patented a number of phosphonic and phosphinic acids as industrial cleaning agents in 1964, subsequently released sulfosate (glyphosate-trimesium) for development in 1980.

The search for the target site of glyphosate action began with the observation by Jaworski (1972) that the control of duckweed (*Lemna* sp.) by glyphosate could be overcome by the addition of aromatic amino acids to the growth medium. However, it was not until 1980 that Steinrücken and Amrhein identified the enzyme 5-enoyl-pyruvyl shikimic acid 3-phosphate (EPSP) synthase (E.C. 2.5.1.19) as being particularly sensitive to glyphosate. This enzyme is involved in the biosynthesis of the aromatic amino acids tryptophan, phenylalanine and tyrosine, and also leads to the synthesis of numerous secondary plant products (Figure 9.5). Approximately 20% of carbon fixed by green plants is routed through the shikimate pathway with an impressive number of significant end products, including vitamins, lignins, alkaloids and a wide array of phenolic compounds such as flavonoids. A common pathway exists from erythrose 4-phosphate, provided from photosynthetic carbon reduction in the chloroplast stroma, to chorismic acid, so inhibition at EPSP

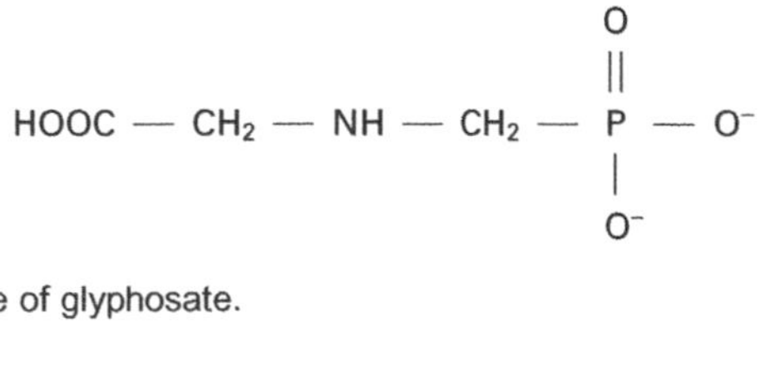

$$HOOC - CH_2 - NH - CH_2 - \overset{\displaystyle O}{\underset{\displaystyle O^-}{\overset{\displaystyle \|}{P}}} - O^-$$

Figure 9.4   The structure of glyphosate.

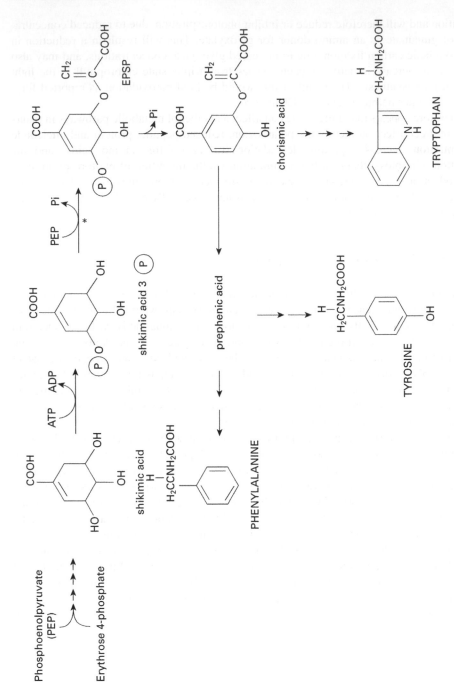

Figure 9.5   Biosynthesis of aromatic amino acids. EPSP, 5-enoylpyruvate shikimic acid 3-phosphate; * , EPSP synthase.

synthase by glyphosate is at a particularly strategic location. Additionally, the enzyme is not found in animals and so the chance of non-target organism toxicity is reduced.

The higher plant enzyme has a molecular weight of 45–50 kDa, found mainly in the chloroplast, which can be reversibly inhibited by glyphosate with an I_{50} of about 10–20 μM. EPSP synthase is synthesised in the cytoplasm before being targeted to the chloroplast. For glyphosate to be active it must enter into the cell and then the chloroplast in sufficient concentration to inhibit the enzyme. Glyphosate is one of only a handful of herbicides for which a carrier protein has been identified that aids in the crossing of plasma membranes. In the case of glyphosate it is a phosphate carrier that is involved (Denis and Delrot, 1993).

Studies indicate that glyphosate acts as a competitive inhibitor with respect to phosphoenolpyruvate (PEP, K_i; 0.1–10 μM), but as a non-competitive inhibitor with respect to shikimic acid 3-phosphate (S3P). Mechanistically, S3P forms a complex with the enzyme to which glyphosate binds before PEP addition. Glyphosate appears to bind away from the active site of EPSP synthase, at a putative allosteric site, and also has very little affinity for the free enzyme. However, it is 'trapped' on the enzyme in the presence of EPSP. The binding of glyphosate appears to prevent the binding of PEP to the enzyme. Observations that even minor changes in the structure of glyphosate result in loss of binding and subsequent herbicidal activity suggest a very specific enzyme herbicide interaction.

A number of studies have attempted to use X-ray crystallography to resolve the order and binding characteristics of the substrates and glyphosate to EPSP synthase (Stilling *et al.*, 1991; Franz *et al.*, 1997; Schöenbrunn *et al.*, 2001). Observations suggest that EPSP has two distinct hemispherical domains that come together in a 'screw-like' movement that causes the active site to be revealed. It appears that the binding of S3P is responsible for causing this conformational change. The active site, although yet to be identified, is postulated to be near the 'hinge' where the two domains of the EPSP synthase enzyme meet. A number of important, conserved, amino acids have been identified that appear to play important roles in active site binding and activity. The cleft formed between the two domains is largely electropositive and this plays a role in the attraction of the anionic ligands that are EPSP substrates (Figure 9.6).

Furthermore, since the enzyme is more active above pH 7, as may be expected in the chloroplast stroma in the light, it is the ionised form of glyphosate which is most likely to be the inhibitory species. Thus, S3P accumulates and the inhibition cannot be reversed by PEP.

Additional proof that EPSP synthase is the target enzyme for glyphosate has come from the elegant studies of Comai *et al.* (1985). These workers isolated a mutant form of EPSP synthase from *Salmonella typhimurium* which was resistant to glyphosate because of a single amino acid substitution, from proline to serine. Furthermore, they isolated the *aroA* gene encoding the resistant enzyme and successfully transferred it to tobacco, making this crop glyphosate resistant. Other studies have cloned the *aroA* gene and an over-production of EPSP synthase in plant cell cultures has also generated tolerance to glyphosate. Subsequent isolation of glyphosate-resistant EPSP synthase and incorporation into a number of crop species is discussed in detail in Chapter 13.

Glyphosate treatment causes growth inhibition, chlorosis, necrosis and subsequent plant death. Plants treated with glyphosate, however, may not show symptoms of treatment for 7–10 days. This slow action in the field probably reflects the time taken for

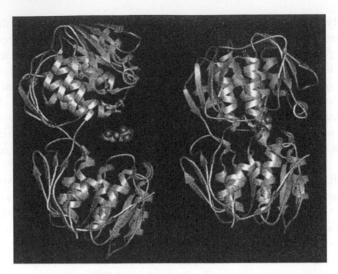

Figure 9.6 The X-ray crystal structure of *E. coli* EPSP synthase. On the left is the open form. On the right is the closed configuration. A modelled glyphosate molecule in the left form is shown, leading to the closed formation on the right (from CaJacob *et al.*, 2004).

the depletion of aromatic amino acid pool sizes to cause decreased rates of protein synthesis. It is now known that glyphosate also has additional effects on phenol and pigment metabolism. Some studies have suggested that glyphosate induces a transient increase in the activity of phenylalanine ammonia lyase, an important enzyme in phenylpropanoid metabolism, so that enhanced concentrations of natural growth inhibitory phenols accumulate. This enzyme activity declines when phenylalanine pools become limiting. Since many phenols are regarded as natural inhibitors of auxin oxidation, then greater auxin metabolism occurs in glyphosate-treated plants with the result that apical dominance is overcome. Consequently, lateral growth of dicots and increased tillering in monocots are often observed in the field.

Relating EPSP synthase inhibition to plant death is not easy owing to the multiple effects this has on a number of metabolic processes. Inhibition of the enzyme results is reduced production of chorisimic acid and a subsequent increase in shikimate and shikimate-3-phosphate due to the blocking of the shikimic acid pathway. In glyphosate-treated tissue, shikimate and shikimate-3-phosphate have been reported as making up 16% of plant dry matter (Schuktz *et al.*, 1990). This causes a reduction in protein synthesis due to depletion of amino acid pools. This alone would result in growth retardation and subsequent plant death. However, blocking the shikimic acid pathway and subsequent reduction in chorisimate causes an increase in flow of carbon compounds out of the photosynthetic carbon reduction cycle. This results in reduction in both photosynthesis and starch production. Reduced carbon compound translocation and, ironically, glyphosate translocation from treated tissue are subsequently observed. In addition, blockage of the shikimic acid pathway results in reduction in the production of auxin growth regulators, lignin, plant defence compounds, UV protectants and photosynthetic pigments, either directly or via reduced synthesis of plastoquinone.

A further metabolic consequence of glyphosate treatment is the development of chlorotic areas on leaves. This may be due to an inhibition of δ-aminolaevulinic acid (δ-ALA) synthetase, an early reaction in the biosynthesis of all porphyrin containing molecules, including the chlorophylls and cytochromes (Chapter 6, section 2). Since these compounds are central to plant metabolism, it would appear that many important biochemical pathways are therefore subject to interference by glyphosate treatment. Indeed, the observation that only 10 µM glyphosate will inhibit EPSP synthase *in vitro* from field rates of greater than 10 000 µM, suggests the involvement of many secondary sites of action which produce an overall herbicidal effect.

Since glyphosate is a non-selective total herbicide, little is known about its metabolism in plants, which is presumed to be either non-existent or ineffectual. However, this herbicide is rapidly biodegraded by soil microorganisms which are able to utilise glyphosate as a sole phosphate source. Specific enzymes operate in a *Pseudomonas* sp. to cleave the phosphate group, and the phosphonomethyl C–N bond is broken to release glycine.

9.5   Inhibition of acetolactate synthase

Since the 1980s five new herbicidal classes have emerged that all share the same site of action. These have proved to be potent, selective, broad-spectrum inhibitors of plant growth at field rates measured in grams rather than kilograms per hectare. The sulphonylureas (SUs), imidazolinones (IMs), triazolopyrimidines (TPs), sulfonylaminocarbonyl-triazolines (often classed as SUs) and pyrimidinyl-oxy-benzoates (pyrimidinyl-carboxy herbicides; PCs) are chemically different (Table 9.1), yet all share the same site of action, namely acetolactate synthase (ALS, E.C. 4.1.3.18, also known as acetohydroxyacid synthase, AHAS), a key enzyme in the biosynthesis of the branched-chain amino acids leucine, isoleucine and valine. In each case, growth inhibition may be overcome by the addition of these amino acids. The efficacy and potency of the ALS inhibitors has ensured the continued success of these herbicides, which have rapidly challenged, and in some instances replaced, traditional products, especially in cereals and soybeans. SUs and TPs are active at field rates of 10–100 g ha^{-1} whereas IMs are required at 100–1000 g ha^{-1} to give a similar degree of weed control. Currently ALS inhibitors represent the second biggest class of herbicidal active ingredients. A major research effort has progressed to understand their mode(s) of action and to develop new products to inhibit the synthesis of branched-chain amino acids. Further information on this class of herbicide can be found in reviews by Babczinski and Zelinski (1991) and Duggleby and Pang (2000).

The highest activity obtained with the SUs is when the aryl group has an *ortho* substituent. Thiophene, furan, pyrimidine and naphthalene groups are also active herbicides when replacing the aryl group, but the *ortho* substitution is still essential with respect to the sulphonylurea bridge. The heterocycle configurations for optimal activity appear to be a symmetrical pyridine or a symmetrical triazine with low alkyl or alkoxy substituents (Beyer *et al.*, 1987). The sulfonylaminocarbonyl moiety also results in the herbicidally active SUs procarbazone and flucarbazone (Amann *et al.*, 2000; Müller *et al.*, 1992). In the case of the imidazolinones, the highest biological activity has been observed when an imidazolinone ring, ideally sustituted with methyl and isopropyl groups, is attached to an aromatic ring containing a carboxyl group in an *ortho* position (Los,

Table 9.1   Structures of a selection of amino acid biosynthesis inhibitors.

(a) Sulphonylureas

General structure: R_1—SO_2NHCNH—(triazine/pyrimidine ring with R_2, R_3, R_4), with C=O on the bridging carbon.

R_1	R_2	R_3	R_4	Common name (crop/dosage, g ha^{-1})
2-Cl-phenyl	–CH$_3$	N	–OCH$_3$	Chlorsulfuron (cereals/4–26)
2-CO$_2$CH$_3$-phenyl	–CH$_3$	N	–OCH$_3$	Metsulfuron-methyl (cereals/2–8)
2-OCO$_2$CH$_2$Cl-phenyl	–CH$_3$	N	–OCH$_3$	Triasulfuron (cereals/10–40)
2-CO$_2$CH$_3$-phenyl	–OCH$_3$	CH	–OCH$_3$	Bensulfuron-methyl (rice/20–75)
2-CO$_2$C$_2$H$_5$-phenyl	–Cl	CH	–OCH$_3$	Chlorimuron-ethyl (soybean/8–13)
2-CO$_2$CH$_3$-phenyl	–CH$_3$	CH	–CH$_3$	Sulfometuron-methyl (non-crop/70–840)
thiophene-CO$_2$CH$_3$	CH$_3$	N	OCH$_3$	Thifensulfuron (cereals, soybean/17–35)
2-CO$_2$CH$_3$-phenyl	CH$_3$	N	OCH$_3$	Tribenuron (cereals/5–30)
2-(C(=O)OCH$_3$)-phenyl	OHCF$_2$	CH	OCHF$_2$	Primisulfuron (maize/20–40)

Table 9.1 (*Continued*)

R_1	R_2	R_3	R_4	Common name (crop/dosage, g ha^{-1})
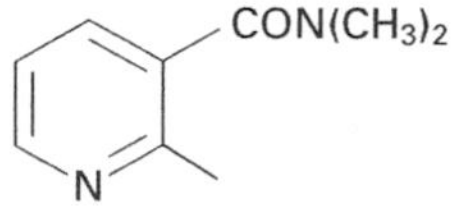 (pyridine with CON(CH$_3$)$_2$)	OCH$_3$	CH	OCH$_3$	Nicosulfuron (maize/40–60)
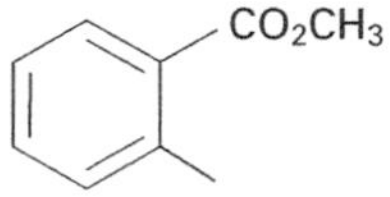 (benzene with CO$_2$CH$_3$)	OCH$_2$CH$_3$	N	NHCH$_3$	Ethametsulfuron (oilseed rape/10–120)
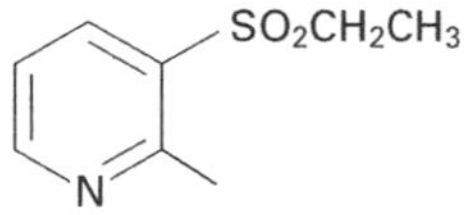 (pyridine with SO$_2$CH$_2$CH$_3$)	OCH$_3$	CH	OCH$_3$	Rimsulfuron (potato/5–15)
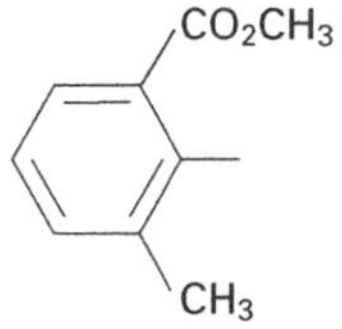 (benzene with CO$_2$CH$_3$, CH$_3$)	N(CH$_3$)$_2$	N	OCH$_2$CF$_3$	Triflusulfuron (sugarbeet/10–25)
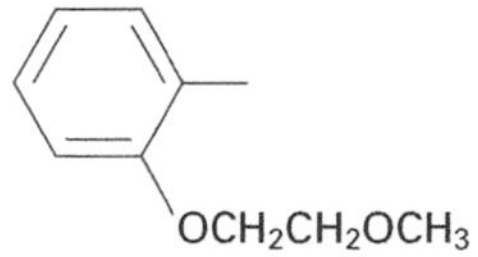 (benzene with OCH$_2$CH$_2$OCH$_3$)	OCH$_3$	N	OCH$_5$	Cinosulfuron (oilseed rape/10–40)
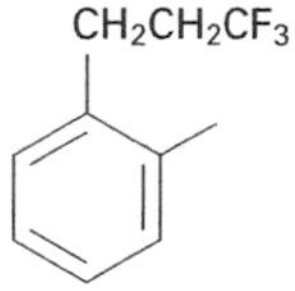 (benzene with CH$_2$CH$_2$CF$_3$)	CH$_3$	N	OCH$_3$	Prosulfuron (cereals/12–40)
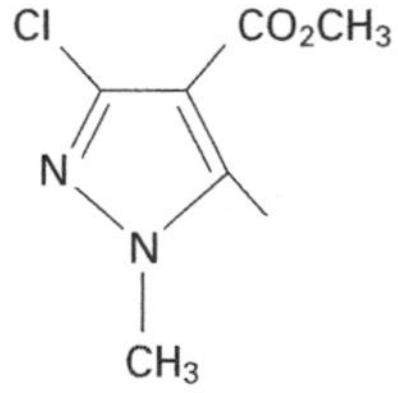 (pyrazole with Cl, CO$_2$CH$_3$, CH$_3$)	OCH$_3$	N	OCH$_3$	Halosulfuron (maize/18–140)
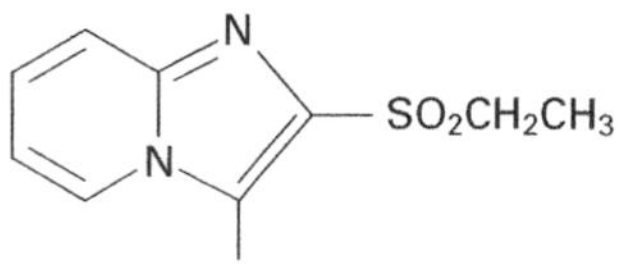 (imidazopyridine with SO$_2$CH$_2$CH$_3$)	OCH$_3$	N	OCH$_3$	Sulfosulfuron (wheat/10–30)

(b) Imidazolinones

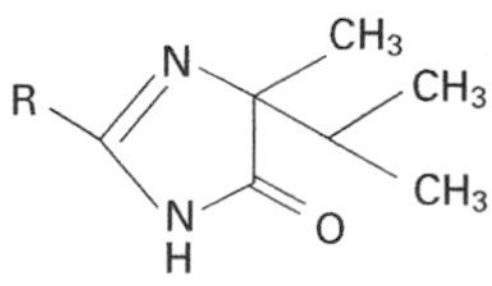

(*Continued*)

Table 9.1 (*Continued*)

R	Common name (crop/dosage g ha^{-1})
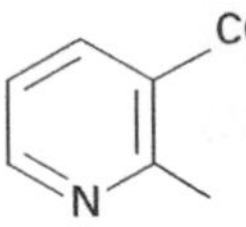	Imazapyr (non-crop/500–2000)
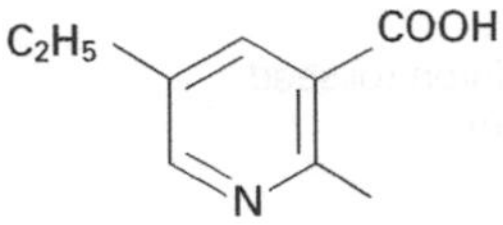	Imazethapyr (legumes/30–150)
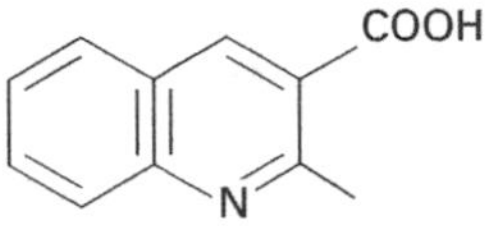	Imazaquin (soybean/140–280)
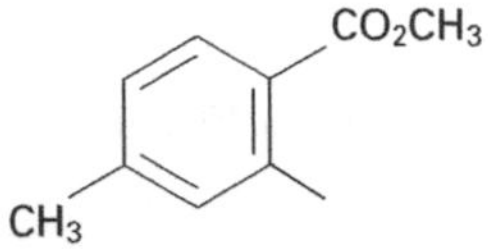	Imazamethabenz (cereals/400–750)
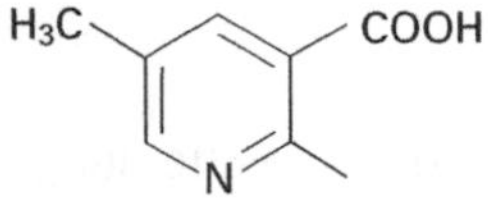	Imazapic (non-crop/50–105)
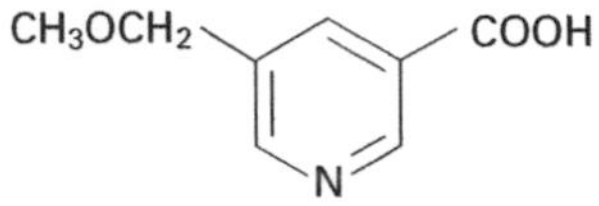	Imazamox (soybean/35–45)

(c) Triazolopyrimidines

	Common name (crop/dosage, g ha^{-1})
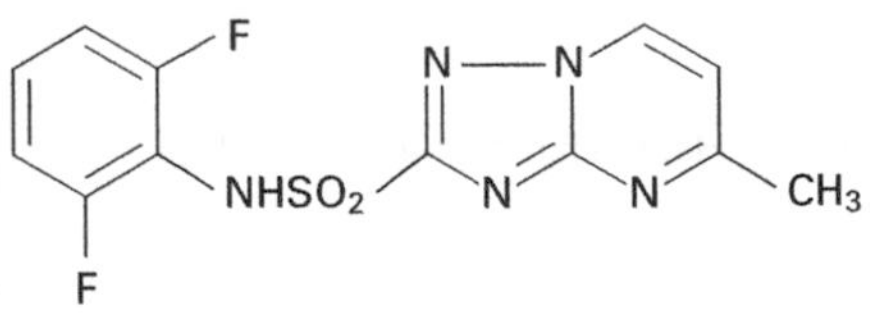	Flumetsulam (soybean/17–70)
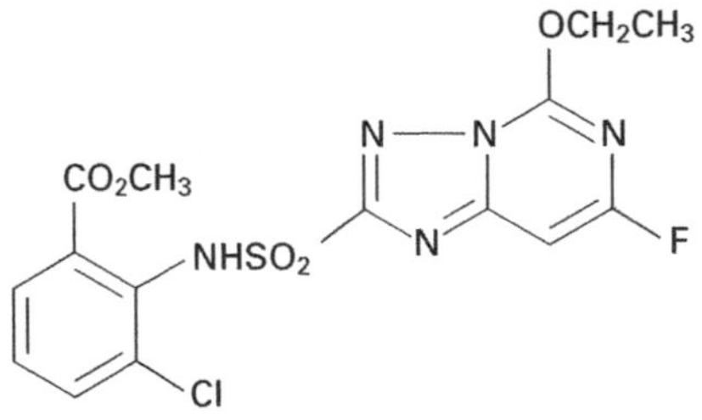	Cloransulam-methyl (soybean/35–44)
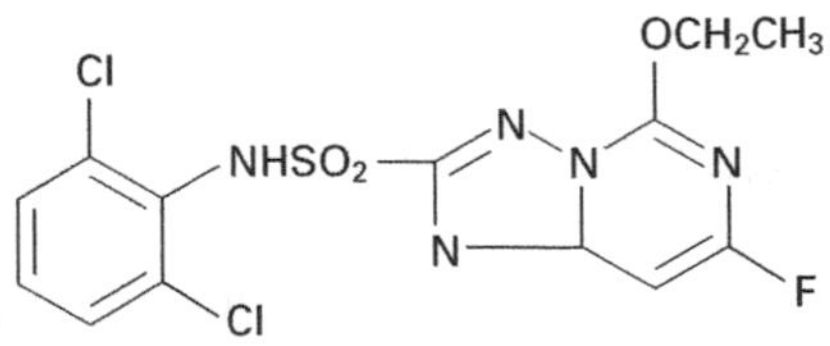	Diclosulam (soybean/26–35)

Table 9.1 (*Continued*)

(d) Sulfonylaminocarbonyltriazoline	
	Common name (crop/dosage, g ha^{-1})
	Flucarbazone-sodium (cereals/30)

(e) Pyrimidinyloxybenzoates	
	Common name (crop/dosage, g ha^{-1})
	Pyrithiobac (cotton/30–100)

1986). The more recently discovered PCs demonstrate activity at similar field rates to SUs and TPs. However, the inhibitory mechanism and cross-resistance patterns appear to be a hybrid between the SUs and the IMs. Elegant studies by Shimizu *et al.* (2002) based upon the synthesis of novel analogues based upon phenoxyphenoxypyrimidine suggest that PCs require both esteric bonding and an appropriate substituted pyrimidine ring in order to demonstrate ALS-inhibiting activity. Highest inhibition was observed when a COOMe group was at the *ortho* position to the pyrimidinyloxy group. A pyrimidine ring imparted greater inhibiting activity than structures containing other N-heterocyclics. Interestingly, the replacement of the O-bridge with an S-bridge reduced ALS-inhibiting activity but did increase crop tolerance in some cases. The presence of an S-bridge also increased herbicide mobility, both via root uptake and by translocation. The base S-containing compound used in these synthesis studies was pyrithiobac-sodium. Studies of this type not only aid in elucidation of structure-function relationships, but also in the discovery of new ALS-inhibiting herbicides.

All five classes of ALS inhibitors possess remarkable herbicidal properties. They are able to control a very wide spectrum of troublesome annual and perennial grass and broadleaf weeds at very low doses. Furthermore, formulations have proved to be both foliar- and soil-active with very low mammalian toxicity. Chlorsulfuron, metsulfuron-methyl, and imazamethabenz give selective weed control in cereals, chlorimuron-ethyl and imazaquin are selective in soybean, imazethapyr is selective in other legumes as well as soybean, bensulfuron-methyl is effective in rice, and sulfometuron-methyl and imazapyr have found industrial and non-crop uses for total vegetation control. Imazapyr is effective in forest management by controlling deciduous trees in conifers, and a coformulation of imazapyr and imazethapyr is being developed as a growth retardant in grassland and turf areas, with an additional control of broadleaf weeds.

ALS, like EPSP synthase, is a nuclear-encoded, chloroplast-localised enzyme in higher plants, and also occupies a strategic location in the biosynthetic pathway of essential amino acids. This pathway has been well studied in microorganisms and is becoming increasingly understood in higher plants. Essentially, synthesis occurs in the stroma from threonine and pyruvate in a common series of reactions (Figure 9.7; Ray, 1989). In isoleucine synthesis, threonine is first deaminated to 2-oxobutyrate by threonine dehydratase, which is controlled by feedback regulation by valine and isoleucine. ALS catalyses the first common step of branched-chain amino acid biosynthesis to yield acetohydroxy acids which undergo oxidation and isomerisation to yield derivatives of valeric acid. Dehydration and transamination then produces isoleucine and valine. 2-Oxoisovalerate reacts with acetyl-CoA to form α-isopropylmaleate which is then isomerised, reduced and transaminated to yield leucine. ALS demonstrates feedback inhibition to leucine, valine and isoleucine.

ALS has been extensively studied in microorganisms and is the subject of increasing scrutiny in plants, although plant ALS is labile and constitutes less than 0.01% of total plant protein. As many as six ALS isozymes have been reported in bacteria to allow carbon flux through this pathway at varying concentrations of pyruvate. However, isozymes are not required for ALS in the chloroplast stroma where more reliable concentrations of substrates are assumed. Study of ALS extracted from pea seedlings identified a 320-kDa ALS that dissociated to a 120-kDa ALS in the absence of flavin adenine dinucleotide (FAD). The larger ALS demonstrated feedback inhibition from valine, leucine and isoleucine whereas the smaller ALS, although still demonstrating enzyme activity, did not exhibit feedback inhibition. This suggests that there are separate, regulatory, subunits that require FAD in order to remain attached to the catalytic subunits of ALS. Current understanding is that the ALS enzyme in pea seedlings consists of at least 4 catalytic and 2 regulatory subunits (Shimizi et al., 2002). Regulation of ALS activity is carried out by leucine, valine and isoleucine (feedback inhibition). This inhibition is also observed for leucine/isoleucine and leucine/valine mixtures but for valine/isoleucine mixtures an antagonism of feedback was observed. This suggests that there are two regulatory sites on the enzyme, one for leucine and one for valine/isoleucine. Two regulatory sites are also suggested by regulatory promoter studies (Hershey et al., 1999). Analysis of the ALS gene from a number of plant species has revealed that it is highly conserved, with very little difference between the rice, maize and barley ALS genes. The isozyme II from S. typhimurium requires FAD, thiamine pyrophosphate (TPP), and Mg^{2+} for complete activation, and the reaction proceeds in a biphasic manner. First, a pyruvate molecule binds to TPP at the active site and is decarboxylated to yield an enzyme–substrate complex plus CO_2. A second pyruvate then reacts with this complex and acetolactate is released. A number of ALS herbicides bind slowly, but tightly, to the enzyme–substrate complex to prevent the addition of the second pyruvate molecule (LaRossa and Schloss, 1984). However, imidazolinones are uncompetitive inhibitors with respect to pyruvate (Shaner et al., 1984), and in the case of sulfonylureas and triazolopyrimides both competitive and non-competitive inhibition is exhibited (mixed-type inhibition) (Subramanian and Gerwick, 1989; Durner et al., 1991). This indicates that the herbicide-binding site for ALS is distinct from the enzyme's active site. Other studies have shown that the herbicide does not bind at the allosteric site either.

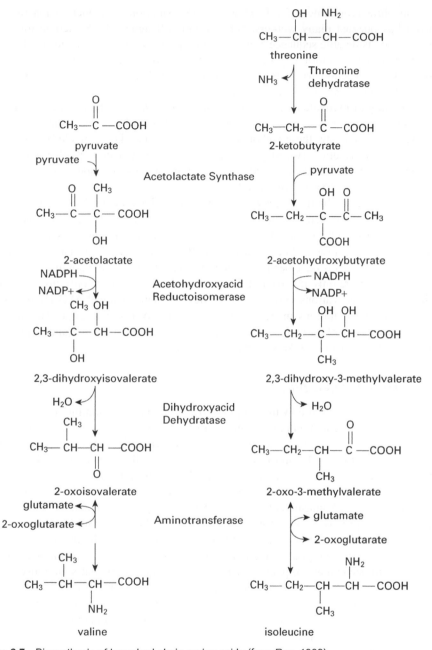

Figure 9.7   Biosynthesis of branched-chain amino acids (from Ray, 1989).

The absolute requirement for FAD where no oxidation or reduction reactions are involved has puzzled many workers. Schloss and colleagues (1988) demonstrated that ALS shows considerable sequence homology with pyruvate oxidase, suggesting that both enzymes may have evolutionary similarities. Indeed, ALS does demonstrate an oxygen-consuming side-reaction when its activity is inhibited (Durner *et al.*, 1994). They have discovered that pyruvate oxidase binds both FAD and a quinone for redox reactions, and that the binding of pyruvate oxidase from *Escherichia coli* with ubiquinone-40 is tightest in the presence of pyruvate. Since the binding of ALS with herbicides is also tightest in the presence of pyruvate, these authors have concluded that the herbicide binding site is derived from an evolutionary vestige of a quinone-binding cofactor site that is no longer functional in ALS. Hence, the herbicide site is extraneous to, or outside of, the ALS active site. McCourt *et al.* (2006) have recently reported that SUs and IMs block a channel in ALS through which substrates access the active site of the enzyme. These studies further reported that SUs approach to within 5 Å of the catalytic centre, whereas IMs bind at least 7 Å from it. This confirms earlier observations of different binding sites for these two classes of herbicide. It seems likely that TPs and PCs will bind to ALS in a similar manner to SUs. Ten amino acids have been identified that are involved in the binding of both SUs and IMs, six that are only involved in SU binding and two that are only involved in IM binding. These observations may lead to the development of novel ALS inhibitors that are unaffected by mutations in ALS genes that currently result in target site-mediated herbicide resistance. Zhou *et al.* (2007) have recently reviewed current understanding of the mechanisms of action of ALS-inhibiting herbicides.

9.5.1 How treated plants die

It is not clearly understood how treated plants die following ALS inhibition. The diminution of the branched-chain amino acid pool will contribute to a cessation of protein synthesis, but although nanomolar concentrations of herbicide can inhibit ALS within minutes *in vitro*, it may take up to two months for the death of intact weeds in the field. Addition of the three branched-chain amino acids to culture media alleviates symptoms of ALS-inhibiting herbicides (Ray, 1984; Shaner and Reider, 1986; Usui *et al.*, 1991; Shimizu *et al.*, 1994; Yamashita *et al.*, 1994). Several physiological and metabolic alterations have been proposed to contribute to weed death. For example, LaRossa and colleagues (cited by Ray, 1989) have found that ALS inhibition results in an accumulation of its substrate, 2-ketobutyrate, which is toxic to *S. typhimurium*. 2-Aminobutyrate has been shown to accumulate in plants treated with ALS inhibitors, but symptoms of ALS inhibition do not appear to be due to 2-aminobutyrate toxicity. Treatment with ALS inhibitors results in a drop in the concentration of valine and isoleucine and a rise in the concentrations of threonine, alanine and norvaline as well as 2-aminobutyrate. The accumulation of abnormal amino acids (e.g. norvaline) due to increased 2-ketobutyrate concentrations cannot, therefore, be discounted as a possible cause of herbicide-related phytotoxicity.

Accumulation of singlet oxygen has also been observed after treatment with ALS inhibitors, possibly as a result of an oxygen-consuming side-reaction exhibited by ALS when its main enzymic process is inhibited (Durner *et al.*, 1994). However, whole plant symptoms differ markedly from other herbicides, the mode of action of which is the production of active oxygen species, suggesting this is not the primary cause of plant death with ALS inhibitors.

One common observation following treatment with ALS inhibitors is a very rapid and potent inhibition of cell division, with the result that an inhibition of elongation of young roots and leaves is evident within 3 h after application. Rost (1984) found that chlorsulfuron blocked the progression of the cell cycle in dividing root cells from peas within 24 h, from G_2 to mitosis (M) and reduced movement from G_1 to DNA synthesis (S). This is a rapid effect on the cell cycle with no direct effect on the mitotic apparatus, which can be overcome by inclusion of isoleucine and valine in the incubation medium. Separate reports that ALS inhibitors may indirectly inhibit DNA synthesis may explain these observations. Furthermore, the involvement of polyamines have been implicated in sulphonylurea action, since the discovery by Giardini and Carosi (1990) that chlorsulfuron causes a reduction in spermidine concentration in *Zea* root tips which could be responsible for this effect in the cell cycle. These findings have created a new interest in cell-cycle research since they imply a possible regulatory role of branched-chain amino acids in the control of plant cell division.

Growth is therefore retarded or inhibited within hours of foliar treatment, but physical symptoms may take days to develop, first appearing as chlorosis and necrosis in young meristematic regions of both shoots and roots. Young leaves appear wilted, and these effects spread to the rest of the plant. Leaf veins typically develop increased anthocyanin formation (reddening), and leaf abscission is commonly observed, both symptoms being typical responses to stress ethylene production. Under optimum growth conditions plant death may follow within ten days, although up to two months may elapse when weed growth is slow. When ALS inhibitors are applied before the crop is planted or has emerged, susceptible weeds will germinate and grow, presumably utilising stored seed reserves. However, further growth of broadleaf weeds stops at the cotyledon stage, and before the two-leaf stage in grasses.

9.5.2 Selectivity

A striking feature of the ALS inhibitors is their highly selective action at low dosage. Indeed, some cereals have been reported to tolerate up to 4000 times more chlorsulfuron than some susceptible broadleaf species. Various studies have shown that this extreme species sensitivity is not due to herbicide uptake, movement or sensitivity to ALS, but is correlated to very rapid rates of metabolism in the tolerant crop. In the tolerant crop soybean the degradation half-life of triazolopyrimidines has been shown to be 49 hours compared to 165.3 hours in the susceptible species pitted morning glory (Swisher *et al.*, 1991). Several sites on the sulphonylurea molecule are locations for enzyme attack and more than one enzyme system has been demonstrated to be active in their detoxification, as summarised in Figure 9.8 (Beyer *et al.*, 1987).

This extreme species sensitivity may have dramatic consequences to a following crop in the same land. For example, since wheat is more than 1000 times more tolerant to chlorsulfuron than the extremely sensitive sugar beet, which may be inhibited by as little as 0.1 part herbicide per billion parts of soil, then low sulphonylurea residues are crucial to the success of this rotational crop. Chlorsulfuron detoxification in the soil is principally due to microbial action which may degrade the herbicide within one to two months under

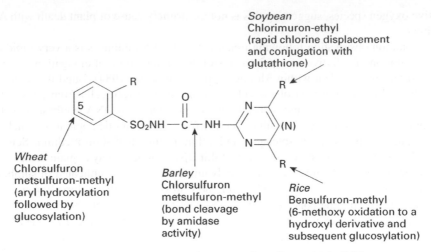

Figure 9.8   Sulphonylurea metabolism in various crops (from Beyer *et al.*, 1987).

optimum conditions. However, when conditions of low temperature, high rainfall, and high soil pH arise, microbial breakdown is greatly reduced and, if prolonged, sufficient residues of chlorsulfuron may remain to cause major damage in succeeding crops for up to a year after the original treatment. The most sensitive crops to chlorsulfuron residues appear to be lentils, sugar beet and onions, but flax, maize, sunflower, mustard, oilseed rape, potatoes and lucerne are also sensitive (Beyer *et al.*, 1987; Blair and Martin 1988). Similar sensitivity to imidazolinone and triazolopyrimidine residues have not yet been reported.

Rapid differential metabolism also provides an explanation for the selectivity of the imidazolinones. Thus, de-esterification of imazamethabenz to phytotoxic acids is observed in susceptible species such as wild oat, and ring methyl hydroxylation followed by glucosylation inactivates the molecule in tolerant maize and wheat (Figure 9.9).

A further aspect of imidazolinone action is noteworthy and remains to be explained. Namely, about 1000 times more imidazolinone is needed to inhibit ALS *in vitro* than an equivalent sulphonylurea, but only five times more herbicide is needed to kill the weed.

Since the sulphonylureas and imidazolinones are such potent inhibitors of ALS and may retain residual activity in soils under certain conditions, it may be predicted that resistant plants will eventually evolve. However, resistance has developed in a surprisingly short time and at present resistance cases to ALS inhibitors are more prevalent than cases to other herbicides that have been used for many more years (Heap, 2007). The use of chlorsulfuron and metsulfuron-methyl in minimum tillage winter wheat monoculture for four to five consecutive years has resulted in resistant biotypes of prickly lettuce (*Lactuca serriola*), Kochia (*Kochia scoparia*), Russian thistle (*Salsola iberica*) and chickweed (*Stellaria media*). Although the degree of resistance varies with both biotype and herbicide, it is clear that cross-resistance to the imidazolinones is evident, and a modified, less sensitive ALS is thought to be present in resistant biotypes (Reed *et al.*, 1989). Furthermore, resistant biotypes of *L. serriola* have been crossed with domestic lettuce (*L. sativa*) by

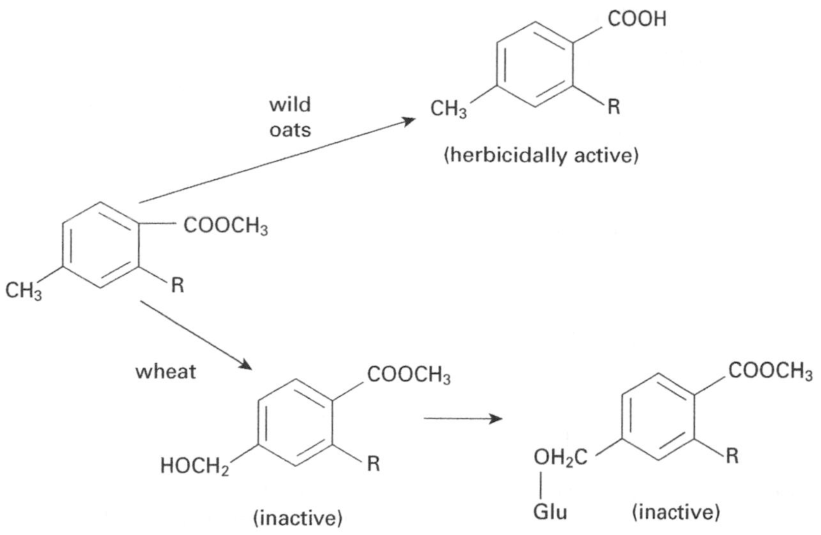

Figure 9.9 Differences in the metabolism of imidazolinones between wheat (tolerant) and wild oats (susceptible).

Mallory-Smith and colleagues (1990), with the result that inherited resistance is controlled by a single gene with incomplete dominance.

9.6   Inhibition of histidine biosynthesis

The biosynthesis of the amino acid histidine has received extensive study in microorganisms and is a highly conserved process among species. It is only in recent years that our understanding of histidine biosynthesis in higher plants has begun to increase (Stepansky and Leustek, 2006), largely due to genetic studies of *Arabidopsis thaliana* mutants (Muralla *et al.,* 2007).

Histidine biosynthesis in plants is summarised in Figure 9.10. Phosphoribosyl-pyrophosphate (PRPP) reacts with ATP to form phosphoribosyl AMP (PRPP-AMP). This is converted to phosphoribulosyl formimino-5-aminoimidazole-4-carboxamide ribonucleotide (PR-FAR), which in turn is converted to imidazole glycerol phosphate (IGP). Histidine is then synthesised from IGP via imidazoleacetol phosphate (IAP), histidinol phosphate (Hol-P) and histidinol (Hol). The enzyme IGP dehydratase (E.C. 4.2.1.19) had long been proposed to be the site of action of the non-selective herbicide 3-amino-1,2,4-triazole (aminotriazole; amitrol) (Figure 9.11). However, observations made by Sandmann and Böger (1992) (among others) that amitrol inhibits ξ-carotene desaturase and lycopene cyclase, in addition to 'bleaching' symptoms in treated plants, have now resulted in aminotriazole being classified as a pigment biosynthesis inhibitor. May and Leaver (1993) have also reported that aminotriazole inhibits both catalase and aconitase. Inhibition of the former results in large increases in reduced glutathione pool size. Aminotriazole is a relatively simple molecule in comparison to many other herbicides

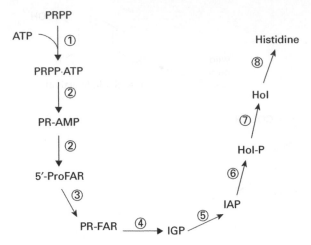

Figure 9.10   Proposed histidine biosynthesis pathway in plants (based upon gene identifications in *Arabidopsis thaliana* by Muralla *et al.*, 2007). For abbreviations see text. Enzymes: 1, phosphorybosyl transferase; 2, cyclohydrolase/pyrophosphohydrolase; 3, BBMII isomerase; 4, PRFAR aminotransferase/cyclase; 5, IGP dehydratase; 6, IAP aminotransferase; 7, histidinol phosphatase; 8, histidinol dehydrogenase.

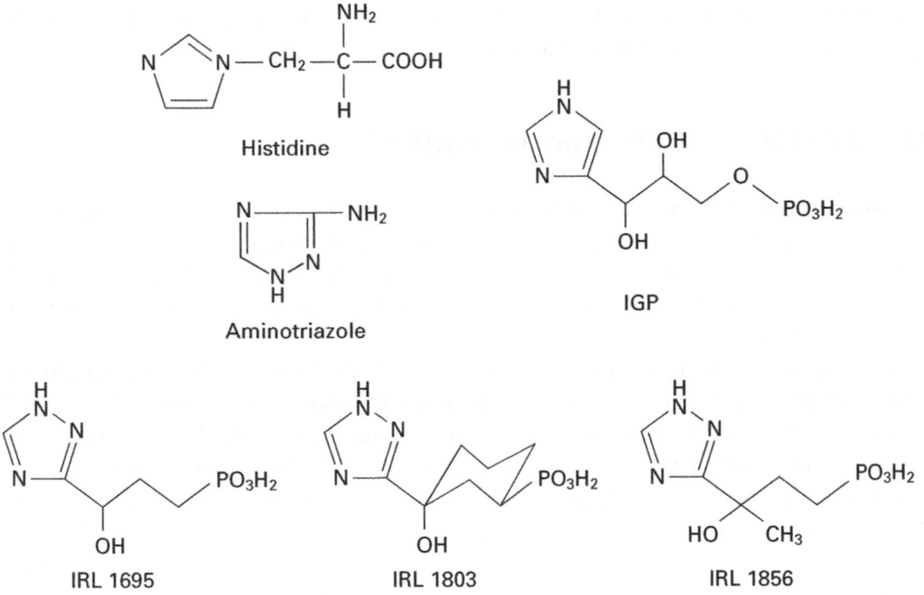

Figure 9.11   The structures of histidine, aminotriazole, IGP and three triazole phosphates reported by Mori *et al.* (1995) to be inhibitors of IGP dehydratase.

and it is perhaps not surprising that it appears to affect a number of metabolic processes in higher plants.

Inhibition of IGP dehydratase has formed the basis for the design of three triazole phosphates that can inhibit histidine biosynthesis in plants (Mori *et al.*, 1995). These compounds were produced by 'rational design', based upon the substrate of IGP dehydratase, namely imidazole glycerol phosphate. All three compounds exhibited herbicidal activity in both cell cultures and intact plants. This activity was attributed to inhibition of histidine biosynthesis rather than other metabolic processes. As such, histidine biosynthesis inhibitors may prove to be a mode of action for a number of commercial herbicides in the future. 'Rational design' of herbicides is a process that has received much attention but has rarely proved successful. The report by Mori and colleagues (1995) is therefore an exciting one, demonstrating that herbicides by design rather than large-scale compound screening can yield positive results.

References

Amann, A., Feucht, D. and Wellmann, A. (2000) A new herbicide for grass control in winter wheat, rye and triticale. *Zeitscrift für Pflanzenkrankheiten und Pflanzenschutz* **17**, 545–553.

Babczinski, P. and Zelinski, T. (1991) Mode of action of herbicidal ALS-inhibitors on acetolactate synthase from green plant cell cultures, yeast and *Escherichia coli. Pesticide Science* **31**, 305–323.

Beyer, E.M., Duffy, M.J., Hay, J.V. and Schlueter, D.D. (1987) Sulfonylureas. In: Kearney, P.C. and Kaufman, D.D. (eds) *Herbicides: Chemistry, Degradation and Mode of Action*, vol. 3, New York: Marcel Dekker, pp. 117–189.

Blair, A.M. and Martin, T.D. (1988) A review of the activity, fate and mode of action of sulfonylurea herbicides. *Pesticide Science* **22**, 195–219.

CaJacob, C.A., Feng, P.C.C., Heck, G.R., Alibhai, M.F., Sammons, R.D. and Padgette, S.R. (2004) Engineering resistance to herbicides. In: Christou, P. and Klee, H. (eds) *Handbook of Plant Biotechnology V.1.* Chichester: John Wiley and Sons, pp. 353–372.

Comai, L., Faciotti, D., Hiatt, W.R., *et al.* (1985) Expression in a plant of a mutant *aroA* gene from *Salmonella typhimurium* confers tolerance to glyphosate. *Nature* **317**, 741–744.

Denis, M.H. and Delrot, S. (1993) Carrier-mediated uptake of glyphosate in broad bean (*Vicia faba*) via a phosphate transporter. *Physiologia Plantarum* **87**, 569–575.

Duggleby, R.G. and Pang, S.S. (2000) Acetohydroxyacid synthase. *Journal of Biochemistry and Molecular Biology* **33**, 1–36.

Durner, J., Gailus, V. and Böger, P. (1991) New aspects of inhibition of plant acetolactate synthase by chlorsulfuron and imazaquin. *Plant Physiology* **95**, 1144–1149.

Durner, J., Gailus, V. and Boger, P. (1994) The oxygenase reaction of acetolactate synthase detected by chemiluminescence. *FEBS Letters* **354**(1), 71–73.

Franz, J.E, Mao, M.K, Sikorski, J.A. (1997) *Glyphosate: a Unique Global Herbicide*. ACS Monograph No. 189. Washington, DC: American Chemical Society.

Giardini, M.C. and Carosi. , S. (1990) Effects of chlorsulfuron on polyamine content in maize seedlings. *Pesticide Biochemistry and Physiology* **36**, 229–236.

Heap, I. (2007) *The International Survey of Herbicide Resistant Weeds*. Available at www.weedscience.com accessed 13 November 2007.

Hershey, H.P., Schwartz, L.J., Gale, J.P. and Abel, L.M. (1999). Cloning and functional expression of the small subunit of acetylacetate synthase from *Nicotiana plumbaginifolia. Plant Molecular Biology* **40**, 795–806.

Jaworski, E.G. (1972) The mode of action of *N*-phosphonomethylglycine: inhibition of aromatic amino acid biosynthesis. *Journal of Agriculture and Food Chemistry* **20**, 1195–1198.

Kishore, G.M. and Shah, D.M. (1988) Amino acid biosynthesis inhibitors as herbicides. *Annual Review of Biochemistry* **37**, 627–634.

LaRossa, R.A. and Schloss, J.V. (1984) The sulphonylurea herbicide sulfometuronmethyl is an extremely potent and selective inhibitor of acetolactate synthase in *Salmonella typhimurium*. *Journal of Biological Chemistry* **259**, 8753–8757.

Leason, M., Cunlife, D., Parkin, D., Lea, P.J. and Miflin, B.J. (1982) Inhibition of pea leaf glutamate synthase by methionine sulfoximine, phosphinothricin and other glutamate analogues. *Phytochemistry* **21**, 855–857.

Los, M. (1986) Synthesis and biology of the imidazolinone herbicides. In: Greenhalgh, P. and Roberts, T.R. (eds) *Pesticide Science and Biotechnology*. Proceedings of the sixth international congress of pesticide chemistry, Ottawa, Canada. Oxford: Blackwell, pp. 35–42.

Mallory-Smith, C., Thill, D.C., Dial, M.J. and Zemetra, R.S. (1990) Inheritance of sulphonylurea herbicide resistance in prickly lettuce (*Lactuca serriola*) and domestic lettuce (*Lactuca sativa*). In: Caseley, J.C., Cussans, G. and Atkin, R. (eds) *Herbicide Resistance in Weeds and Crops*. Guildford, UK: Butterworth, pp. 452–453.

Manderscheid, R. and Wild, A. (1986) Studies on the mechanism of inhibition by phosphinothricin of glutamine synthetase isolated from *Triticum aestivum* L. *Journal of Plant Physiology* **123**, 135–142.

May, M.J. and Leaver, C.J. (1993) Oxidative stimulation of glutathione synthesis in *Arabidopsis thaliana* suspension cultures. *Plant Physiology* **103**, 621–627.

McCourt, J.A., Pang, S.S., King-Scot, J., Guddat, L.W. and Duggleby, R.G. (2006) Herbicide-binding sites revealed in the structure of plant acetohydroxyacid synthase. *Proceedings of the National Academy of Sciences, USA* **103**, 569–573.

Mori, I., Fonné-Pfister, R., Matsunaga, S-I., *et al.* (1995) A new class of herbicides: specific inhibitors of imidazoleglycerol phosphate dehydratase. *Plant Physiology* **107**, 719–723.

Müller, K.H., Koning, K., Kluth, J., Lürssen, K., Santel, H.J. and Schmidt, R.R. (1992) Sulfonylaminocarbonyltriazolinones with oxygen-bound substitutes. EP507171, Bayer AG.

Muralla, R., Sweeney, C., Stepansky, A., Leustek, T. and Meinke, D. (2007) Genetic dissection of histidine biosynthesis in *Arabidopsis*. *Plant Physiology* **144**, 890–903.

Omura, S., Murata, M., Hanaki, H., Hinotozawa, K., Oiwa, R. and Fanaka, H. (1984) Phosalacine, a new herbicidal antibiotic containing phosphinothricin. Fermentation, isolation, biological activity and mechanism of action. *Journal of Antibiotics* **37**, 829–835.

Ray, T.B. (1984) Site of action of chlorsulfuron: inhibition of valine and isoleucine synthesis in plants. *Plant Physiology* **75**, 827–831.

Ray, T. B. (1989) Herbicides as inhibitors of amino acid biosynthesis. In: Böger, P. and Sandmann, G. (eds) *Target Sites of Herbicide Action*. Boca Raton, FL: CRC Press, pp. 105–125.

Reed, W.T., Saladini, J.L., Cotterman, J.C., Primiani, M.M. and Saari, L.L. (1989) Resistance in weeds to sulphonylurea herbicides. *Brighton Crop Protection Conference, Weeds* **1**, 295–300.

Rost, T. (1984) The comparative cell cycle and metabolic effects of chemical treatments on root tip meristems. III. Chlorsulfuron. *Journal of Plant Growth Regulation* **3**, 51–63.

Sandmann, G. and Böger, P. (1992) Chemical structures and activity of herbicidal inhibitors of phytoene desaturase, In: Draber, W. and Fujita, T. (eds) *Rational Approaches to Structure, Activity and Ecotoxicity of Agrochemicals*. Boca Raton, FL: CRC Press, pp. 357–371.

Schloss, J.V., Ciskanik, L.M. and Van Dyk, D.E. (1988) Origin of the herbicide binding site of acetolactate synthase. *Nature* **331**, 360–362.

Schöenbrunn, E., Eschenburg, S., Schuttleworth, W.A., *et al.* (2001) Interaction of the herbicide glyphosate with its target enzyme 5-enolpyruvylshikimate-3-phosphate synthase in atomic detail. *Proceedings of the National Academy of Sciences, USA* **98**, 1376–1380.

Schuktz, A., Munder, T., Holländer-Czytko, H. and Amrhein, N. (1990) Glyphosate transport and early effects on shikimate metabolism and its compartmentation in sink leaves of tomato and spinach plants. *Zeitschrift für Naturforschung* **45**, 529–534.

Shaner, D.L. and Reider, M.L. (1986) Physiological responses of corn (*Zea mays*) to AC 243,997 in combination with valine, leucine and isoleucine. *Pesticide Biochemistry and Physiology* **25**, 248–257.

Shaner, D.L., Anderson, P.C. and Stidham, M.A. (1984) Imidazolinones: potent inhibitors of aceto-hydroxyacid synthase. *Plant Physiology* **76**, 545–546.

Shimizu, T., Kaku, K., Takahashi, S. and Nagayama, K. (1994) Sensitivities of ALS prepared from SU- and IMI-resistant weeds against PC herbicides. *Journal of Weed Science and Technology* **46**, S32–S33.

Shimizu, T., Nakayama, I., Nagayama, K., Miyazawa, T. and Nezu, Y. (2002) Acetolactate synthase inhibitors, In: Boger, P., Wakabayashi, K. and Hirai, K. (eds) *Herbicide Classes in Development.* Berlin: Springer-Verlag, pp. 1–41.

Stepansky, A. and Leustek, T. (2006) Histidine biosynthesis in plants. *Amino Acids* **30**, 127–142.

Stilling, W.C., Abdel-Meguid, S.S., Lim, L.W., *et al.* (1991) Structure and topological symmetry of the glyphosate target 5-enolpyruvylshikimate-3-phosphate synthase – a distinctive protein fold. *Proceedings of the National Academy of Sciences, USA* **88**, 5046–5050.

Subramanian, M.V. and Gerwick, B.C. (1989) Inhibition of acetolactate synthase by triazolopyri-midines: a review of recent developments. In: Whitaker, J.R. and Sonnet, B.E. (eds) *Biocatalysis in Agricultural Biotechnology.* ACS Symposium Series No. 389. Washington, DC: American Chemical Society, pp. 277–288.

Swisher, B.A., de Boar, G.L., Ouse, D., Geselius, C., Jachetta, J.J. and Miner, V.W. (1991) Metabolism of selected triazolo (1,5-c) pyrimidine sulfonamides in plants. Internal report of DowAgro Sci. (Cited in Dorich and Schultz, 1997. *Down to Earth* **52**, 1–10.)

Ullrich, W.R., Ullrich-Eberius, C.I. and Köcher, H. (1990) Uptake of glufosinate and concomitant membrane potential changes in *Lemna gibba* G1. *Pesticide Biochemistry and Physiology* **37**, 1–11.

Usui, K., Suwangwong, S., Watanabe, H. and Ishizuka, K. (1991) Effect of bensulfuron methyl, glyphosate and glufosinate on amino acid and ammonia levels in carrot cells. *Weed Research Japan* **36**, 126–134.

Wallsgrove, R.M. (1989) The biosynthesis of amino acids in plants. In: Copping, L.G., Dalziel, J. and Dodge, A.D. (eds) *Prospects for Amino Acid Biosynthesis Inhibitors in Crop Protection and Pharmaceutical Chemistry.* British Crop Protection Council Monograph. Farnham, Surrey: BCPC, pp. 3–13.

Wild, A. and Ziegler, C. (1989) The effect of bialaphos on ammonia assimilation and photosynthesis. I. Effect on the enzymes of ammonium assimilation. *Zeitschrift für Naturforschung* **44c**, 97–102.

Wild, A., Sauer, H. and Ruehle, W. (1987) The effect of phosphinothricin on photosynthesis. I. Inhibition of photosynthesis and accumulation of ammonia. *Zeitschrift für Naturforschung* **42c**, 263–269.

Yamashita, K., Nagayama, K., Wada, N. and Abe, H. (1994) A novel ALS inhibitor produced by *Streptomyces hygroscopicus. Nippon Nogeikagaka Kaishi* **65**, 658.

Zhou, Q., Liu, W., Zhang, Y. and Liu, K.K. (2007) Action mechanisms of acetolactate synthase-inhibiting herbicides. *Pesticide Biochemistry and Physiology* **89**, 89–96.

Chapter 10
The Disruption of Cell Division

10.1 Introduction

According to Vaughn and Lehnen (1991), approximately 25% of all herbicides that have been marketed affect plant cell division, or mitosis, as a primary mechanism of action. For example, the dinitroanilines have been extensively used for over 30 years to selectively inhibit the growth of many annual grasses and some dicot weeds, especially in cotton, soybeans and wheat. They are applied prior to weed emergence, and herbicide absorption occurs through the roots or shoot tissues growing through the treated soil. In this way, they can provide good control of target weeds, allowing the crop seedlings to establish.

Many cell division inhibitors now have a restricted use or have been replaced by more effective, modern, low-dose herbicides. Furthermore, their continuous use has also led to reports of resistant weeds. They will, however, remain an important chemical weapon for weed control in several crops and have in recent years become resurgent in the control strategies for weeds resistant to the photosynthetic inhibitor herbicides, especially in the control of the important grass weeds *Setaria viridis* and *Eleusine indica*, two of the world's worst weeds.

10.2 The cell cycle

A major difference between plants and animals is that plants contain clearly defined regions, termed meristems, which remain embryonic throughout the life of the plant and are responsible for continued growth. Root and shoot meristems generate axial growth: the cambium is an example of a lateral meristem; the pericycle is a potential meristem for the generation of lateral roots; and grasses possess an additional intercalary meristem at the base of nodes. Cells in a meristematic region undergo a precise sequence of events, known as the cell cycle, which is divided into four discrete periods, each lasting for a time specific to a particular species. In G_1 (first gap phase) each nuclear chromosome is a single chromatid containing one DNA molecule and S phase (DNA synthesis) is characterised by a doubling of the DNA content of the nucleus, so that in G_2 (second gap phase) each chromosome now consists of two identical chromatids with identical DNA molecules. The

Herbicides and Plant Physiology, Second Edition By Andrew H. Cobb and John P.H. Reade
© 2010 A.H. Cobb and J.P.H. Reade

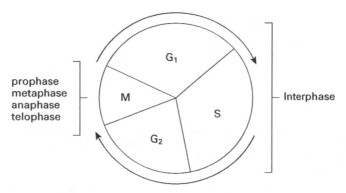

Figure 10.1 The mitotic cell cycle.

gap phases allow the operation of controls that ensure that the previous phase has been accurately completed. The main regulatory steps operate at the G_1/S and G_2/M boundaries, and both are sensitive to environmental conditions. G_1, S and G_2 are collectively termed interphase, to be followed by cell division (mitosis, or M), which is itself subdivided into prophase, metaphase, anaphase and telophase (Figure 10.1). During prophase the nuclear envelope disintegrates, and pairs of distinctly shorter and fatter chromosomes are observed. In metaphase the chromosome pairs move to the centre of the cell along a structure known as the spindle which consists of proteinaceous microtubules extending between the poles of the cell. At anaphase the chromosome pairs separate and migrate to opposite ends of the cell, so that two identical sets of chromosomes are evident at each pole of the spindle. Mitosis is completed in telophase by the formation of a nuclear envelope around each chromosome set, and the creation of a new cell wall by the fusion of the Golgi apparatus at the equatorial region of the cell. A completed cell cycle may take between 12 and 18 hours, and the full metabolic machinery of the cell is required to ensure progression from one stage to the next. It therefore follows that any chemical interference with cell cycle progression will cause growth inhibition and have herbicidal potential. Indeed, herbicides are known that either prevent mitotic entry or disrupt the mitotic sequence.

10.3   Control of the cell cycle

Substantial progress has been made in recent years in our understanding of the molecular mechanisms and control processes of cell division in plants. To ensure that each daughter cell receives the full complement of DNA requires close regulation of the S and M phases. It seems that the basic control mechanisms are highly conserved in all eukaryotes, the main drivers being a class of serine/threonine kinases, known as cyclin-dependent kinases (CDKs). Many regulatory features have evolved in plants that influence CDK activity, including metabolic, hormonal and environmental signals.

To be active, the kinase requires the binding of regulatory proteins, termed cyclins. In plants there are several CDKs, each having a role at different points in the cell cycle. Five types of cyclins exist in plants: the A-type is observed at the beginning of S phase and is destroyed at the G_2/M transition; the B-type appears during G_2 control of the

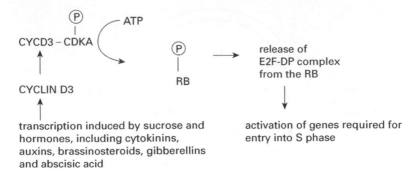

Figure 10.2 Transition from G_1 to S phase. The E2F proteins are transcription factors which, in combination with DP proteins, activate genes involved in DNA replication and other growth and cell cycle processes. For other abbreviations, see text.

G_2/M transition and is destroyed at anaphase; the D-type controls progression through G_1 into S and its presence depends on extracellular signals. If these signals are removed, the cell cycle is blocked in G_1.

Cyclin concentration is regulated by transcription or ubiquitin-mediated destruction at the proteosome (see Chapter 14, section 14.2 for more details).

Phosphorylation of a threonine residue (Thr160) is essential for CDK activity, as it permits proper binding of substrate molecules, and is achieved by a CDK-activating kinase (CAK).

Further regulation of CDK activity occurs by inhibitory phosphorylation of the amino-terminal residues at threonine 14 and tyrosine 15, catalysed by the WEE1 kinase.

Further mechanisms of control also exist by the action of inhibitor proteins, KRPs (kip-related proteins), which bind to both CDK and cyclin. CKS (Cdc kinase subunit) proteins influence the interaction of the substrate with the CDK complex.

How are all these regulatory functions thought to come into play controlling the cell cycle? In essence, the CDK causes the covalent addition of a phosphate group to a substrate protein at a serine or threonine residue, modifying the properties of the substrate. Common substrates are histone proteins in the chromosomes. Other known substrates include cytoskeleton proteins and the retinoblastoma proteins (RBs), whose phosphorylation controls the G/S transition (Figure 10.2).

Figure 10.3 summarises the regulation of cyclin-dependent kinase activity. Readers are referred to the review by Dewitte and Murray (2003) for further details and full references.

10.3.1 Transition from G_2 to M phase

Once the cell has duplicated its DNA in S phase, the next step is to generate the mitotic spindle, disassemble the nuclear envelope, condense the chromosomes and align each pair of sister chromatids appropriately.

The transition is controlled by CDK-CYCB kinase activity. During G_2, the amount of CYCB (cyclin B) increases and becomes associated with both cyclin-dependent protein

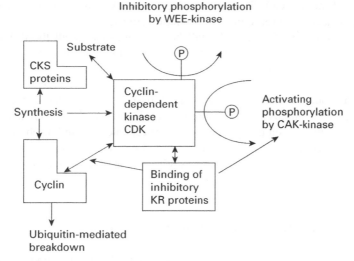

Figure 10.3   Regulation of cyclin-dependent kinase activity.

kinases A (CDKA) and B (CDKB). However, both KRP proteins and inhibitory phosphorylation by WEE kinases render the complex inactive. It is thought that specific protein phosphates reverse the inhibitory protein phophorylation to allow entry into M phase. Thus, cyclin-dependent kinases regulate progression through the cell cycle by reversible phosphorylation.

Progression through mitosis involves the association of the CDK complexes with microtubule and chromatin proteins. CDK activity is therefore likely to play an important part in microtubule dynamics and stability. The kinesins are a class of microtubule-associated proteins that have a motor domain which allows movement along the microtubules. They have key roles in spindle formulation and cell plate dynamics (Vanstraelen *et al.*, 2006). At least 23 kinesins have been implicated in mitosis using the energy of ATP to control microtubule organisation, polymerisation, depolymerisation and chromosome movement. The 23 kinesins are up-regulated during mitosis and their proposed involvement in spindle dynamics is reviewed by Vanstraelen *et al.* (2006).

In essence, progression through the cell cycle requires successive phosphorlyation events and the removal of regulatory and structural proteins by proteolytic degradation.

10.3.2   Hormones and cell division

Plants are adept in their response to changes in their external environment by alterations in physiology redirecting growth and development. Environmental and internal physiological signals are transmitted via hormones to yield appropriate morphogenetic responses. Auxins and cytokinins are the hormones most directly linked to cell division. Auxin increases both CDKA and cyclin mRNA levels but not cell division. Cytokinins and brassinosteroids induce expression of cyclin D3, which promotes entry into the mitotic cell cycle. Cytokinins have also been linked to CDK activation at the G_2/M transition, either by direct activation of a phosphatase or by down-regulation of the WEE1 kinase.

Abscisic acid inhibits cell division by inducing the expression of the CDK inhibitor proteins which cause a decrease in CDKA kinase activity.

A major challenge for future study will be to establish how all the different signal transduction pathways are regulated and integrated in the control of the cell cycle and hence plant growth and development (Stals and Inzé, 2001).

Inhibition of mitotic entry is experimentally observed by an absence or reduction of mitotic chromosomal features in meristematic tissues. Nanomolar concentrations of sulphonylureas and imidazolinones block cell cycle progression at G_1 and G_2 within minutes of application, even though the tissues are capable of S and M phases. This inhibition can be overcome by the addition of branched-chain amino acids and suggests that these compounds are worthy of further study as possible regulators of cell cycle progression.

10.4 Microtubule structure and function

Microtubules appear in electron micrographs as elongated, hollow cylinders occasionally up to 200 µm long and 25 µm in diameter. Chemically, they consist of the dimeric protein tubulin which is composed of similar but distinct subunits of 55 kDa molecular weight. In cross-section they are seen to consist of 13 units composed of the tubulin subunits arranged in a helical fashion. They are found as groups in different parts of the cell and at different times of the cell cycle. They perform a range of cellular functions as four distinct microtubule arrays. Cortical microtubules control the orientation of cellulose microfibrils in the young, developing cell and so determine the final shape of the cell. Preprophase microtubules control tissue morphogenesis, since they determine the planes of new cell divisions. Spindle microtubules, which form the spindle that is visible at metaphase during mitosis, enable the movement of chromosomes. After cell division, phragmoplast microtubules organise the new cell plate between the daughter cells. How these microtubules are formed into distinct arrays, however, remains uncertain. In animal cells, centrioles exist at the poles of cells to control spindle organisation and these are termed microtubule organising centres (MTOCs). Higher plant cells do not possess a centriole. The microtubules commonly originate at the endoplasmic reticulum and the nuclear envelope, but how is unknown.

Microtubule assembly is a very dynamic process. A pool of free tubulin subunits exists in the cytoplasm, which can be reversibly polymerised to form microtubules depending on the stage of the cell cycle. Indeed, the proportion of tubulin assembled into microtubules can vary from 0 to 90%. Mitchison and Kirschner (1984) envisaged the addition of tubulin subunits to the growing end of a microtubule, their loss at the opposite end and the possibility of a loss of the whole structure or parts of it. The biochemistry of tubulin polymerisation and depolymerisation appears complex (Figure 10.4).

The process appears very sensitive *in vitro* to the concentration of calcium and magnesium, and GTP binding appears essential for polymerisation to take place (Gunning and Hardham, 1982; Dawson and Lloyd, 1987). Microtubule-associated proteins (MAPs) have also been identified in plants (Cyr and Palevitz, 1989), which are also believed to influence and interact with microtubule assembly and crosslinking.

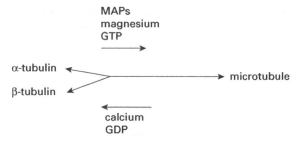

Figure 10.4 Factors affecting tubulin polymerisation and depolymerisation. MAPs, microtubule-associated proteins.

10.5   Herbicidal interference with microtubules

It has been known for some time that members of the dinitroaniline and *N*-phenylcarbamate classes of herbicides interfere with the structure and functions of microtubules. Representative structures of these herbicides are presented in Table 10.1.

Dinitroanilines.   Trifluralin, oryzalin and pendimethalin are widely used pre-emergence herbicides in dicot crops such as cotton and soybean for the control of grasses, and useful selectivity in wheat is evident. At concentrations as low as 1 µM, weeds show characteristic reductions in root length and swollen root tips, an effect identical to those obtained when treating seedlings with the well-known mitotic disruptor colchicine. Ultrastructural analysis reveals cells arrested at prometaphase (Vaughn and Lehnen, 1991) and so no metaphase or later stages are observed. Instead, a nuclear membrane reforms around the chromosomes and the nucleus appears highly lobed. No spindle microtubules are evident in treated cells and cortical microtubules are also affected, with the result that cells appear square-shaped rather than rectangular, explaining the swollen appearance of the root tips.

Recently, Tresch and colleagues (2005) have demonstrated that the cyanoacrylates (CA_1 and CA_2: Figure 10.5) are a new chemical class of herbicides with the same mechanism of action as the dinitroanilines, interacting with the α-tubulin binding site to prevent tubulin polymerisation.

Both dinitroanilines and colchicine inhibit microtubule assembly by forming a tubule–herbicide complex that disrupts polymerisation and hence microtubule assembly. In so doing, the depolymerisation process continues, shortening the microtubules until they are eventually undetectable. Unlike colchicine, the dinitroaniline herbicides do not inhibit tubulin polymerisation in animals, nor disrupt cell division in animal cells. At higher concentrations the dinitroaniline can also inhibit photosynthetic electron flow and oxidative phosphorylation.

These herbicides act by binding to α, β-tubulin dimers. The incorporation of dimer–herbicide complexes into a polymerising tubulin filament is thought to block further polymerisation and to cause microtubule disruption. It follows that any mutation causing changes in the amino acid content of tubulin may alter herbicide binding and could lead to herbicide resistance (Anthony *et al.*, 1998).

The fact that the herbicides display low affinity for animal tubulins renders them relatively environmentally benign.

Table 10.1   Structures of herbicides that interfere with microtubule assembly or function.

(a) dinitroanilines

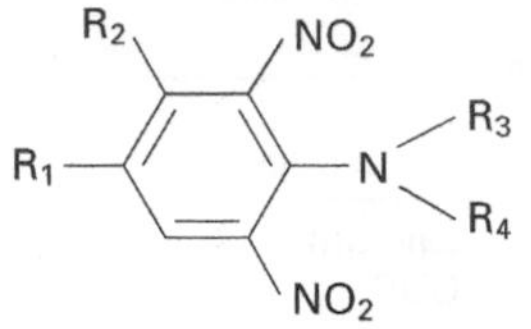

R_1	R_2	R_3	R_4	Common name
CF_3	H	C_3H_7	C_3H_7	trifluralin
SO_2NH_2	H	C_3H_7	C_3H_7	oryzalin
CH_3	CH_3	H	$CH(C_2H_5)_2$	pendimethalin

(b) N-phenylcarbamates

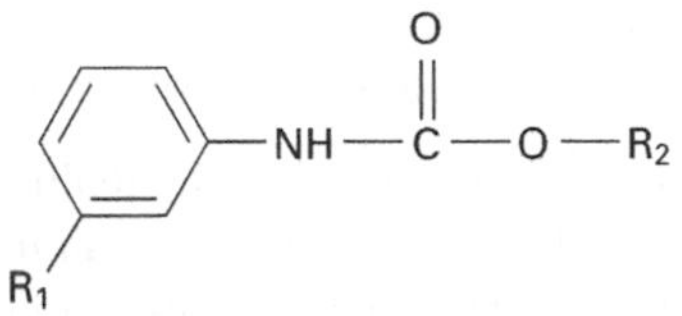

R_1	R_2	Common name
Cl	$CH_2C{=}CCH_2Cl$	barban
H	$CHCH_3CONHC_2H_5$	carbetamide
Cl	$CH(CH_3)_2$	chlorpropham
H	$CH(CH_3)_2$	propham

(c) others

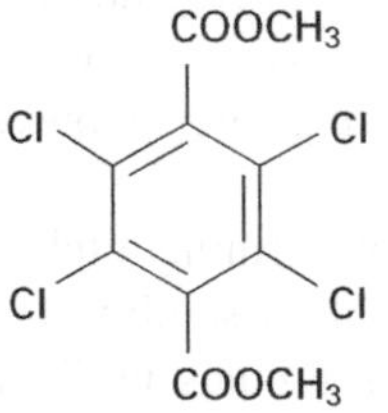

chlorthal-dimethyl(DCPA)

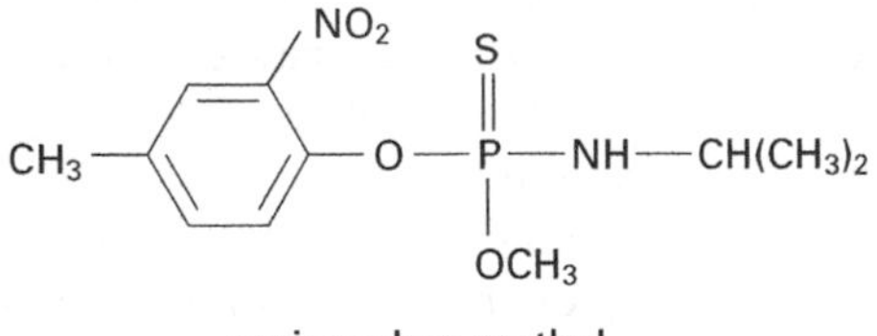

dithiopyr

amiprophos-methyl

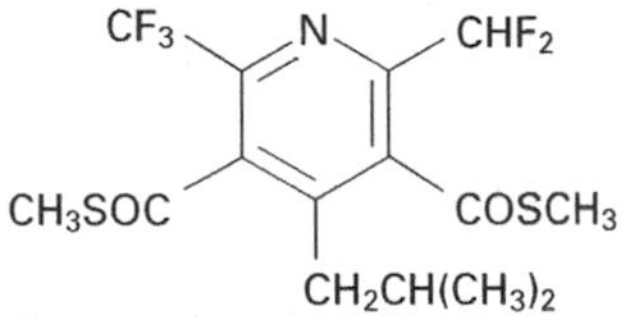

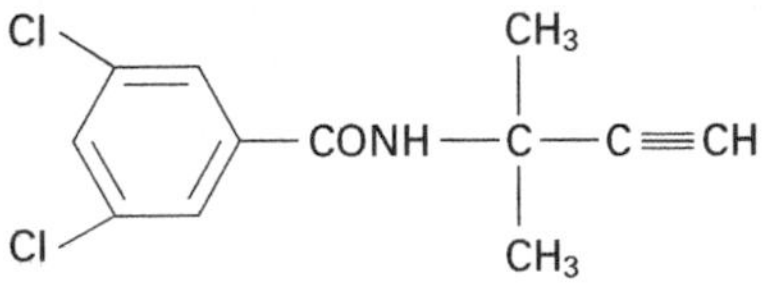

propyzamide

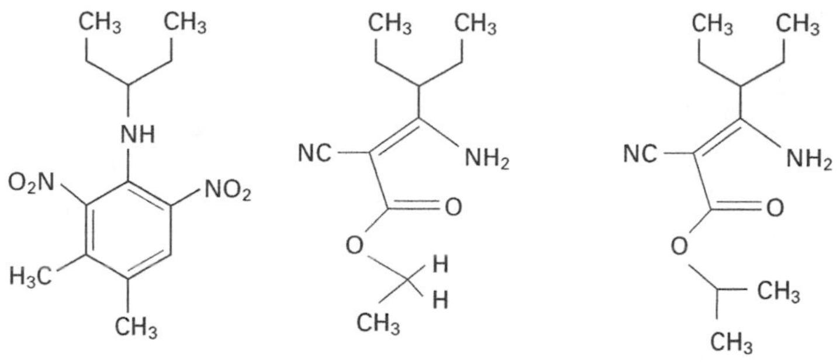

Figure 10.5 The structures of pendimethalin, CA_1 and CA_2.

N-phenylcarbamates. The phenylcarbamates also disrupt mitosis and inhibit photosynthetic electron flow and oxidative phosphorylation at high concentrations. In this case, microscopy reveals intact microtubules but chromosomal abnormalities. Chromosome movement during anaphase normally generates two sites at opposite poles of the cell, but in tissues treated with herbicide, three or more chromosome sites are observed after anaphase. Nuclear membranes then form around each group of chromosomes, and abnormal phragmoplasts organise irregularly shaped, abnormal cell walls. The proposed mechanism is the interference of the spindle microtubule-organising centres, fragmenting them throughout the cell, giving rise to the symptoms of multi-polar cell division. How this is achieved and the sites of action are unknown. Barban, carbetamide, propham and chlorpropham all interfere with mitosis in this manner.

These herbicides have been known since the 1950s and show useful activity against grasses. Nowadays, chlorpropham is principally used for the suppression of sprout growth in stored potato tubers, due to the inhibition of sprout cell division.

Others. Chlorthal-dimethyl (DCPA) is widely used in turf grasses. It appears to block cell plate formation in susceptible species, *via* the disruption of phragmoplast microtubule organisation and production, by an unknown mechanism. There is some indication that CDK phosphatase activity is inhibited by endothal, thus preventing G_2/M transition (Ayaydin *et al.*, 2000). Amiprophos-methyl results in a loss of microtubules and gives similar symptoms to the dinitroanilines. Propyzamide acts in a different fashion to those described so far, such that small microtubules rather than none at all result from treatment with this herbicide. It binds directly to the tubulin, preventing microtubule assembly (Akashi, *et al*, 1988). Dithiopyr acts similarly to propyzamide, but binds to a microtubule-associated protein (MAP) of molecular weight 65 kDa, rather than to tubulin. Vaughn and Lehnen (1991) are of the view that dithiopyr interacts with a MAP involved in microtubule stability, resulting in shortened microtubules.

Clearly, there remain many unanswered questions regarding microtubule assembly and function in plant cell division. The aforementioned herbicides may provide useful probes to study microtubule biochemistry and further work may yield additional novel herbicide targets.

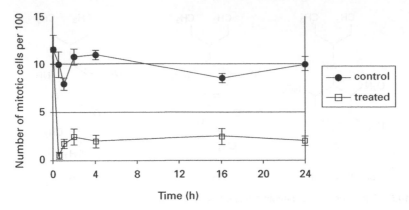

Figure 10.6 The effect of treatment with 0.1 g ai L^{-1} imazethapyr on mitotic divisions of potato (*Solanum tuberosum*) cv. Cara tuber root tips using aerated liquid media. Data are means of 6 replicates ± standard error. Treated and control samples were significantly different at every point ($p < 0.01$) (from Spackman and Cobb, 1999).

Intriguingly, it has also been observed that members of the imidazolinone and sulfonylurea herbicide families inhibit entry into mitotic cell division. Rost (1984) found that chlorsulfuron inhibited the cell cycle progression at G_2 and G_1 transitions and also demonstrated a similar effect of imidazolinones (Rost *et al*, 1990). Both herbicide families inhibit acetolactate synthase (ALS, see Chapter 9) in the synthesis of the branched-chain amino acids (valine, leucine, isoleucine) and cell cycle inhibition can be reversed by the addition of these amino acids. This may imply a link between branched-chain amino acid biosynthesis and cell cycle entry, although the precise nature of this remains to be established.

Spackman and Cobb (1999) observed a rapid inhibition of cell division in potato root tips treated with imazethapyr. Significantly lower numbers of mitotically dividing cells were recorded after only 30 min incubation with 0.35 µM imazethapyr (Figure 10.6). The commercialisation of this response as a novel, low-dose potato-sprout suppressant remains to be demonstrated.

10.6 Selectivity

The dinitroanilines are more phytotoxic to grasses than dicots and it has been proposed that selectivity is based on the lipid content of the seeds (Hilton and Christiansen, 1972). The dinitroanilines are highly lipophilic, while the lipid content of grass seeds is lower than that of dicot seeds. Thus, it is believed that these lipophilic molecules are partitioned into the dicot seed lipid reserves, and so sequestered away from their site of action. These authors also demonstrated that coating the seeds of normally susceptible species with α-tocopherol acetate prior to planting gave some protection to the presence of trifluralin in the growth medium.

Depth protection also offers a more selective action for crops. The dinitroanilines are usually incorporated into the top 10 cm of soil and do not readily leach downwards through the soil column. Hence, large-seeded, deep-sown crops with large lipid reserves can grow through the treated zone, whereas small-seeded, shallow germinating weed seedlings cannot fare as well.

References

Akashi, T., Izumi, K., Nagano, E., Enomoto, E., Mizuno, K. and Shibaoka, H. (1988) Effects of propyzamide on tobacco cell microtubules *in vivo* and *in vitro*. *Plant Cell Physiology* **29**, 1053–1062.

Anthony, R.G., Walden, T.R., Ray, J.A., Bright, S.W.J. and Hussey, P.J. (1998) Herbicide resistance caused by spontaneous mutation of the cytoskeletal protein tubulin. *Nature* **393**, 260–263.

Ayaydin, F., Vissi, E., Meszaros, T. *et al.* (2000) Inhibition of serine/threonine-specific protein phosphatases causes premature activation of cdcMsF kinase at G_2/M transition and early mitotic microtubule organisation in alfalfa. *Plant Journal* **23**, 85–96.

Cyr, R.J. and Palevitz, B.A (1989) Microtubule-binding proteins from carrot. I. Initial characterisation and microtubule bundling. *Planta* **177**, 245–260.

Dawson, P.J. and Lloyd, C.W. (1987) Comparative biochemistry of plant and animal tubulins. In: Davies, D.D. (ed.) *The Biochemistry of Plants. A Comprehensive Treatise*, vol. 12. New York: Academic Press, pp. 3–47.

Dewitte, W and Murray, J.A.H. (2003) The plant cell cycle. *Annual Review of Plant Biology*, **54**, 235–264.

Gunning, B.E.S. and Hardham, A.R. (1982) Microtubules. *Annual Review of Plant Physiology* **33**, 651–698.

Hilton, J.L. and Christiansen, M.N. (1972) Lipid contribution to selective action of trifluralin. *Weed Science* **20**, 290–294.

Mitchison, T.J. and Kirschner, M. (1984) Dynamic instability of microtubule growth. *Nature* **312**, 237–242.

Rost, T.L. (1984) The comparative cell cycle and metabolic effects of chemical treatment on root tip meristems. III. Chlorsulfuron. *Journal of Plant Growth Regulation* **3**, 51–63.

Rost, T.L., Gladdish, D., Steffen, J. and Robbins, J. (1990). Is there a relationship between branched chain amino acid pool size and cell cycle inhibition in roots treated with imidazolinone herbicides? *Journal of Plant Growth Regulation* **9**, 227–232.

Spackman, V.M.T. and Cobb, A.H. (1999) Cell cycle inhibition of potato root tips treated with imazethapyr. *Annals of Applied Biology* **135**, 585–587.

Stals, A. and Inzé, D. (2001) When plant cells decide to divide. *Trends in Plant Science* **8**, 359–364.

Tresch, S., Plath, P. and Grossmann, K. (2005) Herbicidal cyanoacrylates with antimicrotubule mechanism of action. *Pest Management Science* **61**, 1052–1059.

Vanstraelen, M., Inzé, D. and Geelen, D. (2006). Mitosis-specific kinesins in *Arabidopsis*. *Trends in Plant Science* **11**, 167–174.

Vaughn, K. C. and Lehnen Jr., L. P. 1991. Mitotic disrupter herbicides. *Weed Science* **39**, 450–457

Chapter 11
The Inhibition of Cellulose Biosynthesis

11.1 Introduction

The plant cell wall is a very complex yet highly organised network of polysaccharides, proteins and phenylpropanoid polymers (e.g. lignin), set in a slightly acidic solution containing several enzymes and many organic and inorganic substances. It is not a static structure but a dynamic, metabolic compartment of the cell, sharing a molecular continuity with the plasmalemma and the cytoskeleton. It also changes throughout the life of the cell, from cell division to expansion and elongation, expanding its original area by orders of magnitude. Readers are referred elsewhere for a more detailed understanding and review of the cell wall (e.g. Carpita, 1997).

Cellulose is the most abundant plant polysaccharide and provides the framework for the higher plant cell wall. Its biosynthesis is therefore an important target for herbicide development, especially as no animal counterpart exists that could, perhaps, lead to issues of toxicity.

While we know that cellulose biosynthesis takes place at the plasma membrane, cell-free synthesis has not yet been demonstrated. It is thought to be catalysed by multimeric terminal complexes that have been visualised by freeze-fracture images of the plasmalemma. These are, however, irreversibly disrupted and lost on membrane isolation. UDP-glucose is considered to be the primary substrate and progress is awaited on identifying the genes and enzymes involved. *Arabidopsis* mutants have been generated and their further characterisation should shed light on this important process.

Cellulose, the fundamental unit characterising the plant cell wall, consists of microfibrils that are themselves a bundle of parallel $\beta \rightarrow 4$ glucan chains, hydrogen-bound to form a microcrystalline array. Cellulose synthase genes (CesA) that code for the catalytic subunit of cellulose synthase (E.C. 2.4.1.12) were first isolated from cellulose-producing bacteria in the early 1990s and have since been characterised in other prokaryotes and higher plants. All cellulose synthase-like (CSL) proteins share transmembrane helices and a cytoplasmic loop consisting of four conserved regions involved in substrate binding and catalysis. In *Arabidopsis* 10 CesA genes have been identified. Eckardt (2003) has suggested that three CesA proteins (A_1, A_3 and A_6) interact as subunits within the cellulose

Herbicides and Plant Physiology, Second Edition By Andrew H. Cobb and John P.H. Reade
© 2010 A.H. Cobb and J.P.H. Reade

synthase complex in primary cell wall biosynthesis, whereas $CesA_4$, A_7 and A_8 form the complex in secondary cell walls.

The transmembrane nature of the CSL proteins is supported by freeze-facture electron microscopy studies, which show hexagonal arrays of nascent cellulose microfibrils and terminal complexes, termed rosettes, on the cell wall face of the plasma membrane. More conclusive evidence that these rosettes are involved in cellulose synthesis has come from Kimura *et al.* (1999) who labelled freeze-facture relicas of mung bean rosettes with anti-bodies raised against a CesA gene product. It is now generally accepted that the rosettes are functional cellulose synthase complexes.

The Golgi apparatus is the known cell site for the synthesis and processing of glyco-proteins and targetting them to the plasma membrane. Glycans and pectins are also poly-merised at this location, along with the synthesis of hexose- and pentose-containing polysaccharides. It appears that all non-cellulosic polysaccharides are synthesised at the Golgi apparatus.

The cell wall also contains four major structural proteins; extensins, the repetitive proline-rich protein, the glycine-rich protein and the threonine-hydroxyproline-rich protein. Extensins contain repeating $Ser-(Hyp)_4$ and Tyr-Lys-Tyr sequences that are important for wall structure. They are rod-shaped and may cross-link to other extensins. The 33-kDa proline-rich proteins are similar to extensins and are more highly expressed in later stages of cell development. The glycine-rich protein may have over 70% glycine residues and is thought to form a plate-like structure at the interface of the plasma mem-brane and the cell wall, which is thought to serve as an initiation site for lignification. The threonine-hydroxyproline-rich proteins predominate in grasses and are thought to be related to extensins. They accumulate early in the cell cycle and become insoluble during cell elongation and differentiation.

11.2   Cellulose biosynthesis inhibitors

The cellulose biosynthesis inhibitors (CBIs) are a small group of chemically unrelated compounds including the herbicides dichlobenil, isoxaben and flupoxam (Figure 11.1). They are all effective CBIs, and because this site of action is not shared with mammals, they may be ideal choices for the purposes of registration and selective herbicidal activity. A lack of observed resistance in the field may enable them to become useful agents in the control of resistant weeds. Furthermore, they have become valuable probes in our under-standing of cellulose biosynthesis and cell wall formation in plants.

Cellulose is an integral part of the plant cell wall and hence cellulose synthesis is vital in order for the cell wall to be synthesised. The cell wall determines to some extent both the morphology and the function of a cell, but even more importantly it controls the degree to which a cell can expand. It is suggested that if cellulose biosynthesis is inhibited then weakened cell walls will result, causing expansion of the cell and disruption of cellular processes. This in turn leads to abnormal or restricted growth and subsequent plant death. As most CBIs are used pre-emergence then seedling growth is inhibited. Post-emergence use of CBIs leads to stunted growth and swollen roots in treated plants.

It appears that CBIs do not have a common site of action in the synthesis of cellulose, which is a process carried out by multi-enzyme complexes situated at the cell's plasma

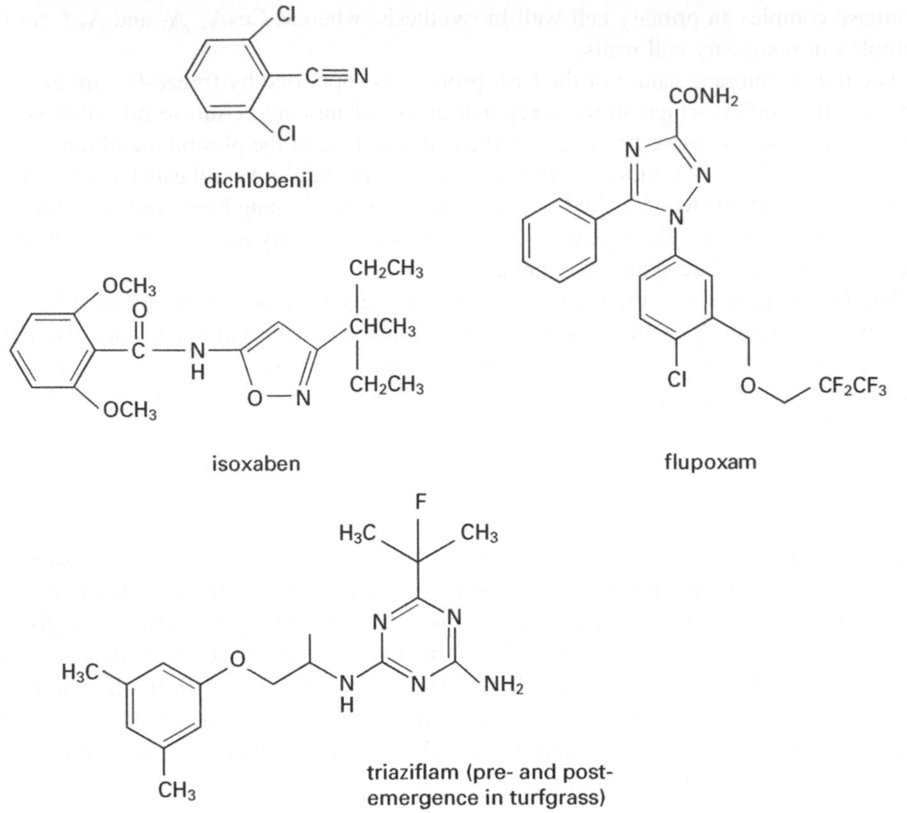

Figure 11.1   Chemical structures of the four major cellulose biosynthesis inhibitor herbicides.

membrane. Research to date has identified at least two separate points at which CBIs inhibit. Dichlobenil analogues have been shown to bind to an 18-kDa polypeptide associated with the multi-enzyme complex. It is postulated that this may be a regulator subunit associated with cellulose synthase, as it appears to be too small to be the enzyme itself. Reported effects of dichlobenil treatment include synthesis of callose in place of cellulose. Callose is a β-1, 3 glucan, often formed in plants as a wounding response. It is not usually present in the cell walls of undamaged cells. It is postulated that dichlobenil inhibits the incorporation of UDP-glucose into cellulose and this is therefore shunted into callose (and xyloglucan) synthesis. Isoxaben has been demonstrated to inhibit the incorporation of ^{14}C-labelled glucose into the cell wall. Other studies have revealed that the synthesis of both cellulose and callose is reduced by this herbicide. This suggests an inhibition of cellulose synthesis at an earlier stage than dichlobenil, preventing incorporation of glucose into both cellulose and callose. It has been postulated that the step where isoxaben acts may be at the point where UDP glucose is formed from sucrose. Flupoxam appears to inhibit at a similar point, although less information on this herbicide's mode of action is available. Quinclorac, an auxin-type herbicide for broad-leaf weed control, also appears to inhibit cellulose biosynthesis in susceptible grasses, although the mechanism by which

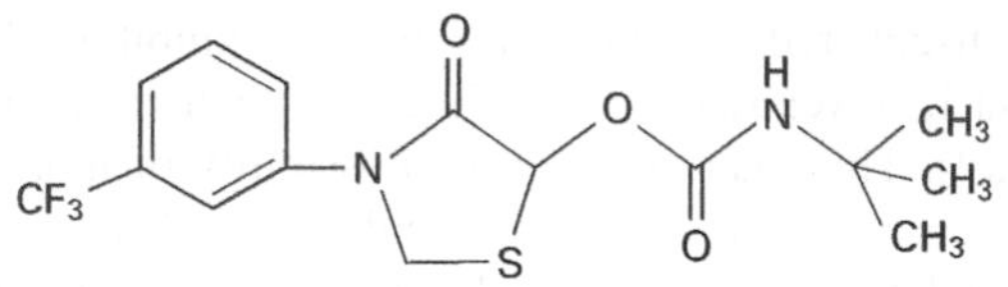

Figure 11.2   5-*tert*-butyl-carbamoyloxy-3-(3-trifluoromethyl) phenyl-4-thiazolidonone (Compound 1).

it accomplishes this has yet to be deduced. This implies that quinclorac has a second site of action in monocots, unrelated to its primary mechanism in dicots (Koo *et al.*, 1997). A new herbicide, 5-*tert*-butyl-carbamoyloxy-3-(3-trifluoromethyl)-phenyl-4-thiazolidinone (Figure 11.2), also appears to inhibit cellulose biosynthesis at the same place as isoxaben (Sharples *et al.*,1998).

This is a representative of a novel class of *N*-phenyl-lactam-carbamate herbicides patented by Syngenta in 1994. It shows potential for the pre-emergence control of a range of grass weeds (including *Setaria viridis*, *Echinochloa crus-galli* and *Sorghum halepense*) and small seeded broad-leaf weeds (including *Amaranthus retroflexus* and *Chenopodium album*) in soybean, at an application rate of approximately 125–250 g ha^{-1}. Root growth was completely inhibited at 2 µM and visual symptoms typically included stunted and swollen roots, with no lateral branches. It was shown by Sharples *et al.* (1998) to be a potent inhibitor (IC$_{50}$ of 50 µM) of the incorporation of ^{3}H-labelled glucose into the acid-insoluble polysaccharide cell wall fraction of *Zea mays* roots – which is assumed to be cellulose. Similar inhibition was observed by isoxaben and dichlobenil, and cross-resistance was demonstrated to an isoxaben-resistant *Arabidopsis* mutant (Ixr1-1) which implies a shared molecular binding site with isoxaben. These authors conclude that the final identification of this site will require a combination of molecular approaches and traditional biochemistry.

The formation of cell plates in the presence and absence of CBIs has received detailed scrutiny in both light and electron microscope studies. Cell plates are the cell walls formed to separate daughter nuclei after cell division. In the presence of dichlobenil, cell plates develop more undulated and thicker regions than in untreated tissues, resulting in cell walls that are incomplete or attached to only one existing cell wall. Treated cell walls also are enriched in callose compared with controls. This may imply that the inhibition of cellulose biosynthesis has increased the amount of UDP-glucose available for callose or xyloglucan synthesis. Conversely, treatment with either isoxaben or flupoxam results in the production of a thin cell plate, with no callose nor xyloglucan enrichment.

Another useful experimental system for the study of CBIs (Vaughn, 2002) is the developing cotton fibre, whose secondary wall is composed almost exclusively of cellulose. Fibre development is severely inhibited by the CBIs. The cell wall of control fibres consists of two layers, an outer wall area enriched in pectin and an inner layer enriched in cellulose–xyloglucan. After treatment with diclobenil, the inner layer became enriched in callose and cellulose was absent. On the other hand, treatment with isoxaben and flupoxam generated relatively more outer wall, highly enriched in de-esterified pectin.

In summary, dichlobenil appears to divert cellulose biosynthesis into callose, while isoxaben and flupoxam inhibit the production of both cellulose and callose in both cell plates and cotton fibres. The further identification of the enzymes involved in the presence of CBIs is awaited with interest.

The most recently discovered cellulose biosynthesis inhibitors, flupoxam and triaziflam, are active at much lower rates ($100–250\,g\,ha^{-1}$) than their predecessors, such as dichlobenil ($3,000\,g\,ha^{-1}$). This may open up the possibility of more imitative chemistry in future to lower the dosage used in this herbicide class. According to Grossman and colleagues (2001), triaziflam may also affect photosynthetic electron transport and microtubule formation. This multiple activity may be a positive feature and may delay the onset of herbicide-resistant weeds.

The relationship between cellulose microfibrils and cortical microtubules has long been debated. Paradez and colleagues (2006) have fluorescently labelled cellulose synthases that assemble into functional cellulose synthase complexes in *Arabidopsis* hypocotyl cells. They have demonstrated that the cellulose synthase complexes move in the plasma membrane along tracks that appear to be defined by co-labelled microtubules. It appears that the force for the movement of the cellulose synthase complex is derived from cellulose microfibril production itself. According to Emons *et al.* (2007), this breakthrough discovery provides the experimental tools that may allow us to answer the following questions:

- How do the cellulose synthase complexes enter the plasma membrane, via exocytosis?
- Where are they inserted into the plasma membrane?
- How are they activated, for how long and how are they regulated?
- Which gene products influence microfibril production?
- How does the network of molecules involved in cellulose production function?

Note that in these studies oryzalin was added to depolymerise the cortical microtubules, showing the value of this herbicide as a tool in unravelling this complex process.

11.3   Selectivity

While mitotic disrupters are most effective on small-seeded monocotyledonous weeds, the CBIs are most effective on dicotyledonous plants. Symptoms following treatment with CBIs are similar to those observed following application of cell division inhibitors, that is, radial swelling of root tips.

Various studies on the metabolic fate of isoxaben in both cell cultures and *Arabidopsis* mutants have implied that selectivity is not due to altered uptake or metabolism, but instead results from an alteration at the target site. Furthermore, the observation (Sharples *et al.*, 1998) that Compound 1 is metabolised very slowly *in vivo* also implies that the selectivity of this molecule in soybean is due to a species-dependent variability at the target site. This is presumed to be the multimeric complex, cellulose synthase. Further details of this enzyme complex are eagerly awaited.

References

Carpita, N. (1997) Structure and biosynthesis of plant cell walls. In: Dennis, D.T., Turpin, D.H., Lefebvre, D.B. and Layzell, D.D. (eds) *Plant Metabolism*, 2nd edn. Harlow, UK: Longman, Ch. 9, pp. 124–147.

Eckardt, N.A. (2003) Cellulose synthesis takes the CesA train. *Plant Cell* **15**, 1685–1687.

Emons, A.M.C., Höfte, H. and Mulder, B.M. (2007) Microtubules and cellulose microfibrils: how intimate is their relationship? *Trends in Plant Science* **12**, 279–281.

Grossman, K., Tresch, S. and Plath P. (2001) Triaziflam and diamino-triazine derivatives affect enantioselectivity multiple herbicide target sites. *Zeitschrift für Naturforschung* **56c**, 559–569.

Kimura, A., Laosinchai, W., Itoh, T., Cui, X., Linder, C. R. and Brown, R.M. (1999) Immunogold labelling of rosette terminal cellulose-synthesising complexes in the vascular plant *Vigna angularis*. *Plant Cell* **11**, 2075–2086.

Koo, S. J., Neil, J.C. and Ditomaso, J.M. (1997) Mechanism of action of quinclorac in grass roots. *Pesticide Biochemistry and Physiology* **57**, 44–53.

Paradez, A.R. *et al.* (2006) Visualisation of cellulose synthase demonstrates functional association with microtubules. *Science* **312**, 1491–1495.

Sharples, K.R., Hawkes, T.R., Mitchell G., *et al.* (1998) A novel thiazolidinone herbicide is a potent inhibitor of glucose incorporation into cell wall material. *Pesticide Science* **54**, 368–376.

Vaughn, K.C. (2002) Cellulose biosynthesis inhibitor herbicides. In: Böger, P., Wakabayashi, K. and Harai, K. *Herbicide Classes in Development*. Berlin and Heidelberg: Springer-Verlag, pp. 139–150.

Chapter 12
Herbicide Resistance

12.1 Introduction

As of summer 2010, herbicide resistance had been reported in 194 plant species (114 dicots and 80 monocots) in over 340,000 fields worldwide (Heap, 2010). Resistance has been reported towards 17 of the 22 Herbicide Resistance Action Committee's herbicide Mode of Action groups. In some species, for example *Alopecurus myosuroides* in the UK, resistance is so widespread that it is now described as being endemic. The number of cases of herbicide resistance will undoubtedly continue to rise. Indeed, resistance to active ingredients that have yet to be commercialised will already be present within existing weed populations. Presently the gene(s) conferring this resistance will offer little or no advantage to the individuals possessing them. However, once a new herbicide is commercialised and applied to weed populations the gene(s) will offer a distinct advantage, as only individuals possessing them will survive herbicide treatment and be in a position to grow and ultimately reproduce. As long as herbicides form a part of a weed management programme, then resistance will remain a problem, although integration of herbicide usage with non-chemical management practices can vastly reduce this risk.

Herbicide Resistance is the inheritable ability of a plant to survive a herbicide dose that would be lethal to a member of a normal population of that plant species.

Inherent in this definition are three important points:

1 Herbicide resistance is inheritable, so it is a characteristic coded for in the plant genome. At least some of the progeny of that plant will either be resistant or will carry the resistance trait. This distinguishes herbicide resistance from other causes of poor herbicide efficacy, perhaps caused by environmental factors including their affect on spray effectiveness, plant physiology and biochemistry.
2 Herbicide-resistant plants survive herbicide treatment and can successfully complete their life cycle by flowering and producing seed. This does not mean that resistant individuals will not show symptoms of herbicide damage, but that they are not killed by herbicides. In many cases some herbicide damage may be observed but it does not lead to plant death.
3 A normal population is a population of the species that when treated with an optimum dose of a herbicide, under ideal conditions, all individuals within it are killed. This

---

Herbicides and Plant Physiology, Second Edition By Andrew H. Cobb and John P.H. Reade
© 2010 A.H. Cobb and J.P.H. Reade

population will be one that has never been exposed to a herbicide. Such 'wild-type' populations are not always available to the researcher, so populations that have demonstrated 100% susceptibility are often used in research, regardless of their field history. In order to determine baseline sensitivity for a particular herbicide acting on a distinct species, several normal populations are ideally used.

Resistance should not be confused with tolerance, which is the term used to describe an individual species that is not controlled by a particular herbicide. For example, cleavers (*Galium aparine*) is described as tolerant to isoproturon because the species as a whole is naturally tolerant to this active ingredient.

12.2 Mechanisms of herbicide resistance

12.2.1 Target site resistance

As detailed elsewhere, herbicides have distinct target sites where they act to disrupt biochemical processes leading to cell, tissue and plant death. The majority of target sites are enzymes and the interaction between herbicide and target site can be disrupted if there is a change in the primary structure of the enzyme protein molecule. Where this occurs, the herbicide may no longer be effective in blocking the action of the target site and the plant will not die, but exhibit herbicide resistance (Figure 12.1).

Target site resistance is often the result of a single point mutation in the gene coding for the target site, so is the result of a change in one nucleic acid in the

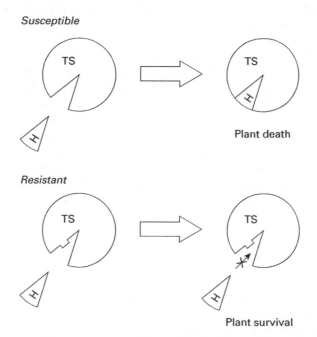

Figure 12.1 Diagrammatic representation of Target Site Resistance. TS, target site; H, herbicide.

target site gene, resulting in one amino acid change in the final target site protein. Examples are given below.

12.2.1.1   Target site-mediated resistance to ACCase-inhibiting herbicides

A number of mutations in the acetyl-CoA carboxylase (ACCase) gene have been reported in grasses, which result in target site resistance to ACCase-inhibiting herbicides. Different cross-resistance patterns are noted for each mutation (Devine, 1997; Devine and Shukla, 2000; Délye, 2005). An isoleucine-to-leucine substitution within the carboxytransferase region of ACCase confers resistance in *Avena fatua* (Christoffers *et al.*, 2002), *Lolium rigidum* Gaud. (Zagnitko *et al.*, 2001; Délye *et al.*, 2002a; Tal and Rubin, 2004), *Setaria viridis* (L.) Beauv. (Délye *et al.*, 2002b) and *Alopecurus myosuroides* Huds (Délye *et al.*, 2002a). In *A. myosuroides* this mutation is designated I1781L (Délye *et al.*, 2002a) indicating the position where the amino acid substitution has taken place. In all cases this mutation confers cross-resistance to all fops and most dims. Another mutation (Ile 2041 → Asn), also in the carboxytransferase region of ACCase, in *L. rigidum* and *A. myosuroides* has been reported, which confers resistance to fops but not to dims (Délye *et al.*, 2003). These observations may increase our understanding of the differences in herbicide binding to ACCase between fops and dims.

12.2.1.2   Target site-mediated resistance to ALS-inhibiting herbicides

Many of the cases of herbicide resistance to acetolactate synthase (ALS) inhibitors reported to date are caused by an altered target site rather than due to enhanced metabolism. Five naturally occurring mutations are found in native populations that give rise to resistance. Many others are found in bacteria and yeast systems. Mutations are found in two regions: A; amino acid sequences 124–205 and B; amino acid sequences 574–653 (Gressel, 2002; Table 12.1). This is not the catalytic site of the enzyme but a separate herbicide-binding site. Different patterns of cross-resistance are noted for each of these amino acid substitutions, and the flexibility of substitution while still maintaining enzymic activity has probably contributed to the rapid rise in resistance to this class of herbicide. Resistance to imidazolinones, conferred by Ser670 → Asp or Ala122 → Thr, does not confer very high levels of resistance to the other classes of ALS inhibitor (Sathasivan *et al.*, 1990, 1991; Bernasconi *et al.*, 1995). Studies with ALS-resistant *Arabidopsis thaliana* mutants *csr*1-1 (Pro197 → Ser) and *csr*1-2 (Ser653 → Asn) identified complex cross-resistance patterns with respect to 22 ALS-inhibiting herbicides (Roux *et al.*, 2005). This suggests complex and non-ubiquitous binding characteristics for this herbicide family.

12.2.1.3   Target site-mediated resistance to Photosystem II-inhibiting herbicides

In 1970 Ryan reported the failure to control groundsel (*Senecio vulgaris*) with simazine and atrazine in a nursery in Washington State, USA. These residual herbicides had been used once or twice a year in the nursery from 1958 to 1968. Susceptible plants grown from seeds collected from a location where triazines had not been in continuous use were adequately controlled by 0.56–1.12 kg ha^{-1}, but seedlings from the nursery were barely

Table 12.1 A selection of mutations that confer target site resistance to ALS (modified from Gressel, 2002). In some cases different cross-resistance patterns have been reported for a single mutation by different researchers.

Mutation	Gene name	Selector and cross-resistances	Little or no cross-resistance
$Met_{124} \rightarrow Glu$		SU, IM	
$Met_{124} \rightarrow Ile$		IM	
$Ala_{155} \rightarrow Thr$		SU, IM, TP, PC	SU, TP
$Pro_{197} \rightarrow His$		SU, some IM	
$Pro_{197} \rightarrow Gln$	C3	SU, some IM	
$Pro_{197} \rightarrow Ala$	S4	SU, some IM	
$Pro_{197} \rightarrow Ser$	Crs-1	SU, TP	IM, PC
$Pro_{197} \rightarrow Arg$		SU, some IM	
$Pro_{197} \rightarrow Leu$		SU, some IM	
$Pro_{197} \rightarrow Thr$		SU, some IM	
$Arg_{199} \rightarrow Ala$			
$Arg_{199} \rightarrow Glu$			
$Ala_{205} \rightarrow Asp$			
$Gln_{269} \rightarrow His$		IM, TP, PC, SU	
$Asn_{522} \rightarrow Ser$			
$Tro_{574} \rightarrow Leu$	ALS3	IM, TP, PC, SU	
$Trp_{574} \rightarrow Ser$		IM>SU	
$Trp_{574} \rightarrow Phe$		IM, TP, SU	
$Ser_{653} \rightarrow Asn$	Csr I-2, imr	IM, PC	SU, TP
$Ser_{653} \rightarrow Thr$		IM	SU
$Ser_{653} \rightarrow Phe$		IM>SU	

SU, sulfonylurea; TP, triazolopyrimidine sulfonamide; IM, imidazolinone; PC, pyrimidyloxybenzoate.

affected, even by doses as high as $17.92 \, kg \, ha^{-1}$. Triazine resistance is now known to have evolved independently throughout the world due to persistent and prolonged use in monocultures in orchards, vineyards, nurseries and maize crops. Cross-resistance is commonly observed (Table 12.2).

A single mutation in the *psbA* gene coding for a 32-kD protein that forms part of the Photosystem II (PS II) complex in the thylakoids can lead to resistance to the herbicides that inhibit photosynthesis at PS II (Oettmeier, 1999). These single mutations cause resistance to a number of PS II inhibiting herbicides to varying degrees (see Table 5.2). In some cases 1000 times more herbicide is needed to displace Q_B from this site. As this binding site is common to many PS II inhibiting herbicides the cross-resistance observed is not surprising. However, cross-resistance does not always extend to all PS II herbicides as is clearly demonstrated in Table 5.2. The mutation Ser264 $\rightarrow$ Gly results in decreased binding and efficacy of triazines but not other classes of PS II inhibitors. The mutation Ser264 $\rightarrow$ Thr results in a broader spectrum of resistance to phenylureas and the triazine herbicides, for example linuron and atrazine, respectively. Resistance to diuron and metribuzin is conferred by the mutation Val219 $\rightarrow$ Ile. In addition to these mutations of the D1 protein, Ala251 $\rightarrow$ Arg and Val280 $\rightarrow$ Leu are also found in weed populations in the wild. In addition laboratory studies have also identified a further mutation, Ser268 $\rightarrow$ Pro, that also confers resistance. It is interesting to note that there is a fitness price to pay for these mutations to the D1 protein, with reduced CO_2 fixation, quantum

Table 12.2 Photoreduction of DCPIP by thylakoids isolated from atrazine-resistant and susceptible biotypes of oilseed rape (*Brassica napus* L. cv. Candelle) (A. Cobb, unpublished observations).

Herbicide	Resistant (R)	Susceptible (S)	R/S
Atrazine	7.5×10^{-4}	1.0×10^{-8}	75,000
Phenmedipham	1.0×10^{-5}	2.0×10^{-7}	50.00
Diuron	1.75×10^{-6}	6.5×10^{-8}	26.90
Metamitron	9.0×10^{-6}	1.8×10^{-6}	5.00
Bentazone	8.0×10^{-5}	6.0×10^{-5}	1.33
Ioxynil	2.6×10^{-7}	3.5×10^{-7}	0.74
Dinoseb	1.1×10^{-7}	0.8×10^{-6}	0.14

Values are the molar concentration of herbicide needed to inhibit the photoreduction of DCPIP by 50% (I_{50}). The resistance ratio $I_{50}R/I_{50}S$ reflects the degree of resistance measured in the assay.

yield and biomass accumulation, all reported for biotypes possessing a mutation to the D1 protein that results in herbicide resistance. The mechanism of this fitness price has not been satisfactorily elucidated to date, but it has been shown to affect electron flow, chlorophyll a : b ratios and to increase damage in high-intensity light conditions (Holt and Thill, 1994).

12.2.1.4 *Target site-mediated resistance to Photosystem* I-*diverting herbicides*

There have been only limited reports of resistance to this class of herbicide and little is known regarding the resistance mechanisms responsible. No evidence of target-site resistance involvement has yet been found and it is thought that most cases of resistance are a result of enhanced herbicide metabolism or sequestration away from the site of action. In *Lolium perenne* and *Conyza bonariensis* increased levels and activities of enzymes that detoxify active oxygen species have been measured in resistant biotypes. These are superoxide dismutase, ascorbate reductase and glutathione reductase (Shaaltiel and Gressel, 1986). Other studies with *C. bonariensis* have suggested the immobilisation of paraquat in resistant biotypes, possibly by binding to cell wall components, so that less of the herbicide can reach the thylakoid. Alternatively, paraquat uptake and movement may be reduced in resistant biotypes of *Hordeum glaucum*.

12.2.1.5 *Target site-mediated resistance to cell division inhibitors*

Dinitroanilines have been used annually for approaching three decades in the cotton fields of the USA and on oilseeds and small grain cereals in the Canadian prairies. It was therefore not unexpected when resistance was reported in 1984 for *Eleusine indica* (L.) Gaertn. (goosegrass), in 1989 for *Setaria viridis* (green foxtail) and in 1992 for *Sorghum halepense* (Johnsongrass), as detailed by Smeda and Vaughn (1994). In *E. indica* the resistant biotype has shown cross-resistance to all dinitroanilines, dithiopyr and amiprophos-methyl, but sensitivity to propyzamide and chlorthal-dimethyl, and enhanced sensitivity to the carbamates (Vaughn *et al.*, 1987). Furthermore, there seems to be no differences in fitness between the resistant and susceptible biotypes.

In resistant biotypes of *S. viridis* and *E. indica* a specific base change in the sequence of the *TUA1* gene causes an amino acid substitution in α-tubulin from threonine to iso-leucine at position 239. This results in a conformational change to the surface of the α-tubulin molecule, which prevents herbicide binding.

In both species, resistance is controlled by nuclear, recessive genes that are mutant alleles of an α-tubulin gene that causes a substitution of either Thr239 to Ile, or Met268 to Thr in *E. indica* or Leu136 to Phe in *S. viridis*. Thr239 → Ile is reported to give rise to resistance to the dinitroanaline herbicides, whereas Met268 → Thr gives rise to low levels of trifluralin resistance (Vaughn *et al.*, 1990; Yamamoto *et al.*, 1998).

12.2.1.6   Target site-mediated resistance to glyphosate

For such an extensively used herbicide, there have been surprisingly few reports of glyphosate-resistant weed populations. The mode of action of glyphosate is to inhibit the enzyme EPSP synthase. A single mutation in the gene encoding this enzyme is reported to have resulted in glyphosate resistance in *Eleusine indica* in Malaysia (Lee and Ngim, 2000). The same mutation (Pro106 → Thr) has also been reported in the EPSP synthase gene from herbicide-resistant *Lolium rigidum* in Australia (Wakelin and Preston, 2006). A further case of glyphosate resistance appears to be due to an increase in the transcription of EPSP synthase, resulting in twice the amount of enzyme (Gruys *et al.*, 1999). There have been many reports of glyphosate-resistant bacterial EPSP synthase and these have proved useful in the development of glyphosate-tolerant crops (see Chapter 13).

12.2.1.7   Target site-mediated resistance to auxin-type herbicides

Although auxin-type herbicides have been extensively used for over 60 years there are few reported examples of weeds that are resistant to them. This is not too surprising – as these herbicides are generally foliar-applied and non-persistent, it is difficult to envisage the generation of the major selection pressures needed to favour the evolution of resistant weeds. However, in 1985 Lutman and Lovegrove demonstrated an approxi-mate ten-fold difference in sensitivity to mecoprop in two populations of chickweed (*Stellaria media*), the more resistant population being found in grassland where mecoprop was seldom used. In subsequent studies Lutman and colleagues suggested that resistance was not due to differences in uptake or movement of mecoprop, but that more of the herbicide was apparently immobile in resistant plants, presumably bound to structural polymers (Coupland *et al.*, 1990). An alternative hypothesis has been proposed by Barnwell and Cobb (1989) and Cobb *et al.* (1990). Their studies confirmed the original observations of resistance by Lutman and Lovegrove (1985) and reported reduced vigour in resistant plants (Figure 12.2). Furthermore, an investigation of H^+-efflux induced by mecoprop in etiolated *S. media* stems revealed that 170,000 times more mecoprop was needed in resistant plants to induce an amount of efflux equivalent to that obtained from susceptible plants. Such differences in mecoprop binding could imply an alteration in the auxin receptor in resistant plants that may also contribute in mecoprop resistance in this important weed species.

Further examples of increased tolerance of weeds to auxin-type herbicides have been observed in New Zealand and reported by Popay and colleagues (1990). These include

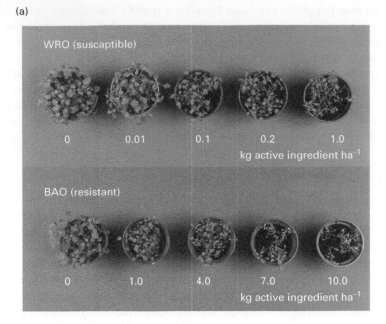

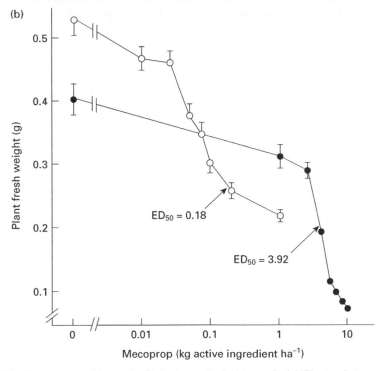

Figure 12.2 Mecoprop resistance in *Stellaria media* (chickweed). (a) Plants photographed from above. (b) ED_{50} graph. ○, WRO (susceptible); ●, BAO (resistant). (After Barnwell and Cobb, 1989.)

populations of nodding thistle (*Caruns nutans*) requiring 5 to 30 times more MCPA and 2,4-D than expected, and populations of giant buttercup (*Ranunculus aris*) being tolerant of 4.8 times the dose of MCPA compared with susceptible populations. Since uptake, translocation and enhanced metabolism have been eliminated as resistance mechanisms in some cases of resistance to auxin analogues, then it seems a distinct possibility that mutations in the auxin-binding protein might be involved in the resistance (Peniuk *et al.*, 1992). However, mutation in the auxin-binding protein has only been confirmed in one case, *Sinapis arvensis*. In this case the auxin receptor did not bind auxin-type herbicides. However, it also did not bind endogenous auxins either and this represents a severe fitness price for possessing herbicide resistance. Plants containing the insensitive auxin receptor lacked apical dominance and, in the absence of auxin-type herbicides, were very uncompetitive due to their heavily branched growth habit and their lack of vertical growth (Deshpande and Hall, 2000; Webb and Hall, 1999).

12.2.1.8   *Target site-mediated resistance to cellulose biosynthesis inhibitors*

There have been no reports of resistance in the field to the cellulose biosynthesis inhibitors (CBIs). Equally, the generation of resistant mutants of *Arabidopsis* in the laboratory has been difficult (Vaughn, 2002). Selection from extensive mutant screens eventually generated isoxaben- and dichlobenil-resistant plants, which were not cross-resistant to either flupoxam or isoxaben, indicating that the three herbicides did not share a common site. Furthermore, none of the mutants exhibited enhanced herbicide metabolism. Vaughan (2002) reports that at least one of the isoxaben-resistant mutants has an alteration in a cellulose synthase gene. Further confirmation of this finding at the gene sequence is awaited.

12.2.2   *Enhanced metabolism resistance*

Plants possess a host of enzymes for the metabolism of xenobiotics and unwanted substances. It is these detoxifying enzymes that modify or break down herbicides once they enter a plant cell. The rate at which these enzymes carry out this task will determine whether a plant lives or dies, and is the main contributor to herbicide selectivity between crops and weed species (see Chapter 4 for more information about the enzyme systems involved).

If an individual weed biotype within a population has the ability to metabolise a herbicide at an increased rate, then it may survive a herbicide treatment. Such biotypes are described as possessing enhanced metabolism resistance (Figure 12.3). This type of resistance has recently been reviewed by Yuan *et al.* (2006).

The main enzymes involved in metabolism of xenobiotics in plants are the cytochrome P450s and glutathione *S*-transferases (GSTs). These enzyme families have both been implicated in herbicide metabolism in tolerant crop and weed species and also in biotypes possessing enhanced metabolism resistance (Table 12.3). Enhanced GST activity has been correlated with resistance to isoproturon, clodinafop-propargyl and fenoxaprop-*p*-ethyl in *A. myosuroides* populations and offers a possible screening test for resistance in this and other species (Reade and Cobb, 2002). Increased GST and P450 activity may be due to a more active enzyme, increased transcription of gene(s) coding for the enzyme, or presence of more copies of the gene(s).

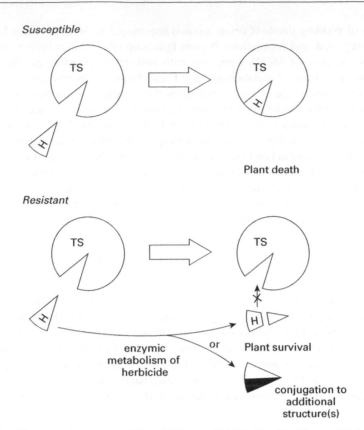

Figure 12.3 Diagrammatic representation of Enhanced Metabolism Resistance. TS, target site; H, herbicide.

Cytochrome P450 monooxygenases are a large and ubiquitous family of enzymes that carry out a number of reactions (see Chapter 4). Enhanced P450 activity has been reported in some weed biotypes that demonstrate herbicide resistance, implying that increased herbicide metabolism by P450s is the mechanism of herbicide resistance (Katagi and Mikami, 2000; Mougin et al., 2001). Inhibitors of P450 activity (e.g. 1-aminobenzotria-zole) have been shown to reduce herbicide metabolism and also levels of herbicide resistance in some resistant populations (e.g. Singh et al., 1998). It appears that in most cases the P450 is not induced in resistant biotypes by herbicide treatment but is consti-tutively expressed. This raises the interesting possibility that the protection mechanisms usually induced by herbicide application – often too late in the case of susceptible bio-types – are already being expressed in the resistant biotypes. Where P450s are responsible for resistance, then wide-ranging and unpredictable levels of cross-resistance can be encountered. This will be determined by the means by which a herbicide is metabolised rather than by its mode of action.

Glutathione S-transferases (GSTs, E.C. 2.5.1.18) are a superfamily of enzymes which catalyse the conjugation of a wide range of substrates, including some herbicides and

Table 12.3   Weed species for which GSTs and/or P450s have been implicated in herbicide resistance (from Devine and Preston, 2000).

Weed species	Herbicide(s)[a]	Proposed enzymatic system
Alopecurus myosuroides	Chlorotoluron Pendimethalin Diclofop-methyl Fenoxaprop-P-ethyl Propaquizafop Chlorsulfuron	Cytochrome P450 monooxygenases/ glutathione S-transferase
Abutilon theophrasti	Atrazine	Glutathione S-transferase
Avena sterilis	Diclofop-methyl	Cytochrome P450 monooxygenases
Avena fatua	Triallate	Cytochrome P450?
Digitaria sanguinalis	Fluazifop-P-butyl	Unknown
Echinochloa colona	Propanil	Aryl acylamidase
Echinochloa crus-galli	Propanil	Aryl acylamidase
Hordeum leporinum	Fluazifop-P-butyl	Unknown
Lolium rigidum	Simazine Diclofop-methyl Fluazifop-P-butyl Tralkoxydim Chlorsulfuron Metribuzin Chlorotoluron	Cytochrome P450 monooxygenases/ unknown
Phalaris minor	Isoproturon	Cytochrome P450 monooxygenases
Stellaria media	Mecoprop	Cytochrome P450 monooxygenases

[a] 'P' in this column refers to herbicidally active isomer.

xenobiotics, to the tripeptide glutathione (GSH, γ-Glu-Cys-Gly; see Chapter 4). Researchers have suggested a role for GSTs in the metabolism of atrazine (Anderson and Gronwald, 1991; Gray *et al.*, 1996), alachlor, metolachlor and fluorodifen (Hatton *et al.*, 1996). GST activity in a herbicide-resistant black-grass biotype has been reported to be approximately double that of herbicide-susceptible biotypes (Reade *et al.*, 1997). GSTs may also play a role in protecting plants from damage from active oxygen species that result from the action of certain herbicides. In black-grass a GST possessing glutathione peroxidase activity has been identified which may undertake this role in the metabolism of organic hydroperoxides (Cummins *et al.*, 1999). This raises the possibility of GSTs not only aiding in the metabolism of certain herbicides, but also in protecting the plant from the damage resulting from herbicide treatment. The diverse roles of GSTs in toxin metabolism, stress responses and secondary metabolism have been reviewed by Marrs (1996).

Resistance to propanil in a biotype of *Echinocloa colona* is due to raised activities of aryl acylamidase. Inhibition of this enzyme in resistant biotypes increases their susceptibility and, as some of these inhibitors are also herbicides, have proved useful in control of resistant *E. colona* populations.

12.2.3   *Enhanced compartmentalisation*

Little evidence exists to indicate enhanced compartmentalisation (Figure 12.4) as a resistance mechanism and what is known suggests it may act in conjunction with enhanced metabolism mechanisms. An example of this is the identification of an ATP-binding cassette (ABC) transporter which moves xenobiotic substances from the cytoplasm into the cell vacuole once they have been conjugated to glutathione. One such ATP-dependent glutathione *S*-conjugate pump has been identified in barley (*Hordeum sativum*) (Martinola *et al.*, 1993). Conjugation of the herbicide metolachlor to glutathione in this system results it its movement across the tonoplast into the vacuole. Similar ABC transporters have been reported to be responsible for the movement of chloroacetanilide and triazine herbicides into vacuoles (Ishikawa *et al.*, 1997). Where conjugation to glutathione forms part of the metabolism of a herbicide, then enhanced activity or the presence of such an ABC transporter would remove herbicides from their site of action quicker, perhaps resulting in plant survival. Alternatively, an increase in the number of ABC transporters might also result in resistance. Similar transport systems for substances that have been conjugated to sugars have also been postulated as a means by which herbicides are removed from the cytosol. Yuan *et al.* (2006) have recently reviewed the role of enhanced compartmentalisation as a mechanism of herbicide resistance.

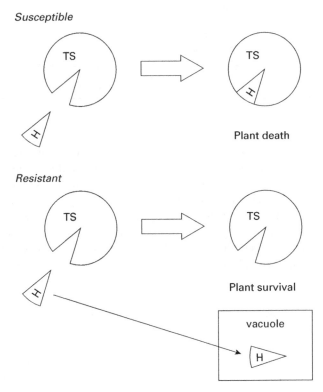

Figure 12.4   Diagrammatic representation of Enhanced Compartmentalisation Resistance. TS, target site; H, herbicide.

12.2.4   *Cross-resistance*

Cross-resistance refers to a plant's resistance to a number of herbicides due to the presence of only one resistance mechanism. Where a resistant plant has an altered target site, it is likely that resistance to many, if not all, herbicides that act at that target site will be found. In the case of target site resistance to PS II-inhibiting herbicides, three distinct herbicide groups are identified for which cross-resistance occurs. Cross-resistance also occurs where resistance is due to enhanced metabolism. It is less easy to predict which herbicides will be included in such cross-resistance situations, as a detailed knowledge of herbicide metabolism is generally lacking.

12.2.5   *Multiple resistance*

Multiple resistance is a plant's resistance to two or more herbicides due to the presence of more than one resistance mechanism. An individual biotype with target site resistance to ACCase inhibitors and ALS inhibitors would be deemed as possessing multiple resistance, as would a biotype possessing both target site and enhanced metabolism resistance. In theory, a biotype could contain many distinct resistance mechanisms, and quite clearly in such cases, predicting which herbicides would still work in controlling such a biotype becomes very difficult. An example of multiple resistance, in *Lolium rigidum*, is presented by Tan *et al.* (2007).

12.3   How resistance occurs

Herbicide resistance occurs through imposing selection pressure on a weed population. The selection pressure is the application of herbicides, which select for survival of members of the population that contain a mechanism imparting resistance to them. These individuals will survive and reproduce, so year upon year – as long as the same herbicides are used – the percentage of such individuals in a population will increase (Figure 12.5). In practice, population size plays a very important part in the occurrence of herbicide resistance. If selection pressure is exerted, but a population is kept small by other weed management methods, then the number of resistant individuals in the population may increase but the population size will remain small, resulting in less of a problem. Conversely, where population size remains large the number of surviving resistant individuals will be large, and so greater problems are encountered. Therefore, resistance problems in practice are a result of repeated imposition of selection pressure coupled with agricultural practices that result in large weed populations.

Factors affecting the speed at which resistance develops include:

- Initial frequency of resistant individuals in the population.
- Size of seed population in the seed bank and the length of time the seeds remain viable.
- The susceptibility of susceptible individuals.
- Repeated use of a single herbicide active ingredient or single mode of action, year upon year.
- Use of herbicides with long periods of residual activity in the soil.
- Use of herbicides with highly specific single modes of action.

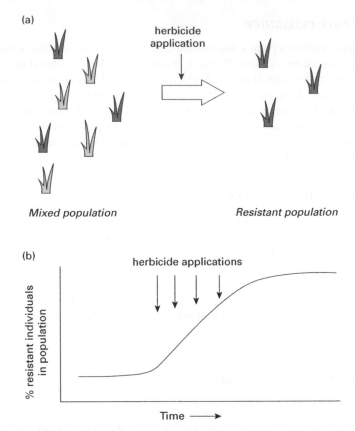

Figure 12.5 How resistance occurs. (a) Herbicide application imposing a selection pressure on a mixed population. Light grey = herbicide sensitive; dark grey = herbicide resistant. (b) Repeated use of the same herbicide will repeat the selection process. Susceptible individuals may enter the population from the soil seed bank, from seed from outside the field and from susceptible plants surviving due to poor herbicide efficacy.

12.4 Chronology of herbicide resistance

The first reported case of herbicide resistance is generally held to be that of resistance to triazines in *Senecio vulgaris* in the USA. Ryan (1970) reported that resistance to simazine and atrazine was detected in this species in 1968. Undoubtedly, herbicide resistance had occurred prior to this date and some earlier reports do highlight poor control that can, in hindsight, be interpreted as possible detection of resistant biotypes. Since the first confirmed case of herbicide resistance, numbers of cases have continued to rise year upon year (Figure 12.6). By summer 2010, 346 resistant biotypes had been confirmed from 194 species (114 dicots and 80 monocots) (Heap, 2010). Resistance has been found to many classes of herbicide and is most prevalent against ACCase inhibitors, triazines and ALS inhibitors. The current 'worst weeds' with respect to herbicide resistance are shown in Table 12.4. Although only introduced to the market in 1982, ALS inhibition is the most widespread

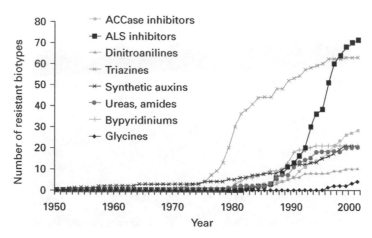

Figure 12.6 The development of herbicide resistance over the last 40 years (from Heap, 2009).

Table 12.4 The most important herbicide-resistant weed species worldwide, as stated by Heap, 2009. Ranking is based upon number of cited references.

1	Rigid ryegrass	*Lolium rigidum*
2	Wild oat	*Avena fatua*
3	Redroot pigweed	*Amaranthus retroflexus*
4	Fat hen	*Chenopodium album*
5	Green foxtail	*Setaria viridis*
6	Barnyardgrass	*Echinochloa crus-galli*
7	Goose grass	*Eleusine indica*
8	Kochia	*Kochia scoparia*
9	Horseweed	*Conyza canadensis*
10	Smooth pigweed	*Amaranthus hybridus*

mode of action for which resistance is reported. This is probably a combined result of the extensive use of these compounds, the number of different active ingredients of this type on the market and the specificity of the site of action. Additionally, the natural occurrence of a number of mutations of the target enzyme (ALS) that are less sensitive to herbicide, but are still enzymically active, has undoubtedly added to the problem.

Resistance is likely to continue to increase, especially where the number of available herbicides with different modes of action is reduced, where production costs reduce herbicide choice and where genetically modified, herbicide-tolerant (GM-HT) crops encourage the repeated use of a single herbicide in cropping systems.

12.5 Herbicide resistance case study – black-grass (*Alopecurus myosuroides* Huds)

Black-grass is an autumn-germinating annual grass that is a major weed of winter cereal crops in many countries (Figure 12.7). Its presence can result in yield reductions of up to

Figure 12.7 Black-grass in a field of wheat. On the left side the field is free of black-grass. On the right the black-grass population is approximately 500 plants m^{-2}. Yield will be seriously reduced in the part of this trial where black-grass has been allowed to remain (photograph courtesy of J.P.H. Reade).

44% (Moss 1987) and in some cases results in total crop loss. Black-grass is capable of producing a very high number of seeds that remain viable in the soil for up to seven years. In the early 1980s poor herbicidal control of black-grass was reported at a farm in Peldon, Essex, UK where the PS II-inhibiting herbicide chlorotoluron (CTU) was routinely used for weed control. Subsequent studies demonstrated that the Peldon biotype was resistant to CTU (Moss and Cussans, 1985). Little or no reduction in photosynthetic rate was noted when field-rate CTU (3,500 g a.i. ha^{-1}) was applied to the Peldon biotype *in vivo* (Sharples *et al.*, 1997; Figure 12.8).

Since the early 1980s many further cases of herbicide resistance to different herbicides have been reported in black-grass. By 1999, as many as 750 herbicide-resistant populations had been identified in the UK alone (Moss *et al.*, 1999). Herbicide resistance in black-grass has also been reported in Belgium, France, Germany, Israel, The Netherlands, Spain, Switzerland and Turkey (Heap, 2009). Many of these cases involve cross- or multiple resistance, and in the UK herbicide resistance due to enhanced metabolism is considered endemic.

A number of studies have attempted to identify the mechanisms responsible for herbicide resistance in black-grass. GST activity in the herbicide-resistant biotype Peldon is approximately double that in herbicide-susceptible biotypes (Reade *et al.*, 1997). Study of further biotypes demonstrating resistance to fenoxaprop-ethyl and clodinafop-propargyl showed correlation between the level of herbicide resistance and level of GST activity (Reade *et al.*, 1997). GST activities were found to be raised in a number of black-grass biotypes that had previously been characterised as possessing enhanced metabolism resistance, but not in biotypes that had been characterised as possessing target site resistance or that were herbicide susceptible (Cocker *et al.*, 1999; Reade and Cobb, 2002). Field studies have further demonstrated that subpopulations of black-grass that have survived herbicide treatment have higher GST activity and abundance than the untreated 'parent' populations (Reade *et al.*, 1999; Figure 12.9), which implies that the individuals with lower GST activity and abundance have been removed by the herbicide treatment. Studies

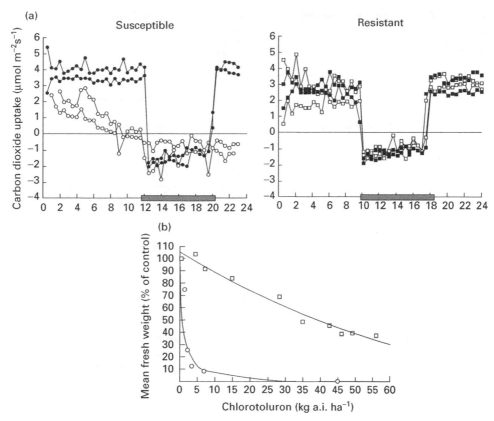

Figure 12.8 (a) CO_2 uptake of susceptible (○, •) and resistant (□, ■) black-grass over a 24-h time period. Plants were initially treated with distilled water (•, ■) or chlorotoluron added at the amount recommended in the field (3.5 kg a.i. ha^{-1}; ○, □). A dark period is indicated by the shaded area on the x-axis. (b) The effect of increasing amounts of chlorotoluron (kg a.i. ha^{-1}) on the fresh weight of susceptible (○) and resistant (□) black-grass, 2 weeks after herbicide treatment. The calculated doses to cause a 50% reduction in fresh weight were 0.93 and 39.3 kg ha^{-1}, respectively, giving a resistance factor of 42. Reproduced from Sharples *et al.* (1997, with permission from the Annals of Botany and Oxford University Press).

suggest that the increased GST activity in resistant biotypes is not due solely to increased abundance of a GST present in both susceptible and resistant plants, but is also due, in part, to the presence of a GST not detected in susceptible biotypes. Cummins *et al.* (1997) identified GSTs with molecular weights of 27 kDa and 28 kDa in addition to a 25-kDa GST in resistant biotypes: the susceptible biotype only had the 25-kDa GST. Reade and Cobb (1999) purified a 30-kDa GST from resistant black-grass that was absent from susceptible black-grass. This GST, the identity of which was confirmed by polypeptide sequencing (Reade and Cobb, 2002), showed different kinetic properties to the one found in both susceptible and resistant biotypes.

CTU did not induce GST activity in either biotype (Sharples *et al.*, 1995) but treatment with the herbicide safeners benoxcor, flurazole and oxabetrinil did increase GST activity

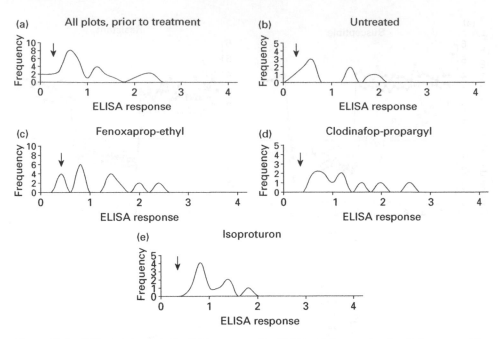

Figure 12.9 GST abundance in the field, measured by ELISA using monoclonal anti-*A. myosuroides* GST antiserum. (a) Prior to herbicide treatment; (b–e) 25 days after treatment. Arrow indicates plants giving a low ELISA response (b) that are absent from treated plots (c,d,e). (Reproduced from Reade and Cobb, 2002.)

in susceptible but not in resistant biotypes (Sharples, 1996). This is interesting, as studies in wheat *(Triticum aestivum* L.) have indicated that some herbicide safeners do increase glutathione conjugation of herbicides (Tal *et al.,* 1993, 1995; Edwards and Cole, 1996). GST activity in black-grass increases from winter to spring, which corresponds to a drop in the efficacy of the herbicides fenoxaprop-ethyl and clodinafop-propargyl with respect to black-grass control (Milner *et al.,* 1999, 2001).

Cytochrome P450s have also been implicated in herbicide resistance in black-grass. Kemp and Caseley (1987) and Kemp *et al.* (1988) used the P450 inhibitor 1-aminobenzotriazole to demonstrate that P450s played a role in metabolism of both chlortoluron and isoproturon in a resistant black-grass biotype. Hall *et al.* (1995) also showed that inhibitors of P450 activity reduced tolerance to chlortoluron and reduced herbicide metabolism in a resistant black-grass biotype, although this was not the case for fenoxaprop (Hall *et al.* 1997). Cell culture studies by Menendez and De Prado (1997) also suggest a role for P450s in chlortoluron metabolism. It is tempting, therefore, to interpret these findings as demonstrating a role for P450s in chlortoluron metabolism, but not in fenoxaprop metabolism, in black-grass. GST activity in resistant black-grass biotypes has been identified against fenoxaprop, fluorodifen and metolachlor (Cummins *et al.,* 1997).

Interestingly, Brazier *et al.* (2002) have also identified *O*-glucosyltransferase in a resistant black-grass biotype, raising the possibility of a third mechanism of enhanced metabolism resistance in this species.

12.6   The future development of herbicide resistance

The appearance of herbicide resistance over the last 40 years has undoubtedly encouraged weed control to be viewed as an integrated process in which chemical means of control are used alongside non-chemical means. Using such approaches should reduce the selection pressure for the survival of herbicide-resistant biotypes by using, among other methods, rotations, stale seedbeds and delayed drilling to control weeds. Unfortunately, as crop production costs rise and as the number of active ingredients available continues to fall (or, at least, the number of herbicide target sites remain limited), then resistance is likely to increase for the foreseeable future. The use of GM-HT crops will also increase selection pressure for the survival of resistant biotypes, as well as posing the risk of genes conferring resistance being transferred from crop to weed. It is alarming to consider that even if herbicides with new sites of action are developed and released to market, then the weeds that will be resistant to them are already present in existing weed populations. They simply require the appropriate selection pressure to be applied, which will ensure their survival advantage. Thus, weed populations are in a constant state of flux. Those that succeed are best adapted to the biotic and abiotic environments presented to them.

References

Anderson, M. and Gronwald, J. (1991) Atrazine resistance in a velvetleaf (*Abutilon theophrasti*) biotype due to enhanced glutathione *S*-transferase activity. *Plant Physiology* **96**, 104–109.

Barnwell, P. and Cobb, A.H. (1989) Physiological studies of mecoprop-resistance in chickweed (*Stellaria media* L.). *Weed Research* **29**, 135–140.

Bernasconi, P., Woodworth, A.R., Rosen, B.A., Subramanian, M.V. and Siehl, D.L. (1995) A naturally occurring point mutation confers broad range tolerance to herbicides that target acetolactate synthase. *Journal of Biological Chemistry* **270**, 17381–17385.

Brazier, M., Cole, D.J. and Edwards, R. (2002) *O*-glucosyltransferase activities towards phenolic natural products and xenobiotics in wheat and herbicide-resistant and herbicide-susceptible blackgrass (*Alopecurus myosuroides*). *Phytochemistry* **59**, 149–156.

Christoffers, M.J., Berg, M.L. and Messersmith, C.G. (2002) An isoleucine to leucine mutation in acetyl-CoA carboxylase confers herbicide resistance in wild oat. *Genome* **45**, 1049–1056.

Cobb, A.H., Early, C. and Barnwell, P. (1990) Is mecoprop-resistance in chickweed due to altered auxin-sensitivity? In: Caseley, J.C., Cussans, G. and Atkin, R. (eds) *Herbicide Resistance in Weeds and Crops*. Guildford, UK: Butterworth, pp. 435–436.

Cocker, K.M., Moss, S.R. and Coleman, J.O.D. (1999) Multiple mechanisms of resistance to fenoxaprop-P-ethyl in United Kingdom and other European populations of herbicide-resistant *Alopecurus myosuroides* (black-grass). *Pesticide Biochemistry and Physiology* **65**, 169–180.

Coupland, D., Lutman, P.J.W. and Health, C. (1990) Uptake, translocation and metabolism of mecoprop in a sensitive and a resistant biotype of *Stellaria media*. *Pesticide Biochemistry and Physiology* **36**, 61–67.

Cummins, I., Moss, S., Cole, D.J. and Edwards, R. (1997) Glutathione transferases in herbicide-resistant and herbicide-susceptible black-grass (*Alopecurus myosuroides*). *Pesticide Science* **51**, 244–250.

Cummins, I., Cole, D.J. and Edwards, R. (1999) A role for glutathione transferases functioning as glutathione peroxidases in resistance to multiple herbicides in black-grass. *The Plant Journal* **18**, 285–292.

Délye, C. (2005) Weed resistance to acetyl coenzyme A carboxylase inhibitors: an update [Review]. *Weed Science* **53**, 728–746.

Délye, C., Matejicek, A. and Gasquez J. (2002a) PCR-based detection of resistance to acetyl-CoA carboxylase inhibiting herbicides in black-grass (*Alopecurus myosuroides*) and ryegrass (*Lolium rigidum* Gaud.). *Pest Management Science* **58**, 474–478.

Délye, C, Wang, T. and Darmency, H. (2002b) An isoleucine–leucine substitution in chloroplastic acetyl-Co A carboxylase from green foxtail (*Setaria viridis* (L.) Beauv.) is responsible for resistance to the cyclohexanedione herbicide setoxydim. *Planta* **214**, 421–427.

Délye, C., Zhang, X.Q., Chalopin, C., Michel, S. and Powles, S.B. (2003) An isoleucine residue within the carboxyl-transferase domain of multidomain acetyl-coenzyme A carboxylase is a major determinant of sensitivity to aryloxyphenoxypropionate but not to cyclohexanedione inhibitors. *Plant Physiology* **132**, 1716–1723.

Deshpande, S. and Hall, J.C. (2000) Auxinic herbicide resistance may be modulated at the auxin-binding site in wild mustard (*Sinapis arvensis* L.): a light scattering study. *Pesticide Biochemistry and Physiology* **66**, 41–48.

Devine, M.D. and Shukla, A. (2000) Altered target sites as a mechanism of herbicide resistance. *Crop Protection* **19**, 881–889.

Devine, M.D. (1997) Mechanisms of resistance to acetyl-coenzyme A carboxylase inhibitors: a review. *Pesticide Science* **51**, 259–264.

Devine, M.D. and Preston, C. (2000). The molecular basis of herbicide resistance. In: Cobb, A.H. and Kirkwood, R.C. (eds) *Herbicides and their Mechanisms of Action.* Sheffield, UK and Boca Raton, FL: Sheffield Academic Press/CRC Press, pp. 72–84.

Edwards, R. and Cole, D.J. (1996) Glutathione transferases in wheat (*Triticum*) species with activity towards fenoxaprop-ethyl and other herbicides. *Pesticide Biochemistry and Physiology* **54**, 96–104.

Gray, J.A., Stoltenberg, D.E. and Balke, N.E. (1996) Increased glutathione conjugation of atrazine confers resistance in a Wisconsin velvetleaf (*Abutilon theophrasti*) biotype. *Pesticide Biochemistry and Physiology* **55**, 157–171.

Gressel, J. (2002) *Molecular Biology of Weed Control.* London and New York: Taylor & Francis Group.

Gruys, K.J., Biest-Taylor, N.A., Feng, P.C.C., *et al.* (1999) Resistance to glyphosate in annual ryegrass (*Lolium rigidum*) II. Biochemical and molecular analysis. *Weed Science Society of America Abstracts* **39**, 163.

Hall, L.M., Moss, S.R. and Powles, S.B. (1995) Mechanism of resistance to chlortoluron in two biotypes of the grass weed *Alopecurus myosuroides. Pesticide Biochemistry and Physiology* **53**, 180–192.

Hall, L.M., Moss, S.R. and Powles, S.B. (1997) Mechanisms of resistance to aryloxyphenoxypropi-onate herbicides in two resistant biotypes of *Alopecurus myosuroides* (blackgrass): herbicide metabolism as a cross-resistance mechanism. *Pesticide Biochemistry and Physiology* **57**, 87–98.

Hatton, P.J., Dixon, D., Cole, D.J. and Edwards, R. (1996) Glutathione transferase activity and herbicide selectivity in maize and associated weed species. *Pesticide Science* **46**, 267–275.

Heap, I. (2009) *The International Survey of Herbicide Resistant Weeds.* Available online at www.weedscience.com accessed 28 July 2009.

Heap, I. (2010) *The International Survey of Herbicide Resistant Weeds.* Available online at www.weedscience.com accessed 1 July 2010.

Holt, J.S. and Thill, D.C. (1994) Growth and productivity of resistant plants. In: Powles, S.B. and Holtum, J.A.M. (eds) *Herbicide Resistance in Plants: Biology and Biochemistry.* Boca Raton, FL: Lewis Publishers, pp. 299–316.

Ishikawa, T., Li, Z-S., Lu, Y-P., and Rea, P.A. (1997) The GS-X pump in plant, yeast, and animal cells: structure, function, and gene expression. *Bioscience Reports* **17**, 189–207.

Katagi, T. and Mikami, N. (2000) Primary metabolism of agrochemicals in plants. In: Roberts, T.R. (ed.) *Metabolism of Agrochemicals in Plants*, Chichester, UK: John Wiley, pp. 43–106.

Kemp, M.S. and Caseley, J.C. (1987) Synergistic effects of 1-aminobenzotriazole on the phytotoxicity of chlorotoluron and isoproturon in a resistant population of black-grass (*Alopecurus myosuroides*). *Proceedings of the British Crop Protection Conference – Weeds*, vol. 1. Farham, Surrey: BCPC, pp. 895-899.

Kemp, M.S., Newton, L.V. and Caseley, J.C. (1988) Synergistic effects of some P450 oxidase inhibitors on the phytotoxicity of chlorotoluron in a resistant population of black-grass (*Alopecurus myosuroides*). *Proceedings of the EWRS Symposium*. Wageningen, Netherlands: European Weed Research Society, pp. 121–126.

Lee, L.J. and Ngim, J. (2000) A first report of glyphosate resistant goosegrass (*Eleysine indica*) in Malaysia. *Pest Management Science* **56**, 336–339.

Lutman, P.J. and Lovegrove A.W. (1985) Variations to the tolerance of *Galium aparine* (cleavers) and *Stellaria media* (chickweed) to mecoprop. *British Crop Protection Conference, Weeds* **2**, 411–418.

Marrs, K. (1996) The functions and regulation of glutathione *S*-transferases in plants. *Annual Review of Plant Physiology and Plant Molecular Biology* **47**, 127–158.

Martinola, E., Grill, E., Tommasini, R., Kreuz, K. and Amrhein, N. (1993) ATP-dependent glutathione *S*-conjugate 'export' pump in the vacuolar membrane of plants. *Nature* **364**, 247–249.

Menendez, J. and De Prado, R. (1997) Detoxification of chlorotoluron in a chlorotoluron resistant biotype of *Alopecurus myosuroides*. Comparison between cell cultures and whole plants. *Physiologia Plantarum* **99**, 97–104.

Milner, L.J., Reade, J.P.H. and Cobb, A.H. (1999) An investigation of glutathione *S*-transferase activity in *Alopecurus myosuroides* Huds. (black-grass) in the field. *Proceedings of the 1999 Brighton Conference – Weeds*, vol. **1**. Farnham, Surrey: BCPC, pp. 173–178.

Milner, L.J.M., Reade, J.P.H and Cobb, A.H. (2001) Developmental changes in glutathione *S*-transferase activity in herbicide-resistant populations of *Alopecurus myosuroides* Huds. (black-grass) in the field. *Pest Management Science* **57**, 1100–1106.

Moss, S.R. (1987) Competition between black-grass (*Alopecurus myosuroides*) and winter wheat. *Proceedings of the British Crop Protection Conference – Weeds*, vol. 2. Farnham, Surrey: BCPC, pp. 365–374.

Moss, S.R. and Cussans, G.W. (1985) Variability in the susceptibility of *Alopecurus myosuroides* (black-grass) to chlorotoluron and isoproturon. *Aspects of Applied Biology* **9**, 91–98.

Moss, S.R., Clarke, J.H., Blair, A.H., *et al.* (1999) The occurrence of herbicide-resistant grass-weeds in the United Kingdom and a new system for designating resistance in screening assays. *Proceedings of the 1999 Brighton Conference – Weeds*, vol. 1. Farnham, Surrey: BCPC, pp. 179–184.

Mougin, C.P., Corio-Costet, M.F. and Werck-Reichhart D. (2001) *Plant and fungal cytochrome P-450s: their role in pesticide transformation*. In: Hall, J.C. *et al.* (eds) *Pesticide Biotransformations in Plants and Microorganisms: Similarities and Divergences*. Washington, DC: American Chemical Society, pp. 166-181.

Oettmeier, W. (1999) Herbicide resistance and supersensitivity in photosystem II. *Cellular and Molecular Life Sciences* **55**, 1255–1277.

Peniuk, M.G, Romano, M.L. and Hall, J.C. (1992) Absorption, translocation and metabolism are not the basis for differential selectivity of wild mustard (*Sinapis arvensis* L.) to auxinic herbicides. *Weed Science* **32**, 165.

Popay, A.I , Bourdot, G.W, Harrington, K.C and Rahman, A. 1990. Herbicide resistance in weeds in New Zealand. In: Caseley, J.C., Cussans, G. and Atkins, R. (eds) *Herbicide Resistance in Weeds and Crops*. Guildford, UK: Butterworth, pp. 470–471.

Reade, J.P.H. and Cobb, A.H. (1999) Purification, characterisation and comparison of glutathione *S*-transferases from black-grass (*Alopecurus myosuroides* Huds) biotypes. *Pesticide Science* **55**, 993–999.

Reade, J.P.H. and Cobb, A.H. (2002) New, quick tests for herbicide resistance in black-grass (*Alopecurus myosuroides* Huds) based on increased glutathione *S*-transferase activity and abundance. *Pest Management Science* **58**, 26–32.

Reade, J.P.H., Hull, M.R. and Cobb, A.H. (1997) A role for glutathione-*S*-transferase in herbicide resistance in black-grass (*Alopecurus myosuroides*). *Proceedings of the 1997 Brighton Crop Protection Conference – Weeds*, vol. 2. Farham, Surrey: BCPC, pp. 777–782.

Reade, J.P.H., Belfield, J.L. and Cobb, A.H. (1999) Rapid tests for herbicide resistance in black-grass based on elevated glutathione *S*-transferase activity and abundance. *Proceedings of the 1999 Brighton Crop Protection Conference – Weeds*, vol. 2. Farnham, Surrey: BCPC, pp. 185–190.

Roux, F., Matejicek, A. and Reboud, X. (2005) Response of *Arabidopsis thaliana* to 22 ALS inhibitors: baseline toxicity and cross-resistance of *csr*1-1 and *csr*1-2 resistant mutants. *Weed Research* **45**, 220–227.

Ryan, G.F. (1970) Resistance of common groundsel to simazine and atrazine. *Weed Science* **18**, 614–616.

Sathasivan, K., Haughn, G.W. and Murai, N. (1990) Nucleotide sequence of a mutant acetolactate synthase gene from an imidazolinone-resistant *Arabidopsis thaliana* var. Columbia. *Nucleic Acids Research* **18**, 2188.

Sathasivan, K., Haughn, G.W. and Murai, N. (1991) Molecular basis of imidazolinone herbicide resistance in *Arabidopsis thaliana* var. Columbia. *Plant Physiology* **97**, 1044–1050.

Shaaltiel, Y. and Gressel, J. (1986) Multienzyme oxygen radical detoxifying system correlated with paraquat resistance in *Conyza bonariensis*. *Pesticide Biochemistry and Physiology*, **26**, 22–28.

Sharples, C.R. (1996) An investigation of herbicide resistance in black-grass using safeners and synergists. PhD Thesis, The Nottingham Trent University, UK.

Sharples, C.R., Hull, M.R. and Cobb, A.H. (1995) The effect of herbicide safeners on chlorotoluron susceptible and resistant black-grass (*Alopecurus myosuroides*). *Proceedings of the 1995 Brighton Crop Protection Conference – Weeds*, vol. 1. Farnham, Surrey: BCPC, pp. 359–360.

Sharples, C.R, Hull, M.R. and Cobb, A.H. (1997) Growth and photosynthetic characteristics of two biotypes of the weed black-grass (*Alopecurus myosuroides* Huds) resistant and susceptible to the herbicide chlorotoluron. *Annals of Botany* **79**, 455–461.

Singh, S.S., Kirkwood, R.C. and Marshall, G. (1998) Effect of ABT on the activity and rate of degradation of isoproturon in susceptible and resistant biotypes of *Phalaris minor* and in wheat. *Pesticide Science* **53**, 123–132.

Smeda, R. J. and Vaughn, K.C. (1994) Resistance to dinitroaniline herbicides. In: Powles, S.B. and Holtum, J. (eds) *Herbicide Resistance in Plants: Biochemistry and Physiology*. Boca Raton, FL: CRC Press, pp. 215–228.

Tal, A., Romano, M.L., Stephenson, G.R., Schwann, A.L. and Hall, J.C. (1993) Glutathione conjugation: a detoxification pathway for fenoxaprop-ethyl in barley, crabgrass, oat and wheat. *Pesticide Biochemistry and Physiology* **46**, 190–199.

Tal, J.A., Hall J.C., Stephenson G.R. (1995) Nonenzymatic conjugation of fenoxaprop-ethyl with glutathione and cysteine in several grass species. *Weed Research* **35**, 133–139.

Tal, A. and Rubin, B. (2004) Molecular characteristics and inheritance of resistance to ACCase-inhibiting herbicides in *Lolium rigidum*. *Pest Management Science* **60**, 1013–1018.

Tan, M-K., Preston, C. and Wang, G-X. (2007) Molecular basis of multiple resistance to ACCase-inhibiting and ALS-inhibiting herbicides in *Lolium rigidum*. *Weed Research* **47**, 534–541.

Vaughn, K.C, Marks, M.D. and Weeks, D.P. (1987) A dinitroaniline-resistant mutant of *Eleusine indica* exhibits cross-resistance and supersensitivity to antimicrotubule herbicides and drugs. *Plant Physiology* **83**, 956–964.

Vaughn, K., Vaughan, M. and Gossett, B.(1990) A biotype of goosegrass (*Eleusine indica*) with an intermediate level of dinitroaniline herbicide resistance. *Weed Technology* **4**, 157–162.

Vaughn, K.C. (2002) Cellulose biosynthesis inhibitor herbicides. In: Böger, P., Wakabayashi, K. and Harai, K. *Herbicide Classes in Development*. Berlin and Heidelberg: Springer-Verlag, pp. 139–150.

Wakelin, A.M. and Preston, C. (2006) A target-site mutation is present in a glyphosate-resistant *Lolium rigidum* population. *Weed Research* **46**, 432–440.

Webb, S.R. and Hall, J.C. (1999) Indole-3-acetic acid binding characteristics in susceptible and resistant biotypes of wild mustard (*Sinapis arvensis* L.). *Weed Science Society of America Abstracts* **33**, 196.

Yamamoto, W., Zeng, L.H. and Baird, W.V. (1998) Alpha-tubulin mis-sense mutations correlate with anti-microtubulin drug resistance in *Eleusine indica*. *Plant Cell* **10**, 297–308.

Yuan, S.J., Tranel, P.J. and Stewart, C.N. (2006) Non-target-site herbicide resistance: a family business. *Trends in Plant Science* **12**, 6–13.

Zagnitko, O., Jelenska, J., Tevzadze, G., Haselkorn, R. and Garnicki, P. (2001) An isoleucine/leucine residue in the carboxyl-transferase domain of acetyl-CoA carboxylase is critical for interaction with aryloxyphenoxypropionate and cyclohexanedione inhibitors. *Proceedings of the National Academy of Sciences, USA* **98**, 6617–6622.

Chapter 13
Herbicide-Tolerant Crops

13.1   Introduction

The development and use of genetically modified, herbicide-tolerant (GM-HT) crops represents the single greatest change in weed control technology since the commercial arrival of selective herbicides in the mid-1940s. Herbicides that up until now have only been used when a crop was not present (or, at least, when a crop had yet to emerge) may now be used to perform selective weed control in emerged crops. In addition, herbicides that were not permitted in certain crops, for reasons of crop damage, may now be used safely in those crops. In practical terms, this offers growers simpler and more effective chemical weed control strategies that incorporate lower numbers of sprays and therefore less energy inputs. GM-HT crops, by allowing more efficient weed control may, however, be more likely to reduce biodiversity in crops and, through gene escape and the repeated use of a single type of herbicide, might give rise to the spread of herbicide resistance in weeds. GM-HT crops continue to be grown commercially in an increasing number of countries and at increasing hectarage. The debate regarding their use, and the use of genetic manipulation technologies in general, continues unabated, especially in Europe. In addition to GM-HT crops, a number of crops have also been produced by non-GM techniques that exhibit herbicide tolerances which are of use in the field. These crops have many of the same benefits and drawbacks as GM-HT crops, but are arguably more acceptable to the general public as they do not contain genes sourced from other organisms. In this chapter the term Genetically Modified (GM) is used to denote the transfer of genes from one organism to another or the alteration of gene expression. Although the use of mutagens can cause the modification of genes and resultant protein sequences, they have been used in crop improvement for many years and this is not included in definition of Genetic Modification within the main body of this chapter and is covered separately, in section 13.12.

13.2   History of genetically modified, herbicide-tolerant crops

Herbicide tolerance is the trait that has received the most commercialisation of all genetically modified (GM) crops to date. One of the first herbicide-tolerant crops grown

Herbicides and Plant Physiology, Second Edition By Andrew H. Cobb and John P.H. Reade
© 2010 A.H. Cobb and J.P.H. Reade

commercially was soybean (USA, 1996) followed by cotton (USA, 1997), maize (USA, 1998) and oilseed rape (USA, 1999) (Duke, 2005). The soybean was modified to be resistant to the broad-spectrum herbicide glyphosate (Roundup ™) and is known as Roundup-Ready soybean (Padgette *et al.*, 1995). The gene conferring resistance to glyphosate, encoding *Agrobacterium tumefaciens* EPSP synthase, has been introduced into a number of commercially important crops including oilseed rape, maize and cotton. It has also been used to transform a number of other crops, including sugar beet and wheat, although these have yet to be grown commercially (CaJacob *et al.*, 2004). GM-mediated tolerance to the broad-spectrum herbicide glufosinate has also been commercialised in a number of similar crops: maize (1997), oilseed rape (1999), soybean (1999), cotton (2003), using a gene from the bacterium *Streptomyces hygroscopicus* (Lyndon and Duke, 1999). These crops are commercialised under the Liberty Link™ trade name. GM-mediated tolerance to the herbicide bromoxynil has also been commercialised in oilseed rape under the trade name BXN™ since 1995 (Pallett *et al.*, 1996). Table 13.1 summarises the GM-HT crops commercialised up until 2004. In addition, many other crops have been genetically modified in the laboratory to possess resistance to herbicides, but have yet to be commercialised. It is worth noting that by 2003, 54% of soybean crops grown worldwide were genetically modified and that by 2004, 85% of US-grown soybean was genetically modified (Economic Research Service, 2007). Table 13.2 presents the increase in growth of herbicide tolerant crops worldwide in recent years. Evans (2010) reported that 14 million farmers in 25 countries planted 134 million hectares of GM crops in 2009. GM crops have been more poorly received in Europe than in Asia, North and South America, mainly due to public and consumer pressure against the use of genetic engineering technology. To date, the UK has permitted no commercial work on growth of any genetically engineered crop. However, a three-year Farm-Scale Evaluation of GM-HT oilseed rape, maize and sugar beet has been performed, mainly to establish the effects of such crops on biodiversity, focusing on, among other things, soil seedbanks, non-crop plants, spiders, beetles and butterflies (see Table 13.6). While public perception in the UK of such crops remains negative, their commercialisation will be delayed or postponed.

13.3 How genetically modified crops are produced

The genetic information of an organism, encoded by the sequence of nucleotide bases within its DNA, determines the proteins it can manufacture and, through these, the physical and biological nature of the organism. Protein synthesis can be subdivided into two distinct processes: transcription and translation (Figure 13.1). Transcription is the copying of the information from a segment of DNA, often a gene encoding for a single protein, to a strand of mRNA. Translation, which takes place in association with ribosomes, involves the 'reading' of the nucleotide sequence in the mRNA molecule and the construction of the amino acid chain which forms the primary structure of the protein. Subsequent to this, the amino acid chain will fold, cross-link and may undergo post-translational processing to form the functional protein.

GM crops usually have one or more genes added to their genetic make-up and so produce one or more extra proteins. It is these proteins that cause the crop to have different properties to those of the unmodified crop. In some cases the modification may involve

Table 13.1 Major herbicide-tolerant genetically modified crops worldwide, as of 2004 (adapted from Christou and Klee, 2004).

Crop	Trait (gene)	Country
Soybean	Glyphosate tolerance *(CP4 EPSPS)*	Argentina, Australia, Brazil, Canada, China, Czech Republic, EU, Japan, Korea, Mexico, Russia, Switzerland, South Africa, Taiwan, UK, Uruguay, USA
Soybean	Glufosinate tolerance *(pat)*	Canada, Japan, USA
Soybean	Glufosinate tolerance *(bar)*	USA
Maize	Glyphosate tolerance *(Maize EPSPS)*	Argentina, Australia, Canada, China, EU, Japan, Korea, Philippines, Taiwan, USA
Maize	Glyphosate tolerance *(CP4 EPSPS)*	Argentina, Australia, Canada, EU, Japan, Philippines, Taiwan, South Africa, USA
Maize	Glyphosate tolerance *(CP4 EPSPS, gox)*	Canada
Maize	Glufosinate tolerance *(bar)*	Canada, Japan, Philippines, Taiwan, USA
Maize	Glufosinate tolerance *(pat)*	Argentina, Australia, Canada, European Union, Japan, Philippines, Taiwan, USA
Oilseed rape	Glyphosate tolerance *(CP4 EPSPS, gox)*	Australia, Canada, China, EU, Japan, Philippenes, USA
Oilseed rape	Glufosinate tolerance *(pat)*	Australia, Canada, European Union, Japan, USA
Oilseed rape	Bromoxynil tolerance *(bxn)*	Australia, Canada, Japan, USA
Cotton	Glyphosate tolerance *(CP4 EPSPS)*	Argentina, Australia, Canada, EU, Japan, Philippines, South Africa, USA
Cotton	Bromoxynil tolerance *(bxn)*	Australia, Canada, Japan, USA
Cotton	Sulfonylurea tolerance *(als)*	USA
Sugar beet	Glufosinate tolerance *(pat)*	Canada, Japan, USA
Sugar beet	Glyphosate tolerance *(CP4 EPSPS, gox)*	Australia, Philippines, USA
Sugar beet	Glyphosate tolerance *(CP4 EPSPS)*	Philippines, USA
Wheat	Glyphosate tolerance *(CP4 EPSPS)*	USA
Rice	Glufosinate tolerance *(bar)*	USA
Flax	Sulfonylurea tolerance *(als)*	Canada, USA

Table 13.2 Adoption of GM crops in the European Union (EU) versus worldwide herbicide-tolerant (HR) crops; italics indicate approval for commercial release of new GM crops (from Madsen and Sandøe, 2005).

Year	European Union	HR crops world area
1986	HR marker gene in tobacco field tested in France and USA	
1994	*HR tobacco approved to be used as conventional tobacco*	
1996	GM soybean imported into the EU *HR tobacco, oilseed rape, soybean and chicory approved for breeding/import*	0.6 M ha
1997	*HR maize, oilseed rape and carnations approved to be used as conventional crops*	6.9 M ha
1998	Stop for GM crops under debate in UK Stakeholders in Denmark agree on a voluntary pause for commercial use of GM crops in 1999 *HR oilseed rape, maize and carnation approved as conventional crops*	20.1 M ha
1999	25 June: moratorium (suspension of new approvals for marketing of GMOs)	31.0 M ha
2000		35.9 M ha
2001		40.6 M ha
2002	New directive on the deliberate release of GM crops into the environment in force (EU Directive 2001/18/EC)	44.2 M ha
2003	October, Regulation on the traceability and labelling of GMOs and the traceability of food and feed products produced from GMOs (Regulation [EC] No. 1829/2003 and No. 1830/2003) published	49.7 M ha
2004	April, Regulation on the traceability and labelling of GMOs and the traceability of food and feed products produced from GMOs fully into force	

GMO, genetically modified organism.

over-expressing one of the genes, in which case it will produce more of one of its proteins, or under-expressing one of its genes, in which case the plant will produce less or none of one of its proteins.

Since the early 1980s scientists have developed a number of ways by which genes from other organisms may be added to the genetic material of the plant. The use of the micro-organism *Agrobacterium tumefaciens* has proved useful in modifying a number of commercially important crops (Komari *et al.*, 2004). *A. tumefaciens* is a common soil bacterium that is a plant pathogen, causing galls to be produced on the infected plants. It accomplishes this in nature by transferring some of its genes to plant cells. These genes are transferred in a plasmid, a circular DNA molecule, and in the case of *A. tumefaciens* the plasmid causing gall formation is termed the Ti (tumour inducing) plasmid. Using this naturally occurring system of genetic modification, scientists have successfully transformed a number of crop species (Komari *et al.*, 2004). The gene of interest is first inserted

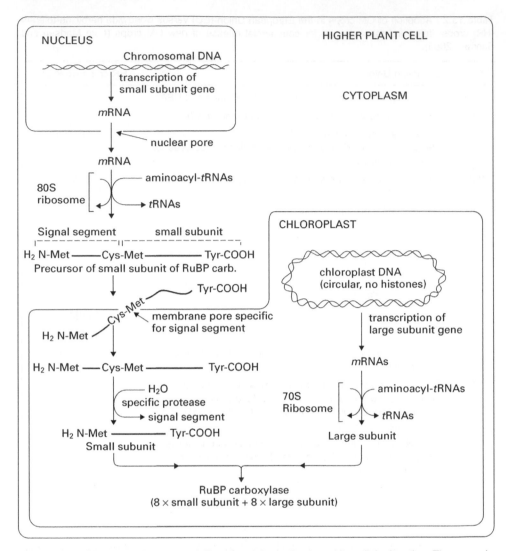

Figure 13.1 Diagrammatic representation of protein synthesis and its cellular location. The example given is for RuBisCo (reproduced from Goodwin and Mercer, 1983).

into the Ti plasmid of *A. tumefaciens* and the transformed bacteria are used to infect pieces of plant tissue. Successful infection will result in the Ti plasmid (and the added gene) being transferred to the plant genome. Due to totipotency, the plant tissue can be regenerated to a whole plant, in which each cell will contain the added gene. In order for this to be successful, the genes in the Ti plasmid that are responsible for tumor initiation are first removed. The introduced genes are integrated into the nuclear DNA of the plant in the same way as takes place when *A. tumefaciens* infects plants in the field.

Although this method has proved very successful for a number of dicotyledonous plant species, until recently it had not been so for monocotyledons, as *A. tumefaciens* would not infect them. However, techniques have now been developed that enable agrobacterium-mediated transformation of rice (Tyagi and Mohanty, 2000), maize (Ishida *et al.*, 1996), barley and wheat (Cheng *et al.*, 1997). It has also proved problematic with soybean, as regeneration from cell cultures is often not possible (McCabe *et al.*, 1988). An alternative method of introducing genes has been used for these species. This involves coating tungsten particles with DNA containing the gene of interest and using a microprojectile gun to shoot the particles and DNA into plant cells. This method, often referred to as the 'Shotgun' method, has allowed transformation of plant species where *A. tumefaciens* is ineffective as a vector. DNA that has entered the cell incorporates into the nuclear DNA and is then passed on as the cell divides.

In addition to these methods, DNA can also be introduced to protoplasts (plant cells that have had their cell wall removed) using electroporation or chemical methods to open up the plasma membrane and allow DNA to enter the cell. Alternatively, microinjection can be used. The protoplast will grow back its cell wall and can be regenerated to a plant whose cells all contain the added gene.

In addition to the gene conferring the required trait a number of other genes and pieces of genetic information are added. These include a promoter (a molecular switch to ensure that the required gene will be transcribed in the transformed plant) and a stop sequence that ensures transcription stops at the end of the required gene. In addition, a marker gene is also added as a way of screening cell cultures to determine if transformation has been successful. For many GM crops the marker gene has been one which confers antibiotic resistance to the successfully transformed plant tissue, allowing easy selection for transformed tissue by culturing it *in vitro* in the presence of antibiotics. A commonly used gene is *NPT II*, which produces the enzyme neomycin phosphotransferase giving resistance to the aminoglycoside antibiotics (Komari *et al.*, 2004). The gene encoding for the enzyme hygromycin phosphotransferase has also been used as a marker (Bilang *et al.*, 1991).

There has been some concern regarding the use of antibiotic resistance genes, especially where they remain in the GM plant when it is at the stage of farm-scale evaluation and commercial production. Alternative methods (for instance, incorporation of the *pmi* gene encoding phosphomannose isomerise, ensuring that on a mannose-only culture medium only transformed tissue survives) have received limited use and may prove more publicly and environmentally acceptable (Joersbo, 2001). The herbicide tolerance gene *bar* has also been used as a marker gene (Vasil, 1996) and, where herbicide tolerance is the trait being introduced, the presence of a marker gene would appear unnecessary, as a simple screen in the presence of the herbicide to which tolerance is introduced should successfully select for transformed tissue by killing untransformed cells. The process of producing a GM-HT crop is summarised in Figure 13.2.

13.4   Genetically engineered herbicide tolerance to glyphosate

Glyphosate is a non-selective herbicide that kills plants by inhibiting their ability to synthesise the aromatic amino acids phenylalanine, tyrosine and tryptophan (see Chapter 9 section 9.4). It accomplishes this by blocking the action of the enzyme EPSP synthase, the penultimate step in the shikimate pathway, vital in the biosynthesis of these three amino acids. Interestingly, glyphosate competes with one substrate (phosphoenol pyruvate, PEP)

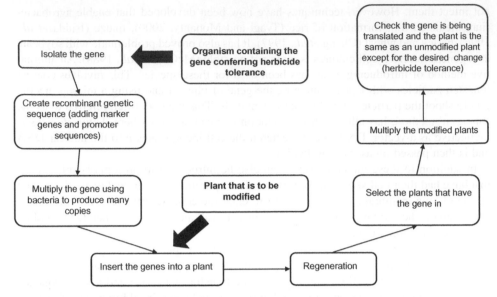

Figure 13.2   An overview of the steps taken in the production of a genetically modified plant. From Bruce and Bruce, 1988.

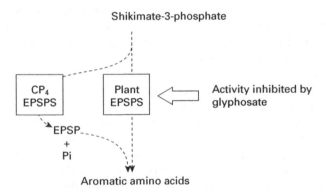

Figure 13.3   Biosynthesis of aromatic amino acids indicating how CP4 EPSPS allows synthesis in the presence of glyphosate. EPSP, 5-enoylpyruvyl shikimic acid 3-phosphate; EPSPS, EPSP synthase.

but forms a very stable herbicide–enzyme complex with the other, resulting in a 'full stop' to the metabolic pathway. This also subsequently reduces the ability of the plant to synthesise a number of vital metabolites including hormones, flavonoids and lignins. GM crops that are tolerant to glyphosate are produced by the insertion of a gene coding for a glyphosate-insensitive EPSP synthase that is obtained from a soil bacterium *Agrobacterium* CP4. In this way, plant EPSP synthase is still inhibited, but the bacterial EPSP synthase is unaffected, allowing the plant to still synthesise aromatic amino acids and the other essential metabolites resulting from the shikimate pathway (Figure 13.3) (for an example of this process, see Vande Berg *et al.*, 2008). The kinetic properties of plant and bacterial wild-type and mutated EPSP synthases are shown in Table 13.3.

Table 13.3 Kinetic properties for selected EPSP synthases (from Dill, 2005). K_i / K_m (PEP) is a measure of the selectivity of EPSPS for PEP over glyphosate. A higher value indicates a greater tolerance to glyphosate while the enzyme still possesses EPSPS activity.

Enzyme source	K_m (PEP) (μM)	K_i (glyphosate) (μM)	K_i / K_m
Petunia (wild type)	5.0	0.4	0.08
G101A	210	2000	9.5
T102I/P106S	10.6	58	5.5
P106S	17	1	0.06
Agrobacterium sp. CP4	12	2720	227

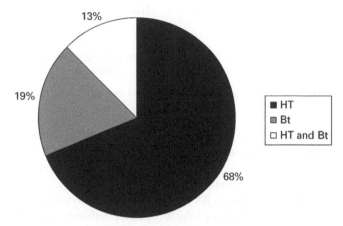

Figure 13.4 The major GM traits that are being commercially exploited (2006). Figures represent the % of total commercial GM plantings worldwide in 2006. HT = herbicide tolerance; Bt = *Bacillus thuringiensis*-mediated insect resistance.

Crops that have been modified by insertion of the EPSP synthase CP4 gene are commercialised under the trade name Roundup-Ready™ and represent the largest proportion of the GM-HT crop market worldwide. In 2006, 68% of GM crops grown were herbicide tolerant (Figure 13.4) and the vast majority of these were Roundup-Ready™. CP4 is a mutant form of EPSP synthase that has an amino acid replacement (Gly100 → Ala) which results in extremely high tolerance to glyphosate (Padgette *et al.*, 1996). The low mammalian toxicity, fast degradation in soil and translocation within plants make glyphosate a popular and effective herbicide. The use of glyphosate-tolerant crops has reduced the number of herbicide applications necessary for a number of crops in addition to reducing fuel costs and land compaction due to reduced machinery use. In addition, these GM crops increase the flexibility of timing of weed control, as glyphosate can be used on weeds at growth stages in excess of those that conventional herbicides can effectively control. Such practices have allowed for the increase in no-tillage and conservation tillage practices alongside GM-HT crops in the USA and Canada.

The effectiveness of glyphosate as a herbicide means that weed control through cultivation is not always necessary. The high level of weed control achieved, however, will

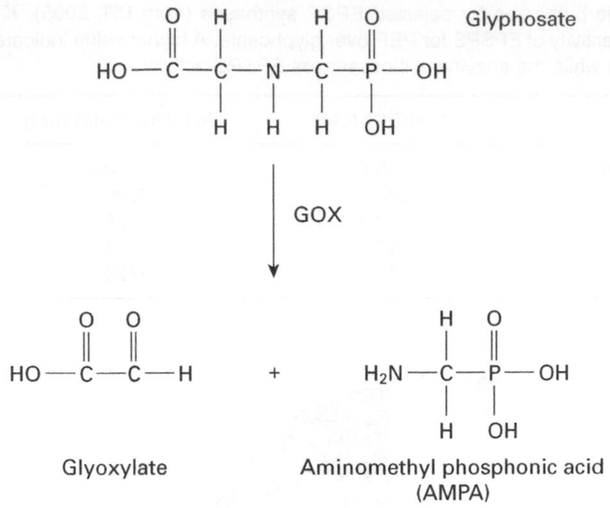

Figure 13.5 The detoxification of glyphosate by glyphosate oxidoreductase (GOX).

clearly impact on biodiversity, as indicated in some of the data collected in the UK during the three-year Farm-Scale Evaluations (see section 13.13). In addition, the overuse of glyphosate as the sole method of weed control may create an unacceptable selection pressure for the survival of naturally occurring populations of glyphosate-resistant weeds. This is a very real concern for the future. As such, GM glyphosate-tolerant crops should be considered alongside GM crops that are tolerant to other herbicides and non-GM crops in rotation in cropping systems. Studies such as those by Westra *et al.* (2008) will also prove important in assessing the effects of changing land management practices under glyphosate-tolerant cropping systems on weed population dynamics.

An alternative method for conferring tolerance to glyphosate involves the insertion of the *gox* gene from *Ochrobactrum anthropi* into crop plants (CaJacob *et al.*, 2004; Reddy *et al.*, 2004). Glyphosate oxidoreductase (GOX) is an enzyme that catalyses the breakdown of glyphosate to AMPA (aminomethylphosphonic acid) and glyoxylic acid (Figure 13.5).

In some oilseed rape lines, both the CP4 EPSP synthase and *gox* genes are inserted and expressed, giving the resultant plant two methods of avoiding damage by glyphosate. The reasons for this gene-stacking have not been made public and the build-up of AMPA (due to the presence of GOX) has been implicated in phytotoxic symptoms in some GM oilseed rape lines (Reddy *et al.*, 2004).

13.5   Genetically modified herbicide tolerance to glufosinate

Glufosinate is a non-selective herbicide that kills plants by inhibiting the enzyme glutamine synthase. This is a key enzyme in the biosynthesis of the amino acid glutamine and also plays a vital role in nitrogen metabolism. Inhibition of this enzyme by glufosinate results in a decrease in glutamine/glutamate pool size and an accumulation of phosphoglycolate,

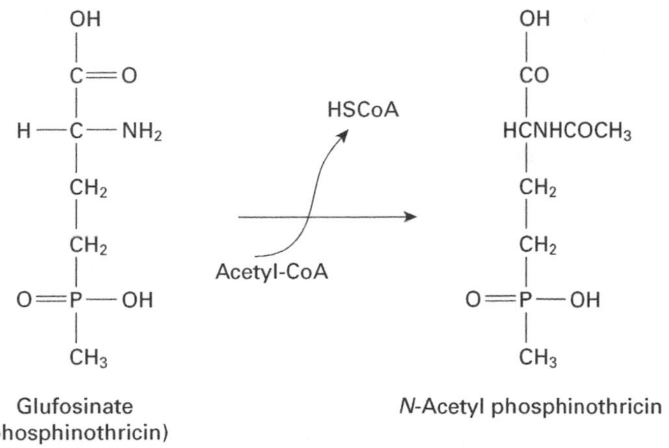

Figure 13.6 The detoxification of glufosinate by phosphinothricin acetyl transferase.

glycolate and glyoxylate. Glyoxylate inhibits the activity of ribulose 1,5 bisphosphate carboxylase-oxygenase (RuBisCo), the key enzyme in photosynthetic carbon reduction. So glufosinate effectively inhibits the ability of a plant to photosynthetically produce carbohydrates and other products (see Chapter 9 section 9.3).

The bacterium *Streptomyces hygroscopicus* possesses the *bar* gene that encodes for the enzyme phosphinothricine acetyl transferase (PAT) that detoxifies glufosinate by acetylation (Lyndon and Duke, 1999). An alternative and very similar gene, *pat*, from *Streptomyces viridichromogenes*, also gives rise to the enzyme PAT and has also been used to modify plants (Lyndon and Duke, 1999) (Figure 13.6). Further details on how glufosinate-degrading bacteria are identified and isolated from soils can be found in Hsiao *et al.* (2007).

Although commercially exploited far less than tolerance to glyphosate, oilseed rape, maize, soybean and cotton have all been commercially grown with inclusion of the *pat* or *bar* gene under the trade name Liberty Link™.

13.6 Genetically modified herbicide tolerance to bromoxynil

Bromoxynil is a contact herbicide that kills plants by inhibiting photosynthesis at photosystem II. It is used for selective weed control in a number of crops including wheat, barley and oats, but is not suitable for use in oilseed rape. GM-HT oilseed rape has been produced using the *oxy* gene from *Klebsiella ozaenae* that encodes a nitrilase enzyme (Pallett *et al.*, 1997). The *K. ozaenae* strain containing this gene, isolated from bromoxynil-contaminated soil, can utilise bromoxynil as its sole carbon source, indicating that the strain contains enzymes that can catabolise bromoxynil. It has subsequently been shown that the nitrilase is responsible for the breakdown of bromoxynil to the non-phytotoxic 3,5-dibromo-4-hydroxybromobenzoic acid (Figure 13.7). To date this method of producing herbicide tolerance has only had limited commercialisation including oilseed rape in Canada (under the trade name BXN). Due to the poor performance of bromoxynil

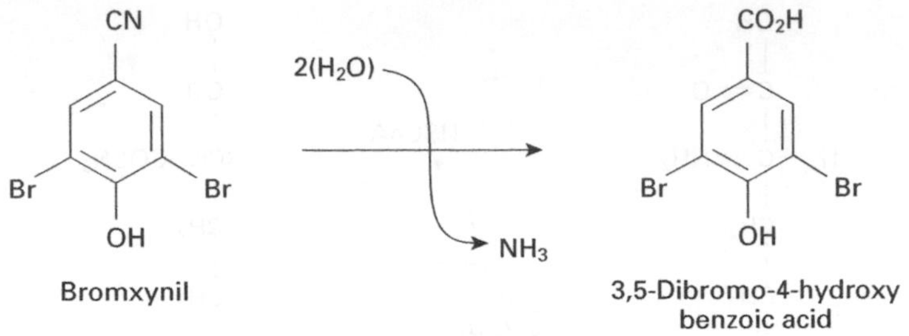

Figure 13.7 The detoxification of bromoxynil by bromoxynil nitrilase.

against grass weeds and certain broad-leaved weeds, it is unlikely that this method will ever prove as popular as those used for creating tolerance to glyphosate or glufosinate.

13.7 Genetically modified herbicide tolerance to sulfonylureas

The identification of genes encoding for acetolactate synthase (ALS) that contain point mutations has allowed the genetic modification of a number of crops so that they possess tolerance to sulfonylurea herbicides. The gene introduced is often from *Arabidopsis* or from tobacco. The most common mutation resulting in tolerance is Pro197 → Ser (Haughn *et al.*, 1988). Many of the commercial sulfonylurea-tolerant crops have been produced by non-GM means and are discussed in section 13.12.

13.8 Genetically modified herbicide tolerance to 2,4-D

Unlike other herbicides discussed in this chapter, the primary site of action of 2,4-D is complex (Chapter 7). Consequently, attempts to date at transferring herbicide tolerance to crops have focused on enhanced breakdown rather than an altered target site. 2.4-D is broken down quickly in soils by a number of microorganisms. One such organism, the bacterium *Alcaligenes eutrophus*, has been shown to be able to utilise 2,4-D as its sole source of carbon (Llewellyn and Last, 1996). It can do this because it possesses a number of genes (called *tfd* genes) that encode for enzymes that metabolise 2,4-D (Llewellyn and Last, 1996, 1999). Of the six identified *tfd* genes, the first (*tfdA*) is of most importance with regard to engineering crop tolerance. 2,4-D (which is phytotoxic) is metabolised to 2,4-DCP (which is not phytotoxic) (Figure 13.8), and *tfdA* encodes for an enzyme, 2,4-D dioxygenase, which carries out the first step in this process.

13.9 Genetically modified herbicide tolerance to fops and dims

Although resistance in weeds to the fops and dims readily occurs and is attributed to an altered target site, enhanced metabolism and possibly altered membrane electrochemical properties, no crops with tolerance to this group of herbicides have yet been produced by

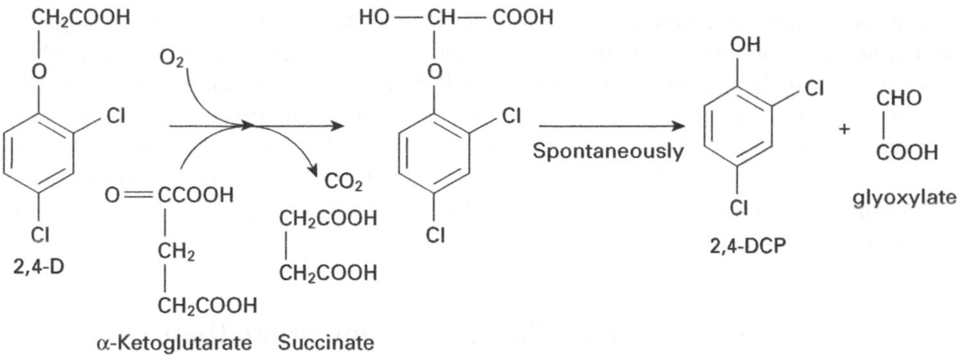

Figure 13.8   The detoxification of 2,4-D by 2,4-D dioxygenase.

GM methods. A few have been produced by non-GM means and are discussed in section 13.12.

13.10   Genetically modified herbicide tolerance to phytoene desaturase

A point mutation in the gene encoding phytoene desaturase (the *pds* gene) results in tolerance to herbicides acting at this target site. Mutations that cause Ser, Cyst or Pro substitution for Arg at position 195 all produce this effect. A herbicide-tolerant phytoene desaturase from *Erwinia uredovora* (encoded by the gene *crtI*) has been used to genetically engineer tobacco to be tolerant to norflurazon and fluridone (Misawa *et al.*, 1993).

13.11   Herbicide tolerance due to engineering of enhanced metabolism

Specific examples of increases in herbicide metabolism as a means of conferring tolerance have already been presented for glufosinate, bromoxynil and the GOX system for glyphosate tolerance. These all involve highly specific enzyme–herbicide interactions and so are not likely to impart a great degree of cross-tolerance to other herbicides. An alternative approach is to engineer the cytochrome P450 and/or glutathione *S*-transferase (GST) enzyme systems that have been implicated in the metabolism of many herbicides and other xenobiotics. In this way, multiple tolerance may be produced within a crop. Certain herbicide safeners, by increasing P450 and GST activities, already aid the crop in metabolising herbicides and so are not damaged by them. An interesting study by Inui *et al.* (1999) introduced human and human/yeast-fused genes encoding for cytochrome P450s into potato plants using standard genetic engineering techniques. The resultant transgenic potato plants showed tolerance to the herbicides chlorotoluron, atrazine and pyriminobac-methyl. Although this method of producing GM-HT crops might been seen as advantageous due to the multiple tolerance that is obtained, it should be considered that there may also be a number of drawbacks. As less is known regarding herbicide metabolism, the herbicides that can be used in such crops – and the control of such crops if they occur as volunteers – will be a lot less predictable. Additionally, the P450 and GST enzyme systems appear to be involved in stress responses as well as protection from xenobiotics, and therefore disruption of these systems may alter the plant's abilities to cope with environmental stressors. Finally, it is highly probable that the consumer, who has already shown some distrust of GM crops, will reject crops that contain human genes.

13.12 Herbicide tolerance through means other than genetic modification

Crop tolerance to herbicides does not necessarily need to be produced by the introduction of genes from other species. A number of methods have been employed, with some success, to produce herbicide-tolerant crops by non-GM means. These include screening of cultivars for tolerance to herbicides during standard breeding programmes, as well as screening for tolerance in cell cultures and in seed germination assays. In addition to these methods, mutations have been induced, using chemicals and/or X-rays, to increase the probability of such screening methods successfully isolating herbicide tolerant lines. Pollen mutagenesis and crossing with wild species that demonstrate herbicide tolerance have also proved successful (Somers, 1996).

In some cases, tolerance to herbicides varies greatly between cultivars of the same species, as is the case for chlorotoluron tolerance in wheat. In this case, the herbicide can only be used on cultivars listed on the product label, as unacceptable levels of phytotoxicity are observed in other cultivars. This could form the basis of breeding programmes to increase tolerance to chlorotoluron, although in reality yield, quality factors and disease resistance are traits considered more important by crop breeders. In addition it is advantageous that the level of tolerance in herbicide-tolerant crops is far greater than is commonly found in naturally occurring cultivars. Selection of ryegrass lines showing resistance to glyphosate did not produce plants with tolerance levels anywhere near those exhibited by genetically engineered crops.

Use of mutagens, often in cell culture, can address this and has been used to produce a number of herbicide-tolerant crops. By causing mutations the probability of selection of a herbicide-tolerant line is increased, but compared to the more certain outcomes of genetic engineering these methods are still somewhat 'hit and miss'.

Glyphosate tolerance due to over-expression of the gene encoding for EPSP synthase in *Petunia* suggested that the low levels of tolerance obtained using this method would not be commercially viable at present. Similar observations have been made in *Salmonella typhimurium* (Ser101 → Pro), *E. coli* (Gly96 → Ala) and can be compared with the mutation in the CP4 EPSP synthase gene (Gly100 → Ala) that has been used for genetic engineering of glyphosate tolerance, as detailed above, especially in Table 13.3. Over-expression of the target site for glufosinate, glutamine synthase, in transgenic tobacco suggested that this method might be of some use in producing glufosinate-tolerant crops, although the genetic engineering methods outlined above have proved more successful to date (Eckes *et al.*, 1989).

Tolerance to triazines in brassica species has been produced simply by cultivar screening and conventional breeding techniques. The physiological basis of this resistance is point mutations in the target site of the triazines, that is the D1 protein. A mutation on the *psbA* gene results in an amino acid replacement (Ser264 → Gly) (Hall *et al.*, 1996) that has a greater tolerance to atrazine. A similar replacement (Ser264 → Thr) results in a greater degree of cross-tolerance to other triazines (Hall *et al.*, 1996). In addition, glutathione *S*-transferases has been implicated in resistance to triazines in the weed species *Abutilon theophrasti* (Anderson and Gronwald, 1991; Gray *et al.*, 1996). The level of this resistance is a lot lower than that imparted by *psbA* gene mutations and has not been exploited in crop tolerance studies.

Tolerance to sulfonylurea (SU) herbicides has been obtained by screening cell cultures in order to select for SU-tolerant mutants in the presence of chlorsulfuron. Using these techniques, SU-tolerant tobacco, sugar beet, oilseed rape, soybean and a host of other crops have been obtained (Saari and Mauvais, 1996; Shaner *et al.*, 1996). The basis of this tolerance is point mutations in the gene encoding for ALS. In *Arabidopsis*, a single mutation results in Pro197 → Ser and increased tolerance to SU herbicides. In tobacco, Pro196 → Ser produces similar results. Substitution of almost any amino acid for Ala (position 117), Pro (position 192) or Try (position 586) results in increased SU tolerance. These multiple sites of mutation that result in viable yet SU tolerant ALS variants may go some way to explaining why resistance in weeds to this family of herbicides appeared so soon after their introduction in the early 1980s and has risen to such a high level today.

Tolerance to imidazolinones, which act at the same target site as SUs, has been obtained in maize (pollen mutagenesis), wheat (seed mutagenesis), oilseed rape (microspore mutagenesis) and in maize (tissue culture in the presence of imazaquin). The latter, which selected for cells containing an altered ALS target site, produced maize plants that were up to 100 times as tolerant as susceptible wild-type maize. Mutations causing Ser653 → Asn give rise to resistance that is specific to imidazolinones. The resulting ALS is still susceptible to SUs and triazolopyrimidines.

Limited research into increasing tolerance to fops and dims in ryegrass (by sexual hybridisation with resistant species), maize and wheat (by selection in tissue cultures) have produced successes, although these have yet to be commercialised (Somers, 1996).

13.13   Genetically modified high-tolerance crops in practice: the UK Farm-Scale Evaluations, 2000–2

GM crops in general have become an integral part of arable agriculture in a number of countries (Table 13.4). Of the traits introduced by GM methods, herbicide tolerance is leading the way (Table 13.5). In other countries, GM crops have either received a limited response or have been actively discouraged. This is particularly true of Northern Europe. In the UK, GM-HT maize, sugar beet and oilseed rape had reached a point by 1999 where most regulatory issues had been satisfied and they were poised for commercial release. However, the UK Government Department for Environment, Food and Rural Affairs (DEFRA), in response to public and pressure group concerns as well as concerns within DEFRA itself, undertook to carry out a three-year Farm-Scale Evaluation (FSE) in order to assess the impact that introduction

Table 13.4 Global area of transgenic crops planted in 2003 by country (from Dill, 2005).

Country	Million hectares	%
USA	42.8	63
Argentina	13.9	21
Canada	4.4	6
Brazil	3.0	4
China	2.8	4
South Africa	0.4	1
Australia	0.1	<1
India	0.1	<1
Romania	<0.1	<1
Uruguay	<0.1	<1
Spain	<0.1	<1
Mexico	<0.1	<1
Philippines	<0.1	<1
Colombia	<0.1	<1
Bulgaria	<0.1	<1
Honduras	<0.1	<1
Germany	<0.1	<1
Indonesia	<0.1	<1
TOTAL	67.7	100

Table 13.5 Dominant GM crops in the world as of 2006 (James, 2006).

Crop and trait(s)	Million hectares
Herbicide-tolerant soybean	58.6
Bt maize	11.1
Bt/herbicide-tolerant maize	9.0
Bt cotton	8.0
Herbicide-tolerant maize	5.0
Herbicide-tolerant oilseed rape	4.8
Bt/herbicide-tolerant cotton	4.1
Herbicide tolerant cotton	1.4
Herbicide-tolerant alfalfa	<0.1
Bt rice	<0.1
TOTAL	

Bt = *Bacillus thuringiensis*-mediated insect resistance. It refers to the introduction of genes coding for the production of the Bt toxin in the plant, which specifically inhibits digestion in the insect gut, leading to insect death.

of these crops would have on biodiversity. No commercial release would be permitted until the completion of the FSE. The FSE focused upon the environmental impact of the introduction of GM-HT maize, sugar beet and oilseed rape when grown in typical UK farming systems.

The evaluations were carried out from 2000 to 2002 and involved maize and oilseed rape engineered to be tolerant to glyphosate, and sugar beet engineered to be tolerant to glufosinate. At each site a randomised block design was used involving two treatments, conventional crop and GM-HT crop. Typically, a field was divided in two, with half being dedicated to each crop. Both the conventional and the herbicide-tolerant crops were managed by commercial farmers using their knowledge and experience of crop production practices. The only difference between the two plots was the farmer's choice to use glyphosate/glufosinate as an in-crop herbicide in the GM-HT half of the field, if they consider it necessary. The choice of herbicide use, including a zero herbicide-use policy, was left completely up to the farmer. For each crop the original aim was to have 60–75 separate field sites. (Practically, the number of sites was less than this for a number of reasons including crop destruction by pressure groups and farmers removing themselves from the trials.)

The primary focus of the FSE was to determine whether the use of these GM-HT crops resulted in differences in biodiversity when compared to conventional management of the same crop. The biodiversity indicators used are listed in Table 13.6. In addition, gene flow from GM-HT crops and the affect of these crops on diversity and abundance of birds was evaluated at a number of sites.

Such an extensive study understandably generated vast amounts of data that have been analysed in various ways by researchers and organisations, sometimes resulting in conflicting conclusions. The general effects of the cultivation of GM-HT sugar beet included reduced weed numbers and a subsequent decrease in bees, butterflies and seed-eating beetles and birds. No overall effects were noted on arthropod bird food, and an increase in springtails and their predators was noted. GM-HT oilseed rape showed broadly similar findings, except that no increase in springtails was observed. GM-HT maize was found to contain more broad-leaved weeds than in conventionally managed maize crops and this led to a rise in bees, broad-leaved weed seeds and a subsequent rise in numbers of beetles and farmland birds that rely on these seeds as a food source. No effects on arthropod bird food or springtails were recorded.

Unsurprisingly, the main conclusion that can be drawn from this is that where weed populations are successfully reduced by herbicide treatment biodiversity is reduced. Where more weeds are present biodiversity is increased. In the case of GM-HT maize, weed control was less successful than conventional methods (that often involved atrazine with its lengthy residual activity). The question is why growers would choose to use GM-HT maize if weed control is less successful? Where GM-HT crops resulted in greater weed control – in sugar beet and oilseed rape – biodiversity indicators were reduced, but in these cases growers were obtaining better weed control than under conventional control measures. It seems that a requirement for better weed control and increased biodiversity might be an impossible 'Catch 22' situation.

Of course, the uptake of GM-HT crops by farmers will also depend on financial savings as much if not more than the efficacy of weed control or environmental impact. May (2003) indicated that if the UK did adopt GM-HT sugar beet then a saving of over £150 ha^{-1} y^{-1} could be made, equating to a UK-wide saving in the region of £23 million y^{-1}. Weed control is expensive in conventional sugar beet crops and £80 of the £150 saving would be due to direct savings on the cost of agrochemicals. Other savings include fuel and labour costs, contractor costs and the indirect cost-saving of reducing soil compaction by carrying out less machinery work on a field.

Table 13.6 Biodiversity indicators measured directly during the DEFRA UK Farm-Scale Evaluations of herbicide-tolerant GM crops (Firbank et al., 2003).

Taxonomic group		Abundance	Diversity	Activity	Other	In-field	Boundary	Following crop	Survey title
Higher plants (excluding crop)	Soil seed bank	X				X		X	Seed bank
	Seedlings	X				X			Seedlings
	Seedlings	X				X		X	Seedlings in following crop
	Adult plants	X			Biomass	X			Weed biomass
	Adult plants	X			Clover, flowers		X		Edge vegetation
		X			Crop growth	X			Crop assessment
	Seed set				Shed seeds	X			Seed rain
	Seed set				Seeds on plants		X		Edge vegetation
Gastropods			X	X		X	X		Within-field gastropods, field margin gastropods
Mites	On crops	X	X			X			Crop pests
Spiders	On soil	X	X			X			Soil-surface arthropods
Aphids	On crops	X	X			X			Crop pests
Bees	Adults	X	X	X		X	X		Bees and butterflies
Beetles	Carabids	X	X	X	Biomass	X	X		Soil-surface arthropods, crop pests, invertebrates on vegetation
	On crops	X	X			X			Crop pests
	Ladybirds	X	X			X			Crop pests
	Soil surface	X				X			Soil-surface arthropods

Table 13.6 Cont'd

Taxonomic group		Abundance	Diversity	Activity	Other	In-field	Boundary	Following crop	Survey title
Bugs	On crops	X	X			X			Crop pests
	On other plants	X	X		Biomass	X	X		Invertebrates on vegetation
Butterflies	Adults		X	X		X	X		Bees and butterflies
	Larvae	X	X		Biomass	X	X		Crop pests, invertebrates on vegetation
Collembola	On soil	X				X			Soil-surface arthropods
	On other plants				Biomass	X	X		Invertebrates on vegetation
Hoverflies	On crops	X				X			Crop pests
Flies	On crops	X	X			X			Crop pests
	Larvae	X			Biomass	X	X		Invertebrates on vegetation
Lacewings	On crops	X	X			X			Crop pests
Moths	Larvae	X	X		Biomass	X			Crop pests, invertebrates on vegetation
Thrips	On crops	X	X			X			Crop pests

13.14   Future developments

The introduction of GM-HT crops has undoubtedly caused significant changes to a number of cropping systems in many countries. These crops have permitted crop production to take place utilising less agrochemical inputs and less fuel due to less need for mechanical cultivations and pesticide applications. These crops suit the minimum and no-tillage systems that are becoming more prevalent owing to their financial and ecological benefits. The effects of such crops on biodiversity, however, have yet to be fully elucidated and farmers are legally prevented from saving a portion of their crop to sow in following seasons. It is also likely that that the global rise in GM-HT crop hectarage has reduced the momentum of the agrochemical industry to research and market herbicides with new target sites, which is vital in the battle against weed resistance. The long-term effect of herbicide-tolerant oilseed rape on agroecosystems has been reviewed by Graef (2009), and such reviews need to play an important role in the monitoring of herbicide-tolerant crops.

It is clear that GM-HT crops will continue to become more widespread globally, and that the technology already exploited in crops such as soybean and cotton will be used in other major world crops in the near future. The prospect of widespread growth of GM-HT wheat, for instance, cannot be far away.

References

Anderson, M.P. and Gronwald, J.W. (1991) Atrazine resistance in a velvetleaf (*Abutilon theophrasti*) biotype due to enhanced glutathione *S*-transferase activity. *Plant Physiology* **96**,104–109.

Bilang, R., Iida, S., Peterhans, A., Potrykus, I. and Paszkowski, J. (1991) The 3′-terminal region of a hygromycin-B-resistance gene is important for its activity in *Escherichia coli* and *Nicotiana tabacum*. *Gene* **100**, 247–250.

Bruce, D. and Bruce, A. (eds) (1998) *Engineering Genesis: the Ethics of Genetic Engineering in Non-human Species*. London: Earthscan Publications.

CaJacob, C.A., Feng, P.C.C., Heck, G.R., Alibhai, M.F., Sammons, R.D. and Padgette, S.R. (2004) Engineering resistance to herbicides. In: Christou, P. and Klee, H. (eds) *Handbook of Plant Biotechnology V.1*. Chichester, UK: John Wiley and Sons, pp. 353–372.

Christou, P. and Klee, H. (eds) (2004) *Handbook of Plant Biotechnology V.1*. Chichester, UK: John Wiley and Sons.

Cheng, M., Fry, J.E., Pang, S., *et al.* (1997) Genetic transformation of wheat mediated by *Agrobacterium tumefaciens*. *Plant Physiology* **115**, 971–980.

Dill, G.M. (2005) Glyphosate-resistant crops: history, status and future. *Pest Management Science* **61**, 219–224.

Duke, S.O. (2005) Taking stock of herbicide-resistant crops ten years after introduction. *Pest Management Science* **61**, 211–218.

Eckes, P., Vijtewaal, B. and Donn, G. (1989) Synthetic gene confers resistance to the broad spectrum herbicide L-phosphinothricin in plants. *Journal of Cell Biochemistry* suppl. 13D.

Economic Research Service. (2007) Adoption of genetically engineered crops in the US, USDA, Washington, DC. Available at http://www.ers.usda.gov/Data/BiotechCrops/ accessed July 2009.

Evans, J. (2010) Food security is all in the genes. *Chemistry and Industry* **9**, 14–15.

Firbank, L.G., Heard, M.S., Woiwod, I.P., *et al.* (2003) An introduction to the farm-scale evaluations of genetically modified herbicide-tolerant crops. *Journal of Applied Ecology* **40**, 2–16.

Goodwin, T.W. and Mercer, E.I. (1983) *Introduction to Plant Biochemistry*, 2nd edn. Oxford: Pergamon Press.

Graef, F. (2009) Agro-environmental effects due to altered cultivation practices with genetically modified herbicide-tolerant oilseed rape and implications for monitoring. A review. *Agronomy for Sustainable Development* **29**(1), 31–42.

Gray, J.A., Balke, N.E. and Stoltenberg, D.E. (1996) Increased glutathione conjugation of atrazine confers resistance in a Wisconsin velvetleaf (*Abutilon theophrasti*) biotype. *Pesticide Biochemistry and Physiology* **55**, 157–171.

Hall, J.C., Donnelly-Vanderloo, M.J. and Hume, D.J. (1996) Triazine-resistant crops: the agronomic impact and physiological consequences of chloroplast mutation. In: Duke, S.O. (ed.) *Herbicide Resistant Crops: Agricultural, Environmental, Economic, Regulatory and Technical Aspects*. Boca Raton, FL: CRC Press, pp. 107–126.

Haughn, G.W., Smith, J., Mazur, B. and Somerville C. (1988) Transformation with mutant *Arabidopsis* acetolactate synthase gene renders tobacco resistant to sulfonylurea herbicides. *Molecular and General Genetics* **211**, 266–271.

Hsiao, C.L., Young, C.C. and Wang, C.Y. (2007) Screening and identification of glufosinate-degrading bacteria from glufosinate-treated soils. *Weed Science* **55**, 631–637.

Inui, H., Ueyama, Y., Shiota, N., Ohkawa, Y. and Ohkawa, H. (1999) Herbicide metabolism and cross-tolerance in transgenic potato plants expressing human CYP1A1. *Pesticide Biochemistry and Physiology* **64**, 33–46.

Ishida, Y., Saito, H., Ohta, S., Hiei, Y., Komari, T., Kumashiro, T. (1996) High efficiency transformation of maize (*Zea mays* L.) mediated by *Agrobacterium tumefaciens*. *Nature Biotechnology* **14**, 745–750.

James, C. (2006) Global Status of Commercialized Biotech/GM crops: 2006. International Service for the Acquisition of Agri-Biotechnology Applications (ISAAA) Briefs No. 35. Ithaca, NY: ISAAA.

Joersbo, M. (2001) Advances in the selection of transgenic plants using non-antibiotic marker genes. *Physiologia Plantarum* **111**, 269–272.

Komari T., Ishida, Y. and Hiei, Y. (2004) Plant transformation technology: Agrobacterium-mediated transformation. In: Christou, P. and Klee, H. (eds) *Handbook of Plant Biotechnology V.1*. Chichester, UK: John Wiley and Sons, pp. 233–261.

Llewellyn, D. and Last, D. (1996) Genetic engineering of crops for tolerance to 2,4-D. In: Duke, S.O. (ed.) *Herbicide Resistant Crops: Agricultural, Environmental, Economic, Regulatory and Technical Aspects*. Boca Raton, FL: CRC Press, pp. 159–174.

Llewellyn, D. and Last, D. (1999) A detoxification gene in transgenic *Nicotinia tabacum* confers 2,4-D tolerance. *Weed Science* **47**, 401–404.

Lyndon, J. and Duke, S.O. (1999) Inhibitors of glutamine biosynthesis. In: Singh, B.K. (ed.) *Plant Amino Acids: Biochemistry and Biotechnology*. New York: Marcel Dekker, pp. 445–464.

Madsen K.H. and SandØe, P. (2005) Ethical reflections on herbicide-resistant crops. *Pest Management Science* **61**, 318–325.

McCabe, D.E., Swain, W.F., Martinell, B.J. and Christou, P. (1988) Stable transformation of soybean (*Glycine max*) by particle acceleration. *Bio/Technology* **6**, 923–926.

May, M.J. (2003) Economic consequences for UK farmers of growing GM herbicide tolerant sugar beet. *Annals of Applied Biology* **142**, 41–48.

Miswawa, N., Yamano, S., Linden, H., *et al.* (1993) Functional expression of the *Erwinia uredovora* carotenoid biosynthesis gene *crtI* in transgenic plants showing an increase of beta-carotenoid biosynthesis activity and resistance to the bleaching herbicide norflurazon. *The Plant Journal* **4**, 833–840.

Padgette, S.R., Kolacz, K.H., Delanny, X., *et al.* (1995) Development and characterisation of a glyphosate-tolerant soybean line. *Crop Science* **35**, 1451–1461.

Padgette, S.R., Re, D.B., Barry, G.F., *et al.* (1996) New weed control opportunities: development of soybeans with Roundup Ready™ gene. In: Duke, S.O. (ed.) *Herbicide-Resistant Crops: Agricultural, Environmental, Economic, Regulatory and Technical Aspects.* Boca Raton, FL: CRC Press, pp. 53–84.

Pallett, K.E., Veerasekaran, P., Freyssinet, M., Pelissier, B., Leroux, B. and Freyssinet, G. (1997) Herbicide tolerance in transgenic plants expressing bacterial detoxification genes. The case of bromoxynil. In: Hatzios, K.K. (ed.) *Regulation of Enzymic Systems Detoxifying Xenobiotics in Plants.* Dordrecht: Kluwer, pp. 337–350.

Reddy, K.N., Duke, S.O. and Rimando A.M. (2004) Aminomethylphosphonic acid, a metabolite of glyphosate, causes injury in glyphosate-treated, glyphosate-resistant soybean. *Journal of Agriculture and Food Chemistry* **52**, 5139–5143.

Saari, L.L. and Mauvais, C.J. (1996) Sulfonylurea herbicide-resistant crops. In: Duke, S.O. (ed.) *Herbicide Resistant Crops: Agricultural, Environmental, Economic, Regulatory and Technical Aspects.* Boca Raton, FL: CRC Press, pp. 127–142.

Shaner, D.L., Bascombe, N.F. and Smith W. (1996). Imidazolinone-resistant crops: selection, characterisation and management. In: Duke, S.O. (ed.) *Herbicide Resistant Crops: Agricultural, Environmental, Economic, Regulatory and Technical Aspects.* Boca Raton, FL: CRC Press, pp. 143–157.

Somers, D.A. (1996) Aryloxyphenoxypropionate and cyclohexanedione resistant crops, In: Duke, S.O. (ed.) *Herbicide Resistant Crops: Agricultural, Environmental, Economic, Regulatory and Technical Aspects.* Boca Raton, FL: CRC Press, pp. 175–188.

Tyagi, A.K. and Mohanty, A. (2000) Rice transformation for crop improvement and functional genomics. *Plant Science* **158**, 1–18.

Vande Berg, B.J., Hammer, P.E., Chun, B.L., *et al.* (2008) Characterization and plant expression of a glyphosate-tolerant enolpyruvylshikimate phosphate synthase. *Pest Management Science* **64**(4), 340–345.

Vasil, I.K. (1996) Phosphinothricin-resistant crops. In: Duke, S.O. (ed.) *Herbicide Resistant Crops: Agricultural, Environmental, Economic, Regulatory and Technical Aspects.* Boca Raton, FL: CRC Press, pp. 85–91.

Westra, P., Wilson, R.G., Miller, S.D., *et al.* (2008) Weed population dynamics after six years under glyphosate- and conventional herbicide-based weed control strategies. *Crop Science* **48**(3), 1170–1177.

Chapter 14
Further Targets For
Herbicide Development

14.1 Introduction

This text is not intended as an exhaustive, nor exhausting, list of herbicides and their proposed sites of action, but instead focuses on the main physiological functions of plants that are sensitive to herbicidal inhibition. The inhibition of photosynthetic electron flow and growth regulation by auxins, lipid biosynthesis and amino acid biosynthesis have been well exploited agrochemically, and much imitative chemistry is still evident in the literature. Nevertheless it is the conference sessions on potential or novel herbicidal targets that always attract the best and the most inquisitive audiences. This is because the rational design of new herbicides is regularly attempted, but invariably lacks success, owing to our incomplete knowledge of plant biochemistry in particular and plant growth in general. Perhaps our increasing understanding of the plant genome may remedy this situation in the years ahead.

Many unsuccessful attempts to design herbicides using biochemical reasoning have been published. In each case a strategic reaction in a key pathway is identified as a suitable target for manipulation, and novel inhibitors are designed on the basis of existing knowledge of the structures of the substrates and the reaction mechanism. The reader is referred to Table 2.8 for potential and patented targets that may lead to new herbicides in the next decade. However, such molecular designs are seldom, if ever, successfully transmitted to the field and commercialised. For example, Kerr and Whitaker (1985) chose phosphoglycollate phosphatase, a key enzyme in photorespiratory carbon oxidation, as a suitable target for herbicide attack. They reasoned that the inhibition of this enzyme would lead to an accumulation of phosphoglycollate, which had been reported to inhibit triose phosphate synthesis and hence photosynthetic carbon reduction. This would also have the effect of preventing photorespiratory carbon recycling that dissipates excess ATP and reducing power generated under high-intensity light conditions. Detailed studies by these workers found that this enzyme could indeed be inhibited *in vitro*, but not in intact leaves, and concluded that 'the effects of an extremely complex chemical on an equally complex biochemical target may be just too subtle for our present understanding'. Indeed, the process of photorespiration has to date defied all attempts at chemical manipulation. It

Herbicides and Plant Physiology, Second Edition By Andrew H. Cobb and John P.H. Reade
© 2010 A.H. Cobb and J.P.H. Reade

therefore follows that if we are not to rely solely on the routine and empirical screening of new chemicals, then only an increased knowledge of plant physiology is sure to generate the new leads, ideas and areas for future agrochemical development. The following sections offer a brief personal choice of three potential areas of plant physiology that deserve further consideration.

14.2   Protein turnover

14.2.1   Introduction

Proteins are constructed from amino acids linked together by peptide bonds. The breakage of these bonds, termed proteolysis, is performed by proteolytic enzymes that are widely referred to as proteases or peptidases. The terms are synonymous, though proteases will be used in this text. Protease research generates thousands of publications each year and the reader is referred to the website of the International Proteolysis Society (www.protease.org) for an overview. The reason for this interest is the realisation in recent years that proteases now appear to be involved in almost all cellular processes in plant growth and development.

The protein content of plant cells is in a state of constant flux, balanced by rates of synthesis and degradation. Alterations in protein content are essential for plant growth and development, and for the most appropriate responses to environmental stimuli. In actively growing leaves the total soluble proteins may turnover every week, while the lifetime of individual proteins is often less than that of a cell. Thus, ACC synthase has a half-life of an hour and the D1 protein may turn over in two hours at high light intensities. This raises the important question: why should a plant expend considerable energy synthesising a protein, only to degrade it within minutes? Clearly, proteolysis regulates metabolism. Protein degradation will release a pool of amino acids to enable the synthesis of new proteins on a different developmental pathway, or simply remove a rate-limiting enzyme. Indeed, many key enzymes are rapidly turned over, indicating metabolic control via proteolysis, controlling their abundance.

Protein mobilisation is therefore essential to the plant cell to:

- supply amino acids for new protein synthesis,
- mobilise storage proteins during germination,
- modify proteins post-translation to their mature forms,
- degrade damaged or misfolded proteins,
- remove enzymes no longer required in metabolism, and
- remove regulatory proteins, such as transcription factors, that are no longer needed.

The general principles governing proteolysis in plants have become better understood in recent times and may offer potential new targets for herbicide development. Unravelling the *Arabidopsis* genome has identified about 1,300 genes involved in the ubiquitin/proteosome pathway and over 600 proteases! Understanding their roles may allow the design of new herbicides to control the accumulation of regulatory proteins and prevent cell division as normal developmental responses to environmental change.

14.2.2   *Proteases*

Proteolysis, or the hydrolytic cleavage of peptide bonds, is highly selective and involves a large family of protease enzymes. In essence, hydrolysis is achieved by nucleophilic attack at the carbonyl carbon, supported by the donation of a proton to the NH group of the peptide bond. The proteases themselves are either endoproteases (cleaving internal peptide bonds) or exoproteases, which progressively cleave peptide bonds from either the C-terminus (carboxypeptidases) or the N-terminus (aminopeptidases).

The endoproteases are further characterised according to the catalytic mechanism. In serine, threonine and cysteine proteases, the hydroxyl or sulphydryl groups of the amino acids at the active site act as the nucleophile during catalysis. On the other hand, aspartic, glutamic and metallo-proteases rely on an activated water molecule as the nucleophile. Further classification in the *MEROPS* database (merops.sanger.ac.uk) is according to sequence similarity. The observation that there are 674 listed for *Arabidopsis* (merops.sanger.ac.uk accessed April 2007) implies considerable structural and presumably functional diversity.

The following brief summary of protease function is taken from Schaller (2004), to whom the reader is referred for further details and references.

1 *Cysteine proteases:* The phytocalpains (calcium-dependent cysteine proteases) regulate numerous development processes, including embryonic pattern formulation, shoot apical meristem formation, cell fate specification in the endosperm and leaf epidermis, and regulation of the balance between cell differentiation and proliferation. Papain-like enzymes are also cysteine proteases involved in protein degradation during seed germination, leaf senescence and programmed cell death. They also contribute to plant resistance against pathogens and insects. The legumains are involved in the processing of seed storage proteins and tissue senescence.

2 *Serine proteases:* This is the largest class of plant proteases, over half being carboxypeptidases and subtilases. The serine carboxypeptidases function in protein turnover and nitrogen mobilisation. Apparently, their abundance and diversity are due to differences in substrate specificity. They have been implicated in programmed cell death, brassinosteroid signalling and seed development. They may also have a function in the formation of plant secondary metabolites. The functions of the subtilases remains uncertain. The study of *Arabidopsis* mutants suggests that they may act as highly specific regulators of plant development, such as the regulation of stomatal density, cuticle formation and embryo development.

3 *Aspartic proteases:* The phytepsins are located in the vacuole of storage tissues (seeds and tubers) and have been implicated in the breakdown of storage proteins (e.g. patatins in potatoes!) during germination and sprout growth. They have roles in organ senescence, cell death, defence against pathogens and insect herbivores. Intriguingly, the nepenthesins are found in the digestive fluids of the insect-trapping organs in carnivorous pitcher plants (*Nepenthes*). They appear to be remarkably stable, being able to function at 50 °C over a wide range of pH and substrates.

4 *Metalloproteases:* These enzymes rely on a divalent cation for activity, commonly Zn, Co or Mn. They are very structurally diverse and few have been characterised to date. The leucine aminopeptidases may be important in the regulation of protein half-life by the cleavage of the N-terminal amino acid exposing or removing destabilising residues.

Most chloroplast proteins are encoded in the nuclear genome and translated in the cytoplasm as large precursor proteins. Import into the chloroplast is governed via the transit peptide, which is then proteolytically removed in the chloroplast stroma by the stromal processing peptidase, a soluble metalloprotease. When the gene for this enzyme was antisensed in *Arabidopsis*, it was shown that altered protein import was vital for chloroplast biogenesis, photosynthesis and consequently plant growth.

14.2.3   The ubiquitin–proteosome pathway

A central mechanism in the degradation of cytoplasmic and nuclear proteins uses the small protein ubiquitin. This globular protein contains 76 amino acids (molecular mass, 8.5 kDa) and is ubiquitous in the eukaryotes. Its structure is highly conserved and very stable in a range of conditions. Its function is to mark proteins for degradation, which it does via two key features, a C-terminal tail (Arg-Gly-Gly) and a lysine residue at position 48. Proteins destined for breakdown are conjugated to ubiquitin in a ligase reaction involving ATP at the carboxy terminal tail. The target protein may then be modified by the binding of several more ubiquitins, resulting in a polyubiquitin chain to be degraded at the proteosome.

Proteosomes are large protein complexes (molecular mass 2,000 kDa) found in the nucleus and cytoplasm of all plant cells, where they function to degrade proteins. They are composed of a Core Particle and two Regulatory Particles (Figure 14.1). The Core Particle (CP) is constructed of 14 different proteins assembled in groups of 7, each group

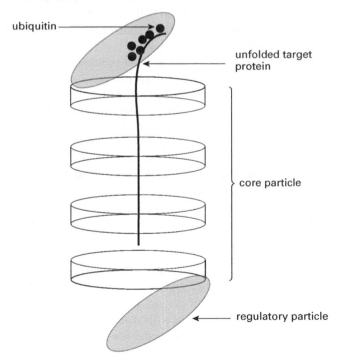

Figure 14.1 Simplied structure of the proteosome.

forming a ring. Four rings are arranged in stacks. There is a Regulatory Particle (RP) at each end of the CP made of 14 different proteins, some of which are ATPases and others that recognise poly-ubiquitin. The target protein is unfolded by the ATPases in the RP and the ubiquitins are released for reuse. The protein moves into the central cavity of the CP where specific proteases on the interior surface degrade the protein releasing peptides of about eight amino acids long.

A good example of how this pathway may control a physiological process involves the role of the cyclins in the cell cycle (Chapter 10). This family of proteins is involved in the cell cycle, where their concentrations are crucial. They accumulate during G_1, S and G_2 phases, and when a critical concentration is reached, they initiate mitosis by binding to and activating the cdc2 protein kinase. During metaphase the cyclins are rapidly degraded, reducing the pool of active cdc2 protein kinase, thereby preventing newly formed daughter cells from entering into another phase of mitosis. Once the cyclin pool declines, degradation ceases, and cyclin concentrations being to rise again, initiating another round of the cell cycle. Clearly, interference with cyclin synthesis and degradation will have a profound effect on the cell cycle and is a potential target for herbicide development.

14.3   Biological control of weeds

14.3.1   Methods of biological control

When plant species are introduced into a new region or country, without their natural and normal pests and pathogens, they invariably flourish and can easily become dominant. These non-native, invasive species can therefore become major problems for weed control by conventional means.

In classical biological control, host-specific natural predators are introduced from the native region where they can reduce the weed population to a manageable level. To date, however, this strategy has not been successfully applied in European agriculture. The Chinese grass carp (*Ctenopharyngodon idella*) has, however, been successfully introduced to control aquatic weeds in waterways, consuming more than its body weight in weeds each day. In the 1980s moths from South Africa were used for the biological control of bracken (*Pteridium aquilinum*). This upland plant is estimated to spread at up to 3% per annum and is toxic to grazing animals. The larvae of noctuid and pyralid moths are able to eat into the plant, allowing secondary invasion by pathogens. Subsequently, the release of the larvae has not been permitted and so bracken continues to spread in the UK uplands. This strategy has not been successfully applied in European agriculture, until now. In 2010 the UK Government approved the use of the non-native psyllid insect, *Aphalara itadori* for the control of the highly invasive, non-native weed Japanese knotweed (*Fallopia japonica*). The psyllid is a natural predator of this plant in Japan and its species specificity has been studied in a UK context. Pilot studies are now taking place to establish its effectiveness in the field. This may represent a significant advance in the control of this invasive species, which is estimated to cost over £1.5 billion per annum to control. Further examples are awaited, perhaps using more native species for biological control.

An alternative is to develop microbial or myco- herbicides. In this strategy, plant pathogenic microorganisms are used as the weed control agent. Greaves (2002) describes the approaches and examples used to date. Candidate microbial herbicides must be host-specific, stable and capable of controlling the target weed in a wide range of environments. They are often applied annually and the challenges of their production, formulation and application are analogous to pesticide use.

The first microbial herbicide, registered in 1981, was Devine, a preparation of *Phytophthora palmivora* that is effective for the control of milkweed vine (*Morrenia odorata*) in citrus groves in Florida. The second, registered in 1982, was Collego, containing the spores of *Colletotrichum gloeosporioides* f.sp. *aeschynomene*, a pathogenic fungus specific to the leguminous weed, Northern joint vetch, *Aeschynomene virginica* in rice and soybean in the southern states of the USA. Camperico, a formulation containing the pathogenic bacterium *Xanthomonas campestris* has been used to control annual meadow grass (*Poa annua*) on golf courses in Japan. Since then considerable research effort has been devoted to the development of new microbial herbicides.

While Greaves (2002) cites many ongoing projects in the 1990s, with several potential pathogens identified, no further commercial breakthroughs have been reported. One reason for this limited progress has been due to formulation and application issues, thus promising activity in the laboratory has not been transferred to the field. New sprayers may be required to reliably deliver microbial herbicides to the target weed and the optimisation of the sprayer systems is essential research that remains to be done (Greaves *et al.*, 1998). The delivery of threshold levels of 10^3–10^5 propagules cm^{-2} of leaf surface is comparable to the early days of herbicide development, when tens of kilograms per hectare were routinely applied. Gressel and Amsellem (2003) see a similar challenge to improve the efficiency of inoculum delivery by several orders of magnitude. These authors consider that enhanced efficiency can be used by transferring virulence factors, such as toxic proteins and secondary metabolites, to the microorganism, and in so doing, tipping the evolutionary balance in favour of the microorganism and more effective weed control. They have engineered genes for auxin over-production and phytotoxic proteins into microbial herbicides to positive effect. The challenge now is to devise failsafe mechanisms to prevent the spread of these genes into the wider environment and to ensure that there is public acceptance of this new approach. Perhaps the future, successful microbial herbicide will be a weed-specific pathogen containing hypervirulence genes, formulated to the target weed at low dose.

14.3.2 Biopesticides

Biopesticides are predicted to have an increasingly important role in the future of crop protection. They are often relatively less toxic than synthetic chemical agents, can be highly specific, have low residues and are relatively inexpensive to develop. They also respond to rising consumer concerns regarding the safety of residues in fruit and vegetable products. On the other hand, just because a product is natural it does not follow that it is safe (strychnine for instance, is a highly toxic natural product)!

While global sales of biopesticides are expected to reach one billion dollars by 2010, they have been mainly associated with the control of insects, by products derived from microorganisms, such as *Bacillus thuringiensis* toxins, or fungal control by plant extracts. However, there is a distinct possibility that herbicides based upon plant-derived allelopathic chemicals will become a more important segment of the agrochemical industry in years to come, as these can be marketed as 'natural' herbicides.

14.4   Natural products as leads for new herbicides

14.4.1   Introduction

In the plant kingdom there is an incredible diversity of what are termed secondary metabolites, that is, those molecules which are not absolutely required for normal cell function. At least 10^5 active plant metabolites are now known and the reader is directed to the NAPRALERT website (www.napralert.org) for a comprehensive database, covering the literature for natural products.

Why do so many natural molecules exist and what is their physiological function in plants? They were first thought to exist as metabolic waste products but, since the metabolic cost to the plant is high, there must be clear benefits to the plant for producing them. It is now thought that secondary metabolism has evolved with the co-evolution of microbial and insect parasites, and animal herbivores, largely as defence and protection mechanisms. Since specialist organisms have evolved that have succeeded in overcoming plant poisons and chemical defences, new metabolic pathways and chemical modifications have evolved and continue to evolve accordingly. The earliest defence molecules were the resins, lignins, condensed tannins and flavonoids. The evolution of the angiosperms and herbaceous plants was marked by the proliferation of further secondary metabolites as products of the acetate–mevalonate pathway.

During steady-state photosynthesis, at least 20% of the carbon fixed in green plants is directed, via triose phosphates, towards synthesis of an impressive array of biologically important end-products. These include lignins, alkaloids, tannins, isoprenoids and a wide range of phenolic compounds, including the 15-carbon flavonoids. Indeed, it is estimated that about 2% of all carbon fixed, equivalent to 10^9 tonnes each year, is converted to flavonoids alone. The precise end products formed depend on the metabolic needs of the plant at any given time, which are highly dependent on the prevailing environmental conditions and growth stage. Furthermore, plants under stress and those with plentiful supplies of nitrogen, are preferentially attacked by insects, and the plant responds by increasing the biosynthesis of defence molecules. During pathogen attack, the concentrations of flavonoids and related compounds greatly increase at the site of infection, to concentrations that are toxic to pathogens in *in vitro* assays. Wounding and feeding by herbivores can induce the biosynthesis of toxic coumarins and tannins, and phenolic acids as precursors to lignins and suberins in wound healing. The accumulation of flavonols, especially kaempferol at wound sites, may also prevent microbial infection. Recent research indicates that stresses such as infection and exposure to ultraviolet light, that are perceived in one part of the plant, can be communicated to the rest of the plant and elicit systemic effects. This communication is thought to be mediated by salicylic acid, which shares part of its biosynthetic pathway with the flavonoids.

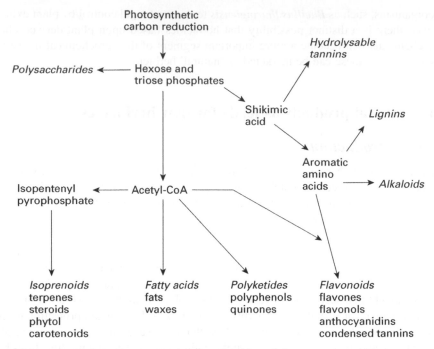

Figure 14.2 A simplified scheme showing the derivation of the main groups of plant secondary metabolites (modified from Duke *et al.*, 2000).

14.4.2 Biosynthesis

In general, most secondary metabolites are synthesised from four starting molecules, each derived from photosynthetic carbon reduction, namely shikimic acid, acetyl-CoA, isopentenyl pyrophosphate and amino acids. A simplified scheme describing how these pathways are integrated is presented in Figure 14.2.

14.4.3 Examples of herbicides derived from natural products

Duke *et al.* (2000) and Wakabayashi and Böger (2004) refer to several case studies where chemical leads for novel sites of action and herbicidal activity have been generated from natural products. Both papers are fully referenced with many examples from the literature. The reader is recommended to mine these valuable seams of information collated by these experts in the field. This section aims to highlight some of the most promising structures that have emerged in recent years (Figure 14.3).

Cinmethylin is a pro-herbicide that resembles the plant monoterpene cineoles and has been developed for the control of monocotyledonous weeds. Following the cleavage of the benzyl ether side chain, it inhibits asparagine synthetase, an activity shared by the natural monoterpene.

Hydantocidin is a nucleoside analogue from *Streptomyces hygroscopicus*. It is also a pro-herbicide that when phosphorylated *in vivo* can inhibit adenylsuccinate synthetase.

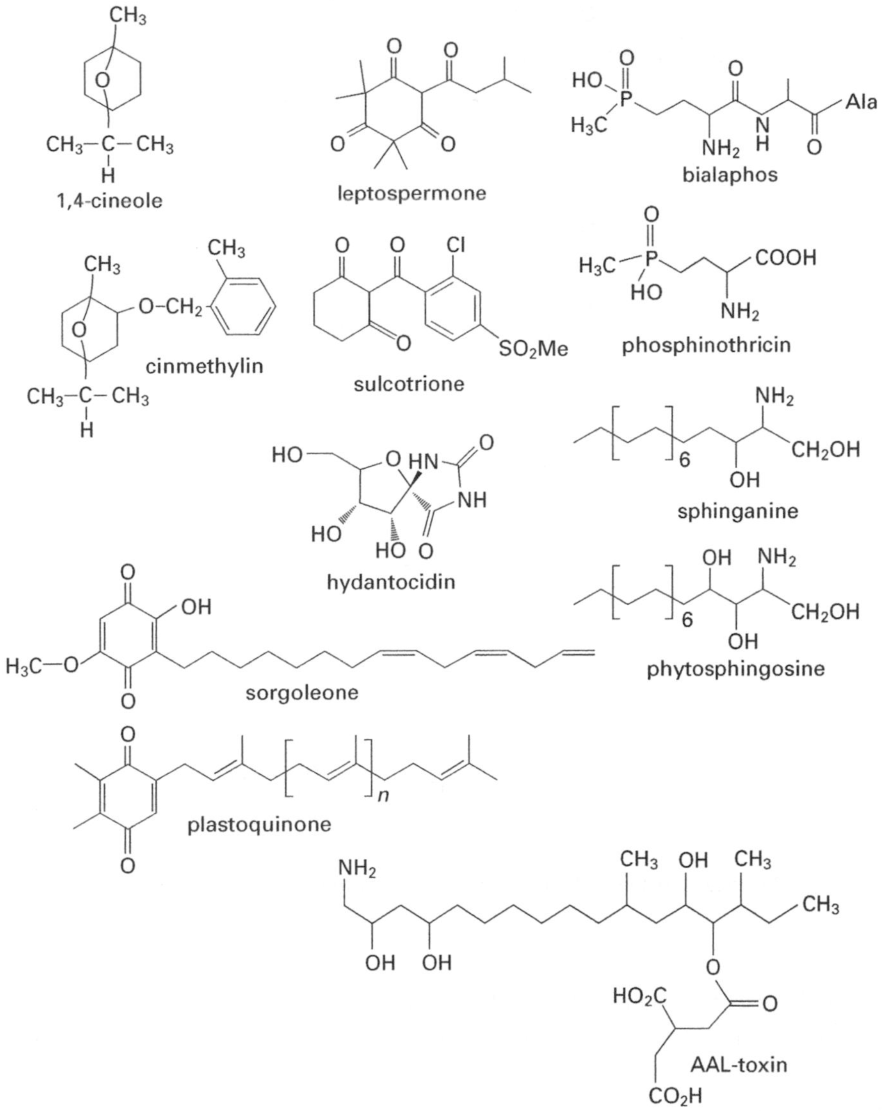

Figure 14.3 Structures of the herbicides mentioned in the text and the natural products from which they were derived.

Phosphinothricin or glufosinate is the breakdown product of bialaphos produced by *Streptomyces viridochromogenes* and *S. hygroscopicus* and inhibits glutamine synthase in a non-selective manner.

Sorgoleone is a natural phytotoxin isolated from root extracts of *Sorghum* spp., which has considerable allelopathic activity towards other plants. It structurally resembles plastoquinone and competes for the Q_B binding site at the D1 protein.

Trikeones, such as sulcotrione, are derivatives of leptospermone found in the bottle-brush plant (*Callistemon* spp.), which inhibit HPPD activity.

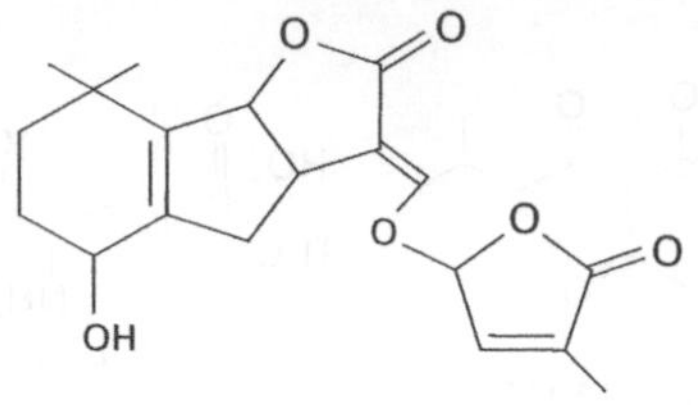

(+)-strigol

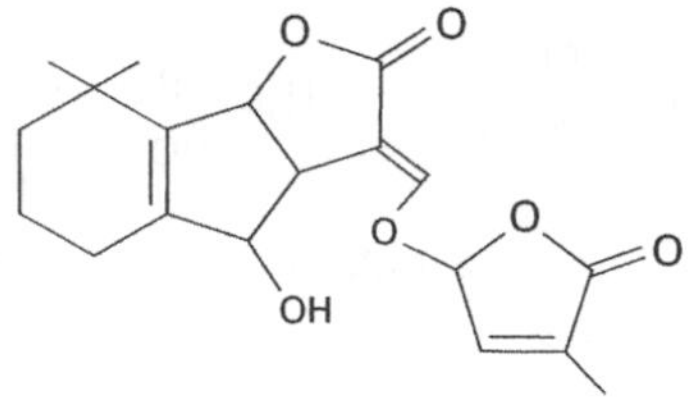

orobanchol

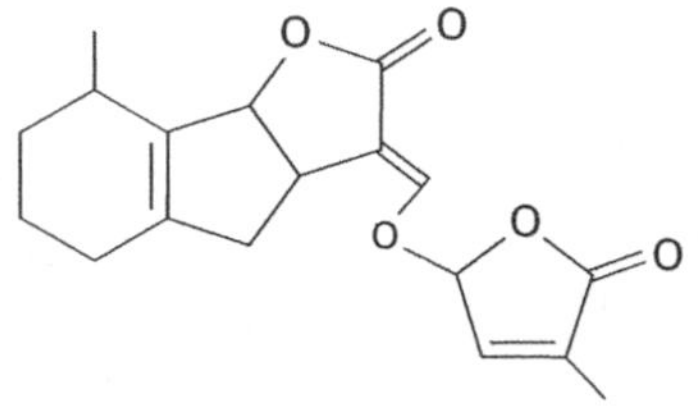

sorgolactone

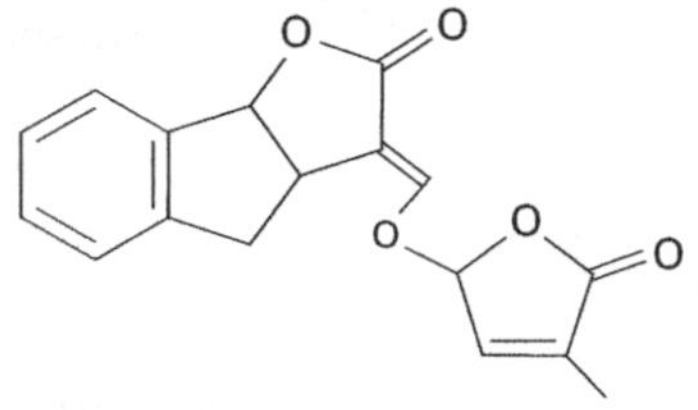

rac-GR24, a synthetic strigolactone

Figure 14.4 Chemical structures of representative strigolactones.

The *AAL-toxin* is a natural herbicide that inhibits ceramide synthase, causing the accumulation of phytosphingosine and sphinganine.

14.4.3.1 Strigolactones (Figure 14.4)

In September 2008 a new group of natural plant hormones was announced, the strigolactones, thought to be involved in the branching process. Mutant pea seedlings that exhibited uncontrolled branching were found not to contain these hormones and that the effect could be overcome by the addition of strigolactones (Gomez-Roldan *et al.*, 2008; Umehara *et al.*, 2008).

These terpenoid lactones are biosynthetically derived from carotenoids (see Figure 14.4 for representative examples of chemical structures). Inhibition of shoot branching can have a profound effect on plant architecture and apical dominance. Thus, cut flowers or ornamental potted plants with either more or less branching may have ornamental appeal, while crop yield might be manipulated.

The strigolactones appear to be synthesised in the roots and move throughout the plant. They are also thought to be involved in the mycorrhzal symbiosis by promoting the growth of mycorrhizal fungi and stimulating the germination of seeds of the parasitic weeds *Striga* and *Orobanche*. It follows that a synthetic analogue which stimulates the germination of these parasites would find a valuable role in the crop protection armoury.

14.4.4   Concluding comment

Swain (1977) has estimated that perhaps as many as 400,000 secondary products are synthesised in the Plant Kingdom, and this surely represents a major source of diverse chemical structures as herbicidal leads. For example, over 700 natural amino acids have been reported in the literature, a significant number of which show promising activity as natural inhibitors of pyridoxal phosphate-dependent enzymes, the syntheses of glutamine and glucosamine, and proteases and peptidases (Jung, 1989). One is therefore forced to conclude that since only a small percentage of plant species have been analysed for natural product chemistry, further investigations will surely prove rewarding.

References

Duke, S.O., Dayan, F.E. and Rimando, A.M. (2000) Natural products and herbicide discovery. In: Cobb, A.H. and Kirkwood, R.C. (eds) *Herbicides and Their Mechanisms of Action*. Sheffield, UK: Sheffield Academic Press, Ch. 5.

Greaves, M.P. (2002) Biological control of weeds. In: Naylor, R.E.L. (ed.) *Weed Management Handbook*, 9th edn. Oxford: Blackwell Science, Ch. 17.

Greaves, M.P., Holloway, P.J. and Auld, B.A. (1998) Formulation of microbial herbicides. In: Burgess, H.D. (ed.) *Formulation of Microbial Biopesticides, Beneficial Organisms and Seed Treatments*. London: Kluwer Academic, pp. 203–234.

Gressel, J. and Amsellem, Z. (2003) Transgenic mycoherbicides for effective, economic weed control. Proceedings of the British Crop Protection Council International Congress. *Crop Science and Technology* 1, 61–68.

Gomez-Roldan, V., Fermas, S., Brewer, P.B., *et al.* (2008) Strigolactone inhibition of shoot branching. *Nature* **455**, 189–194.

Jung, M.J. (1989) Natural amino acids as enzyme inhibitors. In: Copping, L.G., Dalziel, J. and Dodge, A.D. (eds) *Prospects for Amino Acid Biosynthesis Inhibitors in Crop Protection and Pharmaceutical Chemistry*. British Crop Protection Council Monograph No. 42. Farnham, Surrey: BCPC, 15–22.

Kerr, M.W. and Whitaker, D.P. (1985) Energy losses – photorespiration. *Annual Proceedings of the Phytochemical Society of Europe* **26**, 45–57.

Schaller, A. (2004) A cut above the rest: the regulatory function of plant proteases. *Planta* **220**, 183–197.

Swain, T. (1977) Secondary compounds as protective agents. *Annual Review of Plant Physiology* **28**, 479–501.

Umehara, M., Hanada, A., Yoshida, S., *et al.* (2008) Inhibition of shoot branching by new terpenoid plant hormones. *Nature* **455**, 195–200.

Wakabayashi, K. and Böger P. (2004) Phytotoxic sites of action for molecular design of new herbicides (Part 2): Amino acid, lipid and cell wall biosynthesis, and other targets for future herbicides. *Weed Biology and Management* **4**, 59–70.

Glossary

The following glossary provides a short explanation of some relevant terms that are not defined in the text. For further information of a botanical nature, the reader may find the *Penguin Dictionary of Plant Science* (1999) to be of value.

absorption   The process whereby chemicals gain entry into plant tissues. This may be active (against an energy gradient) or passive (no energy expended).

acceptable daily intake   The concentration of **herbicide** or its **residue** that a human may be exposed to on a daily basis that, accordingly to current information, does not appear to create an unacceptable risk to the well-being of the individual. Expressed as milligrams per kilogram body weight per day.

active ingredient   The chemical in a commercial **formulation** that is responsible for the herbicidal effect.

active site   The site at which the substrate(s) of an **enzyme** is bound during catalysis.

acute toxicity   A measure of the amount of chemical, as a single dosage or concentration, required to cause injury or illness in test animals.

adjuvant   A substance added to the formulation or spray tank to modify application characteristics, e.g. to enhance coverage of leaf surfaces.

agroecosystem   The living and nonliving components of an agriculturally active unit area and their interactions within that area.

allelopathy   The release of a chemical by a plant that inhibits the growth of nearby plants and so reduces competition.

amyloplast   A plastid that forms starch grains.

annual   A plant that germinates, grows, flowers, produces seeds and dies within one year.

antagonism   Reduced activity of a herbicide in the presence of another chemical. Opposite is synergism.

Herbicides and Plant Physiology, Second Edition By Andrew H. Cobb and John P.H. Reade
© 2010 A.H. Cobb and J.P.H. Reade

apoplast The continuum of cell walls throughout the plant, important in water movement.

autoradiography A technique for detecting the presence and distribution of radioactive compounds in a biological material. In essence, radioactive emissions darken a photographic plate which, when developed, reveals the location of the isotope as dark patches of silver grains.

auxin A natural (e.g. IAA) or synthetic (e.g. MCPA) plant **growth regulator** involved in cell division, enlargement and differentiation.

band application Treatment to a defined area, such as along a crop row, rather than applied over the entire field.

biennial A plant that completes its life cycle in two years. In the first year vegetative growth predominates and photosynthates are stored over winter. In the following year these products are used for the generation of leaves, flowers and seeds.

bioassay A biological assay. The assessment of the effect of a chemical on an organism by comparison with the effects of standard substances of known concentration.

biological (weed) control The control or suppression of weed growth by the action of one or more organisms.

biorational design The use of biological control agents and analogues of naturally occurring biochemicals for the discovery of new agrochemicals.

biotechnology The use of (usually) microorganisms, or their metabolites and products, in industrial processes.

biotype Subgroup within a species that differs in some respect (e.g. in **herbicide resistance**) from that species.

broadleaf weed Generally dicotyledonous plants with branch-veined leaves and terminal meristems.

carboxylase An enzyme that catalyses the transfer or incorporation of carbon dioxide into a substrate molecule.

carcinogen A substance or agent capable of inducing cancer in animals.

carotenoids Yellow, orange, brown, or red lipophilic pigments that function as accessory photosynthetic pigments.

cell cycle Sequence of events leading to the formation of two daughter cells.

chlorosis Loss of green colour in foliage. Leaves appear typically pale or yellow in colour.

chronic toxicity Toxicity of a single chemical after a prolonged period of exposure, usually from a day to several weeks.

clone A population of genetically identical cells or individuals.

coleoptile A cylindrical sheath of tissue that encloses and protects young shoots of grasses and cereals during growth to the soil surface.

competition The active acquisition of limited resources by an organism, resulting in a reduced growth and yield of other organisms in a shared environment.

contact herbicide A compound that kills on contact rather than relying on **translocation** for activity.

cotyledon Embryonic leaf in seed plants that acts either as a storage organ or in absorbing food reserves from the **endosperm**. Dicotyledonous plants (such as broadleaf weeds) have two cotyledons and monocotyledonous plants (cereals and grasses) have one.

cross-resistance Resistance to more than one herbicide due to the presence of a single resistance mechanism.

cultivar A cultivated variety generated by human selection and so not normally found in natural populations.

cuticle A continuous waxy layer that covers the aerial parts of a plant to prevent excessive water loss.

cutin A water-repellant waxy polymer that is a major component of the **cuticle**.

cytoskeleton An intracellular scaffold of proteins. A dynamic structure that maintains cell shape and function, including intracellular transport and division.

defoliant A chemical that causes leaf loss.

desiccant A chemical that induces rapid desiccation of plant parts.

differentiation A series of changes that occur in cells and tissues during development, resulting in their specialisation.

dormancy An inactive phase during which growth and developmental processes stop.

ED$_{50}$ The chemical dose that produces a desired effect in 50% of the test organisms exposed to the chemical.

emergence The breaking through the soil surface by a seedling or an elongating shoot.

emulsifiable concentrate A formulation containing organic solvent and emulsifier to aid mixing with water.

environment All the organic and inorganic conditions in which an organism lives.

enzyme A protein that catalyses biological reactions.

ephemeral A plant with such a short life cycle that it may be completed many times in one growing season.

epinasty Plant tissue movement to an external stimulus in which increased growth on one side of an organ causes bending of that organ. Commonly observed following application of auxins to young tissues.

ethylene   A gaseous plant hormone that affects growth, development, ripening and senescence.

fatty acid   A long-chain aliphatic carboxylic acid that may be saturated or unsaturated.

formulation   The form in which an agrochemical is prepared for commercial use, including wettable powders, **emulsifiable concentrates**, granules, etc.

gene   The unit of inheritance.

genetic engineering   Isolation of 'useful' **genes** from a donor organism and their functional incorporation into an organism that does not normally possess them.

germination   The process of growth initiation in seeds or spores.

glycosyltransferase   An **enzyme** that transfers a sugar residue from one molecule to another.

graminicide   A **herbicide** which kills grasses.

grass weeds   Generally monocotyledonous plants with parallel-veins of the family Gramineae.

growth regulator   A substance that at very low concentrations may affect growth and differentiation in plants.

herbicide   A chemical that kills plants.

Hill reaction   Transfer of electrons from water to non-physiological acceptors by thylakoids in the presence of light and with the evolution of oxygen.

I_{50}   The concentration required to inhibit the activity of an enzyme or a metabolic pathway by 50%.

induced enzyme   An **enzyme** that is synthesized in response to elevated concentrations of its substrate or a specific inducer.

invasive species   Non-native species that out-compete native species in a defined habitat.

isotopic tracer   A stable, radioactive isotope that can be used to label a **herbicide** or metabolite to monitor its fate within an intact organism and the environment.

isozyme   (or isoenzyme) One of a number of distinct proteins with the same enzyme activity but different kinetic characteristics.

LD_{50}   The dose required to kill 50% of test organisms, usually expressed as milligrams of chemical orally ingested per kilogram body weight.

leaching   The movement of water soluble chemicals down the soil profile.

mechanical weed control   The physical control of weeds using implements such as the hoe or field cultivator.

meristem   A region containing actively dividing cells.

metabolism The sum of the enzymatic reactions taking place in a cell, organ or organism.

metabolite A product of **metabolism** within an organism.

microtubule A cytoplasmic tubule composed of the protein tubulin.

morphology The study of form and shape, particularly with respect to external structure.

multiple resistance Resistance to more than one herbicide due to the presence of more than one resistance mechanism.

mutagen An agent that causes an increased frequency of mutation, i.e. an inherited genetic change.

necrosis Death of a plant cell, group of cells or tissue, while the rest of the plant is still alive. A particularly useful term when dead tissues undergo typical colour changes in comparison to healthy tissues.

no-till The sowing of seeds directly, with minimum soil disturbance.

non-selective herbicide A treatment to kill all vegetation.

noxious weed A plant identified by law as being undesirable, troublesome or difficult to control.

nucleolus The region of the nucleus where ribosomal components are assembled.

organelle A membrane-bound structure in the cytoplasm in which specific but essential processes take place. Examples include photosynthesis in chloroplasts and oxidative phosphorylation in mitochondria.

perennial A plant that lives for many years, surviving as either herbaceous perennials (with underground storage organs) or woody perennials (whose aerial stems persist above ground).

persistence The property of a compound to persist in soil and give prolonged protection to a crop or prevent plant regrowth for extended periods. Persistence is a function of dose, chemical volatility and stability in a given environment.

pest Any organism that damages crop growth or reduces yield.

photosynthate Organic products of photosynthetic carbon reduction.

physiology The study of the processes and functions of life.

phytotoxicity Damage to plants.

plasmid A non-chromosomal, circular, DNA molecule.

plasmodesmata Cytoplasmic channels that span the plasma membrane and cell walls of plants, and form physiological connections between adjacent cells.

plasticity The ability of an organism to change its form in response to varying environmental conditions.

plastids Major organelles found in plant and algal cells.

poisonous plant A plant that can produce toxic substances in sufficient amounts to cause injury or death in animals.

population A community of potentially interbreeding organisms.

post-emergence After a **weed** or crop has emerged.

pre-emergence Before a **weed** or crop has emerged.

proteases Enzymes that break down the peptide bonds of proteins.

protoplast The part of the plant cell internal to the cell wall and bounded by the plasma membrane.

rate Amount of active ingredient applied to a unit area.

recombinant DNA Genetic material that contains novel **gene** sequences using techniques of **genetic engineering**.

registration The process by which a herbicide is legally approved for use.

residue Trace of a pesticide and its metabolites remaining on and in a crop or the environment.

resistance The ability of a plant to withstand exposure to a normally lethal dose of herbicide.

rhizome An underground stem capable of horizontal growth. Buds can play an important role in the vegetative propagation of weeds.

safener A chemical that reduces the **phytotoxicity** of another chemical when used together.

seed dressing A chemical seed treatment to protect against fungal or insect attack.

selective herbicide A herbicide that will kill some species and not others.

signals Internal and external factors that induce changes in cell structure and function.

sink A site in a plant where a demand exists for particular substrates or photosynthate.

soil application Application of a herbicide to the soil surface rather than foliage.

soil incorporation The mechanical mixing of a herbicide into the soil surface.

source A site of production of particular substrates or **photosynthate** for movement to specific **sinks**.

spray drift The movement of airborne droplets of herbicide away from the intended area of treatment.

stem cells Any cells that can proliferate in an undifferentiated state and can give rise to differentiated cells or tissues.

stoma   A moveable pore on aerial plant surfaces to allow gaseous exchange (pl. stomata).

surfactant   An additive to the formulation that modifies the characteristics of the herbicide on the leaf surface.

susceptible species   A plant that can be killed or injured by a herbicide.

symplast   The continuum of cytoplasm throughout the plant, linked by **plasmodesmata**.

synergism   The combined effects of two treatments when acting together greatly exceed the sum of their effects when each acts alone.

systemic   A chemical that is absorbed and translocated throughout the plant body.

tank mixture   Mixture of two or more pesticides in a spray tank at application.

taproot   A persistent, primary root often penetrating to considerable depth and specialised for storage.

target   The organisms, structures, tissues, cells and **enzymes** to be affected by the **herbicide**.

tiller   A shoot that develops at the base of a stem.

tissue culture   The growth of isolated plant cells or pieces of tissue under controlled conditions in a sterile growth medium.

translocation   The transfer of soluble molecules, such as **herbicides**, from one part of the plant to another.

transpiration   The loss of water by evaporation from a plant surface, primarily through open **stomata**.

trichome   An outgrowth from an epidermal cell, variable in form and function, often termed leaf hair.

tuber   A swollen part of a stem or root that is modified for storage.

vacuome   The collective term for all the vacuoles in a cell.

volatility   A measure of the tendency of a substance to vapourise.

weed   A plant growing where it is not wanted, or interferes with the activities or welfare of mankind. A plant whose virtues remain to be discovered!

xenobiotic   A chemical foreign to the organism.

X-ray crystallography   A method of determining the arrangement of atoms within a crystal, allowing the 3D structure of complex molecules, including proteins, to be elucidated.

Index

Printed and bound by CPI Group (UK) Ltd, Croydon, CR0 4YY

Printed and bound by CPI Group (UK) Ltd, Croydon, CR0 4YY

27/10/2024

14580394-0001